Plant Carbohydrate Biochemistry

EXPERIMENTAL BIOLOGY REVIEWS

Series advisors:

D.W. Lawlor
AFRC Institute of Arable Crops Research, Rothamsted Experimental Station, Harpenden, Hertfordshire AL5 2JQ, UK

M. Thorndyke
School of Biological Sciences, Royal Holloway, University of London, Egham, Surrey TW20 0EX, UK

Environmental Stress and Gene Regulation
Sex Determination in Plants
Plant Carbohydrate Biochemistry

Forthcoming titles include:

Biomechanics in Animal Behaviour
Cell Death in Health and Disease
Cambium: the biology of wood formation

Related titles from BIOS in the Environmental Plant Biology *series:*

Abscisic Acid: physiology and biochemistry
Biological Rhythms and Photoperiodism in Plants
Carbon Partitioning: within and between organisms
Embryogenesis: the generation of a plant
Environment and Plant Metabolism: flexibility and acclimation
Pests and Pathogens: plant responses to foliar attack
Photoinhibition of Photosynthesis: from molecular mechanisms to the field
Plant Cuticles: an integrated functional approach
Stable Isotopes: the integration of biological, ecological and geochemical processes
Water Deficits: plant responses from cell to community

Plant Carbohydrate Biochemistry

J.A. BRYANT
School of Biological Sciences, University of Exeter, Exeter EX4 4QG, UK

M.M. BURRELL
Advanced Technologies (Cambridge) Limited, 210 Cambridge Science Park, Cambridge CB4 4WA, UK

N.J. KRUGER
Department of Plant Sciences, University of Oxford, South Parks Road, Oxford OX1 3RB, UK

βIOS
SCIENTIFIC
PUBLISHERS

Oxford • Washington DC

© BIOS Scientific Publishers Limited, 1999

First published in 1999

A CIP catalogue record for this book is available from the British Library.

ISBN 1 85996 112 6 (hardback)
ISBN 1 85996 117 7 (clothbound)

BIOS Scientific Publishers Ltd
9 Newtec Place, Magdalen Road, Oxford OX4 1RE, UK.
Tel. +44 (0) 1865 726286. Fax. +44 (0) 1865 246823
World Wide Web home page: http://www.bios.co.uk/

Production Editor: Jonathan Gunning.
Typeset by Saxon Graphics Ltd, Derby, UK.
Printed by Biddles Ltd, Guildford, UK.

Contents

Contributors

Anderson, L.E., Department of Biological Sciences, University of Illinois, 845 West Taylor Street, Chicago, IL 60607–7060, USA

Andralojc, P.J., Biochemistry and Physiology Department, IACR-Rothamsted, Harpenden, Hertfordshire AL5 2JQ, UK

Athwal, G.S., USDA Plant Science Research, North Carolina State University, Raleigh, NC 27695–7631, USA

Bailey, K.J., Department of Animal and Plant Sciences, University of Sheffield, Sheffield S10 2TN, UK

Ballicora, M., Department of Biochemistry, Michigan State University, East Lansing, MI 48824-1319, USA

Bowsher, C.G., School of Biological Sciences, 3.614 Stopford Building, University of Manchester, Manchester M13 9PT, UK

Bryant, J.A., School of Biological Sciences, University of Exeter, Exeter EX4 4QG, UK

Cairns, A.J., Institute of Grassland and Environmental Research, Plas Gogerddan, Aberystwyth, Ceredigion SY23 3EB, UK

Cornah, J.E., Department of Plant Sciences, University of Cambridge, Downing Street, Cambridge CB2 3EA, UK

Debnam, P.M., School of Biological Sciences, 3.614 Stopford Building, University of Manchester, Manchester M13 9PT, UK

Dennis, D.T., Performance Plants Inc., Bioscience Complex, c/o Queen's University, Kingston, Ontario, K7L 3N6, Canada

Denyer, K., John Innes Centre, Colney Lane, Norwich NR4 7UH, UK

Dever, L.V., Department of Biological Sciences, Lancaster University, Lancaster LA1 4YQ, UK

Douce, R., Laboratoire de Physiologie Cellulaire Végétale, CNRS URA 576, Département de Biologie Moléculaire et Structurale, CEA-Grenoble, F-38054 Grenoble, Cedex 9, France

Edwards, A., John Innes Centre, Colney Lane, Norwich NR4 7UH, UK

Emes, M.J., School of Biological Sciences, 3.614 Stopford Building, University of Manchester, Manchester M13 9PT, UK

Farrar, J.F., School of Biological Sciences, University of Wales Bangor, Bangor, Gwynedd LL57 2UW, UK

Fell, D.A., School of Biological and Molecular Sciences, Oxford Brookes University, Oxford OX3 0BP, UK

Flügge, U.I., Universität zu Köln, Botanisches Institut, Lehrstuhl II, Gyrhofstr. 15, D-50931 Köln, Germany

Fu, Y., Department of Biochemistry, Michigan State University, East Lansing, MI 48824-1319, USA

Gallagher, J., Institute of Grassland and Environmental Research, Plas Gogerddan, Aberystwyth, Ceredigion SY23 3EB, UK

Hanke, G., School of Biological Sciences, 3.614 Stopford Building, University of Manchester, Manchester M13 9PT, UK

Herbers, K., SunGene GmbH & Co.KGaA, Corrensstrasse 3, 06466 Gatersleben, Germany

Huber, J.L., USDA Plant Science Research, North Carolina State University, Raleigh, NC 27695–7631, USA

Huber, S.C., USDA Plant Science Research, North Carolina State University, Raleigh, NC 27695–7631, USA

Ireland, R.J., Biology Department, Mount Allison University, Sackville, New Brunswick, E4L 1G7, Canada

Kaiser, W.M., Lehrstuhl für Moleculare Pflanzenphysiologie und Biophysik, Universität Würzburg, D-97082 Würzburg, Germany

Keeling, P.L., ExSeed Genetics, 1573 Food Science Building, Iowa State University, Ames, IA 50011-1061, USA

Keys, A.J., IACR-Rothamsted, Harpenden, Hertfordshire AL5 2JQ, UK

Kochhar, A., Institute of Grassland and Environmental Research, Plas Gogerddan, Aberystwyth, Ceredigion SY23 3EB, UK

Lea, P.J., Department of Biological Sciences, Lancaster University, Lancaster LA1 4YQ, UK

Leegood, R.C., Robert Hill Institute and Department of Animal and Plant Sciences, University of Sheffield, Sheffield S10 2TN, UK

Loveland, J.E., Biochemistry and Physiology Department, IACR-Rothamsted, Harpenden, Hertfordshire AL5 2JQ, UK

Machado de Carvalho, M.A., Instituto de Botanica, Secao de Fisiologia e Bioquimica, Caixa Postal 4005, 01061–970 Sâo Paulo, SP, Brazil

Martin, C., John Innes Centre, Colney Lane, Norwich NR4 7UH, UK

Maxwell, K., Department of Agricultural and Environmental Sciences, Ridley Building, University of Newcastle, Newcastle upon Tyne NE1 7RU, UK

Möhlmann, T., Universität Osnabrück, Fachbereich Biologie/Chemie, Pflanzen-physiologie, Barbarastr. 11, D-49069 Osnabrück, Germany

Neuhaus, H.E., Universität Osnabrück, Fachbereich Biologie/Chemie, Pflanzen-physiologie, Barbarastr. 11, D-49069 Osnabrück, Germany

Osmond, B., Photobioenergetics Group, Research School of Biological Sciences, Institute of Advanced Studies, Australian National University, Box 475, Canberra, ACT 2601, Australia

Parry, M.A.J., Biochemistry and Physiology Department, IACR-Rothamsted, Harpenden, Hertfordshire AL5 2JQ, UK

Pollock, C.J., Institute of Grassland and Environmental Research, Plas Gogerddan, Aberystwyth, Ceredigion SY23 3EB, UK

Poolman, M.G., School of Biological and Molecular Sciences, Oxford Brookes University, Oxford OX3 0BP, UK

Popp, M., Institute of Plant Physiology, University of Vienna, Althanstr. 14, POB 285, A-1091 Vienna, Austria

Preiss, J., Department of Biochemistry, Michigan State University, East Lansing, MI 48824-1319, USA

Rawsthorne, S., Brassica and Oilseeds Research Department, John Innes Centre, Colney Lane, Norwich NR4 7UH, UK

Rébeillé, F., Laboratoire de Physiologie Cellulaire Végétale, CNRS URA 576, Département de Biologie Moléculaire et Structurale, CEA-Grenoble, F-38054 Grenoble, Cedex 9, France

Robinson, S., Biological Sciences, University of Wollongong, Northfields Avenue, Wollongong, NSW 2522, Australia

Roper, J.M., Institute of Ophthalmology, University College London, 11–43 Bath Street, London EC1V 9EL, UK

Singh, D.P., Department of Plant Sciences, University of Cambridge, Downing Street, Cambridge CB2 3EA, UK

Smirnoff, N., School of Biological Sciences, University of Exeter, Hatherly Laboratories, Prince of Wales Road, Exeter EX4 4PS, UK

Smith, A.G., Department of Plant Sciences, University of Cambridge, Downing Street, Cambridge CB2 3EA, UK

Smith, A.M., John Innes Centre, Colney Lane, Norwich NR4 7UH, UK

Sonnewald, U., Institut für Pflanzengenetik und Kulturpflanzenforschung, Corrensstrasse 3, 06466 Gatersleben, Germany

Stitt, M., Botanisches Institut, Universität Heidelberg, Im Neuenheimer Feld 360, 69120 Heidelberg, Germany

Tetlow, I.J., School of Biological Sciences, 3.614 Stopford Building, University of Manchester, Manchester M13 9PT, UK

Thomas, S., Department of Biology and Biochemistry, Brunel University, Uxbridge, Middlesex UB8 3PH, UK

Tjaden, J., Universität Osnabrück, Fachbereich Biologie/Chemie, Pflanzen-physiologie, Barbarastr. 11, D-49069 Osnabrück, Germany

Toroser, D., USDA Plant Science Research, North Carolina State University, Raleigh, NC 27695–7631, USA

Walker, R.P., Robert Hill Institute and Department of Animal and Plant Sciences, University of Sheffield, Sheffield S10 2TN, UK

Wheeler, G.L., School of Biological Sciences, University of Exeter, Hatherly Laboratories, Prince of Wales Road, Exeter EX4 4PS, UK

Winter, H., Fachbereich Biologie/Pflanzenphysiologie, Universität Osnabrück, Barbarastrasse 11, D-49069 Osnabrück, Germany

Zeeman, S.C., John Innes Centre, Colney Lane, Norwich NR4 7UH, UK

Abbreviations

AAN	aminoacetonitrile
AGPase	ADPglucose pyrophosphorylase
AICAR	5-aminoimidazole-4-carboxamide riboside
ALA	5-aminolaevulinic acid
AOS	active oxygen species
APX	ascorbate-specific peroxidase
BE	branching enzyme
CA	2-carboxy-D-arabinitol
CA1P	2-carboxy-D-arabinitol-1-phosphate
CABP	2-carboxy-D-arabinitol-1,5-bisphosphate
CAM	Crassulacean acid metabolism
CoA	coenzyme A
coprogen	coproporphyrinogen
cytFBPase	cytosolic FBPase
DTT	dithiothreitol
DHA	dehydroascorbate
DHAP	dihydroxyacetone phosphate
DHFR	dihydrofolate reductase
DHFS	dihydrofolate synthase
DHPS	dihydropteroate synthase
DOG	deoxyglucose
ER	endoplasmic reticulum
F6P	fructose-6-phosphate
FAD	flavin adenine dinucleotide
FBP	fructose-1,6-bisphosphate
FBPase	fructose-1,6-bisphosphatase
FFT	fructan: fructan fructosyl transferase
FPGS	folylpolyglutamate synthetase
G1P	glucose-1-phosphate
G6P	glucose-6-phosphate
GAPDH	glyceraldehyde-3-phosphate dehydrogenase
GBSS	granule-bound starch synthase
GDC	glycine decarboxylase complex
GMP	GDPmannose pyrophosphorylase
GPT	glucose phosphate translocator
GSH	glutathione
GT	glucose transporter
GUS	β-glucuronidase
HBP	hamamelose-2^1,5-bisphosphate
HMG-CoA	3-hydroxy-3-methylglutaryl-CoA

HPPK	dihydropterin pyrophosphokinase
KABP	3-ketoarabinitol-1,5-bisphosphate
L-GAL	L-galactono-1,4-lactone
L-GALDH	L-GAL dehydrogenase
L-GUL	L-guluno-1,4-lactone
MDA	monodehydroascorbate
MDH	malate dehydrogenase
ME	malic enzyme
3-O-mG	3-O-methylglucose
NR	nitrate reductase
OPPP	oxidative pentose phosphate pathway
PAGE	polyacrylamide gel electrophoresis
PCK	PEP carboxykinase
PCO	photorespiratory carbon oxidation
PCR	photosynthetic carbon reduction
PDBP	D-*glycero*-2,3-pentodiulose-1,5-bisphosphate
PEP	phosphoenolpyruvate
PEPPT	PEP/phosphate translocator
PFK	phosphofructokinase
PGA	3-phosphoglycerate
PGI	phosphoglucose isomerase
PGK	phosphoglycerate kinase
PGM	phosphoglucomutase
P_i	inorganic phosphate
PMI	phosphomannose isomerase
PPDK	pyruvate P_i dikinase
PP_i	inorganic pyrophosphate
PPT	phosphinothricin
PRK	phosphoribulokinase
protogen	protoporphyrinogen
ROS	reactive oxygen species
Rubisco	ribulose-1,5-bisphosphate carboxylase/oxygenase
RuBP	ribulose-1,5-bisphosphate
SBPase	sedoheptulose-1,7-bisphosphatase
SDS	sodium dodecyl sulphate
SHMT	serine hydroxymethyltransferase
SOD	superoxide dismutase
SPP	sucrose-6-phosphate phosphatase
SPS	sucrose phosphate synthase
SS	starch synthase
SST	sucrose: sucrose fructosyl transferase
SuSy	sucrose synthase
TNC	total non-structural carbohydrate
TPT	triose phosphate translocator
TS	thymidylate synthase
urogen	uroporphyrinogen
XuBP	xylulose-1,5-bisphosphate
ZMP	5-aminoimidazole-4-carboxamide ribonucleiode monophosphate

Preface

All three of us were Ph.D. students in Tom ap Rees' laboratory (indeed, JAB was the first student to join Tom's group in Cambridge) and felt strongly that Tom's retirement, scheduled for 1998, provided an ideal opportunity to mark his immense contribution to plant biochemistry. It seemed to us, as we met for preliminary discussions in 1995, that an international conference and the published proceedings thereof, each with a significant contribution from Tom himself, would be an appropriate tribute. It was our view that the conference should represent the best and most exciting science in the area for which Tom is best known, namely plant carbohydrate biochemistry, and by late summer in 1996 we had started to assemble our cast of speakers. Our planning was interrupted by Tom's tragic death on October 3rd 1996. However, after a pause for reflection and some discussion with Tom's widow, Wendy, we decided to proceed with the conference, and the associated publication, as a memorial to Tom. An appreciation of Tom by his long-term mentor and friend, Harry Beevers, follows this preface. We simply need to say a little more about the book.

For a while, 'conventional' plant biochemistry became unfashionable as the research spotlight turned on genes and molecular biology. There were, however, many in the plant science community who believed that we should continue to study the metabolic pathways themselves. This was for the very good academic reason that there was, and still is, a huge amount that we do not know, and for the more applied reason that a good knowledge of plant metabolism will surely assist in the commercial exploitation of plant genetic manipulation. Thus, several key workers, including Tom ap Rees and the contributors to this book, 'kept the faith'. Their persistence has certainly been rewarded. One of the most exciting developments in the last few years has been not only a renewal of interest in plant metabolism itself but also the increasing application of molecular biological and molecular genetic techniques to the study of metabolism (in addition to the possibilities of actually modifying plant metabolic pathways for applications in agriculture, horticulture and food technology). The marriage of more 'conventional' plant biochemistry to plant molecular biology is amply illustrated throughout this book and it is our hope that the reader will feel the excitement of current research in this important area. Notwithstanding the impressive progress that has been made over the past few years, there is still a lot to do. The recent elucidation of the pathway for synthesis of a major plant product, ascorbate, so long a mystery but described in this book, is just one indication that discoveries remain to be made. So, in addition to presenting the state of the art in an exciting and timely manner, many of our authors also take a look into the future and see a field that will remain vibrant, exciting, rewarding, interesting and challenging. We hope that readers who are at the beginning of their careers will be inspired to meet the challenges offered by this area of plant biology.

In compiling this book, we have received help and support from many quarters. We are grateful to the authors for the way in which they espoused the cause of the conference and of the resultant book. Such was the desire to commemorate Tom's life and work that of the speakers we invited not one refused the invitation. The authors have done a great job for us in conveying the excitement of this important work and we thank them for it. We also thank the staff of BIOS Scientific Publishers, especially Rachel Offord and Jonathan Gunning, for their enthusiasm to publish this volume and for their tactful but persistent and effective efforts in keeping us to our editing timetable. The company has been a pleasure to work with and we are grateful for their desire to publish this book as soon as possible.

We cannot let this opportunity pass without doing something that several authors in the book have done, namely to record our gratitude to Tom ap Rees. As our Ph.D. supervisor he was tireless in his support, guidance, encouragement and enthusiasm. He showed a great interest in our subsequent careers and remained always willing to encourage and advise. We miss him. Nevertheless we are very glad that Wendy ap Rees was able to attend the conference and we thank her for her encouragement to hold the meeting and to publish the proceedings. We hope that she and the plant science community will see this book as a fitting tribute to a great man.

John A. Bryant, Exeter
Michael M. Burrell, Cambridge
Nicholas J. Kruger, Oxford

December 1998

Dedication

This book is dedicated to the memory of Tom ap Rees,
outstanding scientist and inspirational teacher

Tom ap Rees (1930–1996)

Tom ap Rees was working on oxidative enzymes in beech mycorrhiza as a D.Phil. student of my old friend, J.L. Harley, when I met him for the first time in Oxford in 1957. He then did post-doctoral work with me at Purdue University before moving to Australia and subsequently to Cambridge in 1964.

During his career, devoted to different aspects of carbohydrate metabolism in plants, Tom ap Rees established an international reputation as a skilled experimentalist who asked important questions and attacked them with all the tools available. Much of his work, elucidating cyanide-resistant respiration, the regulation of respiratory pathways, the roles of organelles, and the control of starch metabolism, required accurate enzyme assays and he adamantly insisted on first establishing truly optimal conditions for each assay. His papers were refreshing, direct and clear (many began with 'The aim of this work…'), presented unambiguous results, and contained carefully considered conclusions which have stood the test of time. His presentations at scientific meetings and symposia were always outstanding; his lecturing style was superb and the envy of his colleagues.

Over the years, Tom ap Rees was supervisor to a great number of outstanding graduate students and it is quite remarkable that so many of these now hold important and influential positions in England and abroad. Several have contributed to this volume and without exception they hold Tom in the very highest regard and seek to emulate his remarkable qualities as a scientist and a mentor.

I understand too that Tom was eminently successful as Head of the Department in Cambridge. He vigorously introduced new programmes in biology and characteristically took a heavy teaching load. He influenced generations of undergraduate students and has had a permanent impact on biology in Cambridge. In spite of his discipline and devotion to his science he maintained a sunny disposition and was a most delightful companion. He and his wife, Wendy, were wonderful hosts.

It does not surprise me that so many hundreds of Tom's former students, colleagues, admirers and friends assembled to pay their respects to Tom at a memorial service in Cambridge. The terrible accident that so cruelly ended his life came when he was within sight of retirement. His career will live on as a shining example, and his legacy of a devoted group of students will ensure that he will be remembered.

Harry Beevers
University of California
Santa Cruz

The first will be last and the last will be first: non-regulated enzymes call the tune?

Mark Stitt

1. Introduction

Biochemical pathways consist of a sequence of reactions, each catalysed by a specific enzyme. A key question for our general understanding of metabolic regulation, that has been debated for a long time, is whether metabolic flux is regulated at one or a small number of steps that act as bottlenecks and metabolic switchpoints, or whether flux is co-limited by several enzymes.

Many of the reactions in biochemical pathways lie close to their thermodynamic equilibrium. When a reaction is close to equilibrium, the net flux is the difference between the rates of the forward and reverse reactions, both of which will be occurring at an appreciable rate (Rolleston, 1972; Stitt 1994, 1995). Since a shift in the substrate/product ratio will alter the relative rates of the forward and reverse reactions, such steps are readily reversible. It has been assumed that the enzymes that catalyse reversible reactions are present in excess, and are therefore inappropriate sites to alter flux through the pathway. In agreement, enzymes that catalyse readily reversible steps usually, although not always, lack obvious regulatory properties. In contrast, a small number of the reactions in a pathway are removed from thermodynamic equilibrium. At these irreversible steps, the net flux directly depends on the current rate of catalysis, making it plausible that these enzymes regulate flux through the pathway. In support of this view, such enzymes often possess regulatory properties, including allosteric regulation and/or post-translational regulation.

Enzyme activity can be modified via fundamentally different mechanisms. 'Fine' regulation involves perturbation of the activity of the pre-existing protein activity by changes in the levels of substrates, inhibitors or activators and by post-translational modification, whereas 'coarse' regulation involves changes in the amounts of the enzymes, either due to transcriptional regulation or to protein turnover. It is generally

Plant Carbohydrate Biochemistry, edited by J.A. Bryant, M.M. Burrell and N.J. Kruger.
© 1999 BIOS Scientific Publishers Ltd, Oxford.

assumed that 'fine' regulation is important for rapid adjustments of fluxes to short-term perturbations, and 'coarse' regulation is more important during development and gradual adjustments to changes in the environment. However, the numerous reports that a large change in the expression of a target gene leads to remarkably little effect on metabolic fluxes (see Stitt and Sonnewald, 1995, and below for more examples) raise doubts about the importance of 'coarse' regulation of individual enzymes. It is important to know whether coarse control operates at a small number of steps and if these are the same ones where 'fine' control operates, or whether there is co-ordinate 'coarse' regulation of all the steps in a pathway.

Several approaches have been used to identify key control sites and uncover the mechanisms that regulate them (Newsholme and Start, 1973). One approach involves correlative studies of changes in enzyme activities or transcript levels, when the flux through the pathway is changing during a developmental transition or in response to an environmental perturbation. A second widely used approach has been to compare the changes of pathway flux and the changes in the substrate concentration. If the substrate concentration changes reciprocally to the flux, this is a good indication that the enzyme could represent a control point. The *in vitro* properties of the enzyme and the expression pattern can then be characterized to identify the mechanisms that regulate the expression of the corresponding gene or the activity of the enzyme.

Such methods, however, do not unambiguously identify the key sites for control. Correlative investigations of enzyme activities and transcript levels do not show whether they are a functional pre-condition for the change in flux, let alone whether the enzyme investigated is a key site for regulation. This problem is underlined by the large number of studies where quite large changes in the expression of selected enzymes lead to remarkably little effect on metabolic fluxes. The comparison of fluxes and changes in metabolite levels addresses function *in vivo* more directly, especially when information about the *in vitro* properties of the enzymes is combined with *in vivo* information about changes in fluxes, substrate concentrations and levels of effectors, or changes in expression or post-translational regulation. However, this approach still suffers from serious problems. Quite apart from the technical difficulties in accurately measuring metabolite levels and fluxes *in vivo* (see ap Rees and Hill, 1994), redundant or alternative reactions may be present, making it difficult to identify the substrate pool and flux for a given enzyme. Further, the complexity of regulatory networks makes it difficult to distinguish unambiguously between primary and secondary effects: enzymes are regulated for many different reasons, not only to allow the flux to be changed through the pathway, but also to ensure that changes of flux at one site are followed by corresponding changes at other sites in the pathway, or to control the concentrations of key intermediates, or levels of end products. Many of the regulatory features of enzymes may operate to co-ordinate their activity with that of other enzymes in the pathway, rather than to allow the flux through the pathway to be changed. These approaches also have a basic logical flaw: they provide detailed – and important – information about the regulation mechanisms that act on the individual enzymes in a pathway but they do not provide information about how the individual steps interact in the pathway as a system.

2. The application of reversed genetics to identify control steps

These shortcomings have been solved in the last 10 years by combining the techniques of metabolic biochemistry and molecular genetics. The approach involves the production of a set of plants in which the expression of one enzyme is progressively decreased, in as specific a manner a possible. This can be done by using classical mutants to generate a set of plants where the number of functional gene copies is varied (see e.g. Neuhaus and Stitt, 1990; Neuhaus et al., 1989). However, more genes can be addressed and a better 'spread' of activities obtained by transforming plants with antisense or sense constructs to achieve antisense inhibition or co-suppression of the gene in question, and if possible also over-expression. Following careful quantification of the change in gene expression (see ap Rees and Hill, 1994; Haake et al., 1998; Stitt, 1995), the effect on the pathway flux (or any other quantifiable trait) can be measured.

The importance of an enzyme for the control of pathway flux can be visualized by plotting enzyme activity against flux, both normalized on the wild-type value (*Figure 1*). When the enzyme is the sole control point ('limiting' in the strict sense) there is a linear and strictly proportional inhibition of pathway flux as gene expression is inhibited. When the enzyme, together with other enzymes, co-limits flux there will be a curvilinear response with a finite but non-proportional slope in the range corresponding to wild-type plants. When the enzyme is not a significant site for control, the slope will be zero in the wild-type range. At some point, a further decrease of expression may lead to an inhibition of flux, however, the effects of a large inhibition of expression merely tell us whether the enzyme is essential or redundant, and does not mean that the enzyme controls flux in wild-type plants. In some cases, decreased expression of an enzyme, especially one in another competing pathway or one that controls the levels of effector molecules, may increase flux through the pathway of interest. These responses to changes in the amount of enzyme can be expressed in the formalism of control theory as flux control coefficients (ap Rees and Hill, 1994; Stitt, 1994; see Chapter 2). Briefly, $C = (dJ/J) (dE/E)^{-1}$, where J and E are the flux and the amount of enzyme in the wild-type, and dJ is the change in flux that results from dE, a small decrease in the enzyme amount.

In the next sections, I will briefly describe a series of case studies, which illustrate the structure of regulation networks, the meaning (or lack of meaning) of the concept of 'bottleneck' steps, and the close interactions between 'fine' and 'coarse' control.

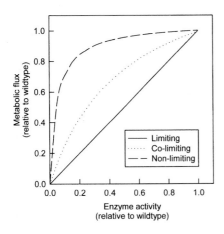

Figure 1. Enzyme activity versus metabolic flux.

3. Nitrate reductase is a highly regulated enzyme, but is expressed in excess

3.1 *Although nitrate reductase is a highly regulated enzyme, decreased expression does not alter the rate of nitrate assimilation*

Nitrate reductase (NR) catalyses the first step in nitrate assimilation and is responsible for the majority of the nitrogen assimilated in most plant species. It catalyses an irreversible reaction, is subject to a sophisticated hierarchy of transcriptional, post-transcriptional and post-translational regulation mechanisms, and is universally thought to be the key site for the regulation of nitrate assimilation (Crawford, 1995; Hoff *et al.*, 1994; Huber *et al.*, 1996).

This view, however, has to be reconciled with the surprisingly small effect that decreased expression of NR has on nitrate assimilation. Tobacco mutants with one or two instead of four functional *NIA* gene copies contain less NR activity but have a similar protein content and grow at the same rate as wild-type plants (Scheible *et al.*, 1997c). Analogous results have been obtained in barley, *Arabidopsis*, and *Nicotiana plumbaginifolia* mutants with a 50–90% decrease of NR activity (see Scheible *et al.*, 1997c for references). The dilemma was aptly stated by Crawford (1995): 'it is a mystery why such a highly regulated enzyme that catalyses a critical step is present in such excess'. We have recently investigated how low-NR mutants compensate for the reduced expression of NR, because we felt this would provide insights into the function of the sophisticated mechanisms that regulate NR, and allow us to solve this dilemma.

3.2 *Decreased expression is compensated via a time-of-day-dependent relaxation of feedback regulation*

The *NIA* transcript level, NR protein and NR activity show a marked diurnal rhythm, with a maximum in the first part of the light period and a decline later in the light period (Hoff *et al.*, 1994; Scheible *et al.*, 1997c). After darkening, NR is further inactivated as a result of post-translational regulation, via phosphorylation and subsequent binding of an inhibitory 14-3-3 protein (Huber *et al.*, 1996; Moorhead *et al.*, 1996). Phosphorylation is also thought to act as a signal for degradation of NR (Kaiser and Huber, 1997; Nausaume *et al.*, 1995;). It is known that nitrate induces and glutamine represses *NIA* (Hoff *et al.*, 1994). Although dark inactivation of NR is linked to changes in sugar levels (Kaiser and Huber, 1994), we have recently shown that dark inactivation is reversed in nitrogen-limited wild-type plants and in low-NR transformants (Scheible *et al.*, 1997c) where glutamine is low, and can be accentuated even in the light by feeding glutamine (Scheible *et al.*, 1997c). The decay of transcript and NR protein during the day therefore probably reflects feedback regulation acting at the level of transcription, post-translational modification and protein stability, as nitrate decreases and glutamine accumulates during the light period as a result of an inbalance between the rate of nitrate reduction and the rate at which nitrate is entering the leaf and glutamine is being further metabolized or exported.

In mutants with a decreased number of functional *NIA* genes and, as a consequence, lower NR activity at the start of the photoperiod, nitrate does not decrease as far and glutamine does not accumulate as much as in wild-type plants, the decay of NR activity and protein during the day is almost completely prevented, and the dark-inactivation of

NR is partially reversed (Scheible *et al.*, 1997c). As a result, even though NR activity is two to three times lower in the mutants compared to wild-type plants during the first part of the light period, it is similar to wild-type plants towards the end of the light period, and is higher than in wild-type plants during the night.

This example illustrates several general points: (i) decreased gene expression may be compensated for by relaxation of feedback mechanisms, (ii) this may be especially effective for a highly regulated enzyme, (iii) although compensation often involves a stimulation of the current activity, it can also (iv) involve changes in the duration or timing of processes, and (v) compensation can nevertheless have side effects, even when the pathway flux is not altered, because it requires changes in the concentrations of metabolites, in this case nitrate and amino acids, which could in turn affect further processes. Several groups are investigating what influence the altered levels of nitrate and glutamine could have for other processes in the plant, including shoot–root allocation (Scheible *et al.*, 1997a), the regulation of carbon metabolism (Scheible *et al.*, 1997b), and nitrate uptake in the roots (Gojon *et al.*, 1998).

3.3 Regulation of NR normally serves to co-ordinate the rate of nitrate assimilation with other processes

With respect to the original dilemma, our results imply that the sophisticated regulation of NR *co-ordinates* the *current* rate of nitrate assimilation with diurnal fluctuations in the influx of nitrate and the export of amino acids and their use in the plant, but does not control the total amount of nitrate assimilated over 24 h. This presumably depends on the amount of nitrate entering the plant, and the rate at which the assimilated nitrogen is used.

3.4 Functional implications of the 'excess' NR during transients and in unfavourable conditions

This result also leads to a further question: why do wild-type plants contain 'excess' NR activity, and down-regulate it for much of the time? A trivial explanation would be that tobacco is a crop plant, and that breeding has led to an inbalance between the expression of *NIA* and that of other genes. However, this can be ruled out, because analogous results have been obtained with *N. plumbaginifolia* (see above) and *Arabidopsis* (M. Stitt and P. Ullrich, unpublished data). One part of the explanation may be that it is counterproductive to decrease NR activity to the point where it restricts the overall rate of nitrate assimilation, because this would mean that the plant would not able to take full advantage of the available nitrate to support growth and would be at a selective disadvantage compared to plants with higher levels of NR. It can nevertheless be asked why there is a 4–10-fold excess in these various species. The second part of the explanation may be that high NR activity may be important in transient conditions, where large amounts of nitrate must be assimilated in a short time.

To test this, we recently carried out two further experiments with the *NIA* dosage mutants. In one experiment, we transferred wild-type plants (four *NIA* gene copies) and a series of mutants (two or one functional *NIA* genes) to nitrate-free conditions for 3 days, and then re-fertilized them with nitrate and investigated how quickly NR activity rose in the following 4 h. There was an almost linear relation between the number of functional *NIA* genes and the speed at which NR activity increased, showing that the gene number has a very direct impact on the rate of transcription and translation in conditions where the feedback regulation (see above) is relaxed (N. Bujard, A. Krapp and M. Stitt, unpublished results). In the second experiment, we grew wild-type tobacco plants and mutants with one instead of four functional *NIA* gene copies in a short day (6 h light/18 h dark) regime. In these conditions, NR activity in wild-type plants continues to rise until the end of the light period. The mutants contained lower levels of NR protein and activity, and grew about 25% more slowly (P. Matt, A. Krapp, U. Schurr and M. Stitt, unpublished data).

4. The contribution of Rubisco to the control of photosynthesis depends on current and past conditions

Rubisco catalyses the unique step in which carbon dioxide is incorporated into organic compounds in the Calvin cycle, contains up to 40% of the protein in the leaf, and represents a key site in the carbon and nitrogen economy of the plant. A set of 'antisense' tobacco plants with decreased expression of ribulose-1,5-bisphosphate carboxylase/ oxygenase (Rubisco) allowed us to examine whether Rubisco limits the rate of photosynthesis (for a review of the following results, see Stitt and Schulze, 1994).

4.1 *The extent to which Rubisco controls the rate of photosynthesis depends on the conditions in which photosynthesis is measured*

When the plants were grown in moderate light, ambient photosynthesis was only slightly inhibited when Rubisco was decreased to about 60% of the wild-type activity, and a control coefficient, C, of 0.05–0.15 was estimated (Quick *et al.*, 1991). However, when plants were grown in low light and the rate of photosynthesis then suddenly increased by increasing the light intensity, there was a near-proportional relation between the amount of Rubisco and the rate of photosynthesis ($C > 0.9$) (Quick *et al.*, 1992). A similar result was obtained if the carbon dioxide concentration was decreased (Stitt *et al.*, 1991). On the other hand, when photosynthesis was measured in 5% carbon dioxide, Rubisco could be decreased by 80% without any effect on the rate of carbon assimilation. The extent to which Rubisco limits photosynthesis clearly depends on the short-term conditions under which the flux is measured.

4.2 *The extent to which Rubisco controls the rate of photosynthesis depends on the past history of the plant*

A further important result was illustrated by comparing plants grown in different conditions. When plants were grown in high light, there was a curvilinear relation between Rubisco activity and photosynthesis with a very small slope in the wild-type range (C was about 0.2), whereas when plants were grown at low light and the light intensity suddenly increased, Rubisco was almost totally limiting for photosynthesis ($C > 0.9$) (Lauerer *et al.*, 1993). When plants were grown on low nitrogen fertilizer, Rubisco decreased and became more limiting for ambient photosynthesis ($C = 0.5$–0.6) (Quick *et al.*, 1992). The extent to which Rubisco limits photosynthesis clearly depends on the history of the plants.

The control exerted by one enzyme on pathway flux therefore varies, depending upon the immediate conditions in which the measurements are carried out, and on the history of the plant. This raises the intriguing question of how these short-term and long-term influences will interact in the field, where the conditions are continually changing. Our experiments have not addressed this important aspect, but some speculative thoughts are offered. Ultimately, the rate of photosynthesis will depend on the availability of light and water, and hence carbon dioxide, on the availability of nutrients to allow the products of photosynthesis to be utilized in growth, and on genetic factors that may restrict the maximum rate of plant growth. A one-sided limitation by Rubisco, as discussed already for NR, would hinder the use of the available resources and represent a selective disadvantage. Our experiments imply that the response to an acute one-sided limitation of photosynthesis by Rubisco is to alter the relative amounts of Rubisco and other proteins, which will allow an escape from or amelioration of the one-sided internal limitation. A potential excess of expression may allow a more rapid 'escape' from short-term limitations that appear after a sudden change of the conditions, for example, a sudden increase of the growth irradiance or of nitrogen fertilization.

5. Control is distributed in a non-fixed manner between the enzymes of the starch synthesis pathway in leaves

The next two examples illustrate how control is distributed between the various enzymes in a linear metabolic pathway. The pathway of starch synthesis in leaves starts with fructose-6-phosphate in the chloroplast stroma. Fructose-6-phosphate (F6P) is converted to glucose-1-phosphate in two steps catalysed by phosphoglucose isomerase (PGI) and phosphoglucomutase (PGM). These enzymes catalyse readily reversible reactions and do not have any known regulatory properties, and would therefore be considered to be irrelevant for understanding the control of pathway flux. The next enzyme, ADPglucose pyrophosphorylase (AGPase), catalyses the conversion of glucose-1-phosphate and ATP to ADPglucose. The reaction is irreversible, because pyrophosphate is hydrolysed by a very active pyrophosphatase in the plastid stroma, AGPase is subject to exquisite allosteric regulation by 3-phosphoglycerate (PGA) and phosphate which act as activator and inhibitor (Preiss, 1988) and to transcriptional regulation (Koch, 1996), and is universally viewed as the step at which starch synthesis is regulated (Preiss, 1988).

The effect of decreased expression of each of these enzymes on the rate of starch synthesis was investigated using dosage mutants of *Clarkia xantiana* and *Arabidopsis* (Kruckeberg *et al.*, 1989; Neuhaus and Stitt, 1990; Neuhaus *et al.*, 1989). When starch synthesis was measured in moderate light, a 50% decrease of plastid PGI or of plastid PGM had no significant effect on starch synthesis, whereas a 50% decrease of AGPase led to a significant but non-proportional inhibition of starch synthesis ($C = 0.28$). When starch synthesis was measured in saturating light and carbon dioxide to increase the flux through the pathway by 2–3-fold, all three enzymes contributed to the control of starch synthesis, with control coefficients of 0.35, 0.26 and 0.56, respectively.

This example illustrates three further general points: (i) the distribution of control between the enzymes in a pathway can vary, (ii) control can be shared between several enzymes and (iii) non-regulated enzymes that catalyse readily reversible steps are not necessarily present in large excess and can exert a partial limitation on flux.

6. Complex patterns of control can develop at branch points

An added layer of complexity is introduced by metabolic branch points. Such branch points constitute a central feature in metabolism, and the way in which flux is partitioned between the two branches is a key feature in regulation. This has been studied by comparing the effect of decreased expression of a cytosolic enzyme in the pathway of sucrose synthesis, and of the corresponding plastid isoform in the pathway of starch synthesis.

6.1 *The patterns of control in the pathways of starch synthesis and sucrose synthesis differ*

Clarkia xantiana mutants with decreased expression of the cytosolic PGI and the plastid PGI were compared to investigate how decreased expression of an enzyme that catalyses the same reaction in two branching pathways affects the fluxes in the two pathways (Kruckeberg *et al.*, 1989; Neuhaus *et al.*, 1989). The results with the cytosolic mutant differed strikingly from those with the plastid mutant. Firstly, whereas a 50% decrease in the expression of the plastid PGI inhibited the rate of starch synthesis in high light but not in low light (see above), the opposite result was obtained for the cytosolic PGI: here a 64% decrease of activity led to an inhibition of sucrose synthesis in low light but not in high light. Secondly, whereas a decreased rate of sucrose synthesis always led to a compensating increase in the rate of starch synthesis and the rate of photosynthesis was not inhibited, decreased expression of plastid PGI led in high light to an inhibition of starch synthesis *and* an inhibition of photosynthesis, with the result that sucrose synthesis was also decreased. These results illustrate that the impact of decreased expression depends on the pathway in which the enzyme is embedded, not the reaction it catalyses.

6.2 *Regulation loops lead to a complex interaction between plastid PGI expression and pathway flux*

The results obtained for cytosolic PGI were surprising, for two reasons: one was that this was the first example we found that a 'non-regulated' enzyme catalysing a reversible reaction was not expressed in large excess; the other was that the impact on flux was largest in conditions where the absolute flux through the pathways was low and disappeared when the flux through the pathway was high. Analyses of metabolites were carried out to try to find an explanation for these results.

The substrate of the cytosolic PGI, F6P, is also the substrate for and the activator of fructose-6-phosphate-2-kinase and is an inhibitor of fructose-2,6-bisphosphatase (Stitt, 1996). These two enzyme activities are responsible for the synthesis and degradation of the regulator metabolite fructose-2,6-bisphosphate, which is a potent competitive inhibitor of the cytosolic fructose-1,6-biphosphatase (FBPase) (Stitt, 1996). A relatively small (*ca* 50%) change in the F6P level, due to decreased expression of cytosolic PGI leads to an amplified (*ca* 100%) increase of the fructose-2,6-bisphosphate level (Neuhaus *et al.*, 1989; Stitt, 1989) and inhibition of the cytosolic FBPase. This explains why a relatively small decrease in cytosolic PGI expression leads to an inhibition of sucrose synthesis in low light. In high light, F6P and fructose-2,6-bisphosphate increase even further in the mutant, but this does not lead to an inhibition of sucrose synthesis because there is also a marked increase of triose phosphates and, presumably, fructose-1,6-bisphosphate which overrides the effect of the competitive inhibitor on the cytosolic FBPase. This example illustrates (i) that quite moderate changes in the levels of the substrates of non-regulated enzymes can have a major 'knock-on' effect if they happen to be a substrate for, or a modulator of, important regulatory loops, and (ii) that interactions between different metabolic pathways can give rise to complex and, until the regulation mechanisms are analysed, counter-intuitive changes in the fluxes through the pathways.

7. 'Non-regulated' enzymes contribute to the control of flux in the Calvin cycle

The Calvin cycle is an even more complex pathway, consisting of a cycle in which the products of the light reactions are consumed at three sites, and carbon is withdrawn at two points. Selected enzymes of the Calvin cycle are subject to exquisite 'fine' regulation involving post-translational regulation by thioredoxin, regulation by pH and magnesium, and regulation by metabolite effectors, allowing up to 100-fold changes in their activity in response to changes in the light intensity (Stitt, 1996).

The Calvin cycle includes four enzymes that catalyse irreversible reactions and are highly regulated: Rubisco, stromal FBPase, sedoheptulose-1,7-bisphosphatase (SBPase) and phosphoribulokinase (PRK). Studies on potato and tobacco transformants in several laboratories have shown that none of the regulated enzymes exerts a one-sided or strong control on the rate of ambient photosynthesis. We have shown (see Section 4.1) that a 40% decrease of Rubisco leads to only a slight inhibition of photosynthesis in ambient conditions. Kossman et al. (1995) have shown that decreased expression of plastid FBPase does not affect photosynthesis until 60% of the wild-type activity is removed. Harrison et al. (1998) have shown that transformants with half of the wild-type SBPase activity show a significant but still non-proportional inhibition of photosynthesis, and Paul et al. (1995) have shown that an 85–95% decrease of PRK activity has no effect on the rate of photosynthesis. Plants have also been investigated with decreased expression of NADP-glyceraldehyde-3-phosphate dehydrogenase (NADP-GAPDH) (Price et al., 1995). Although this enzyme catalyses a near-equilibrium reaction, it is regulated by thioredoxin. Photosynthesis was inhibited when NADP-GAPDH activity was decreased by 65% or more.

We recently, for comparison, investigated the effect of decreased expression of two enzymes that catalyse readily reversible reactions and lack any 'fine' regulatory properties, the plastid aldolase and transketolase. The results confirm for a second, and more complex, pathway that the 2-fold decreases in the expression of non-regulated enzymes can lead to inhibition of pathway flux.

7.1 Plastid aldolase exerts a small but significant control on the rate of photosynthesis

A 30% decrease of aldolase activity in potato transformants led to a small (5–10%) inhibition of ambient photosynthesis, and reduction below 30% of the wild-type activity led to a severe inhibition of photosynthesis in plants growing in the greenhouse in moderate light (about 200 μmol m^{-2} s^{-1}) (Haake et al., 1998; V. Haake, U. Sonnewald and M. Stitt, unpublished). To check that this was not an anomalous result due to the particular growth conditions, we also grew the potato transformants in low light (70 μmol m^{-2} s^{-1}), high light (450 mmol m^{-2} s^{-1}), or high light plus elevated (800 ppm) carbon dioxide (Haake et al., 1999). A small decrease in plastid aldolase expression led to a significant inhibition of ambient photosynthesis in all of these growth conditions, with the inhibition being smallest in low light and highest in high light and elevated carbon dioxide (C of about 0.18 and 0.56, respectively). This non-regulated enzyme evidently exerts a small degree of control over the rate of photosynthesis across a wide range of growth conditions, and exerts a strikingly large degree of control in high light and elevated carbon dioxide.

Analyses of Calvin cycle metabolites and carbohydrates revealed another unexpected result. In low or moderate light, decreased expression of aldolase led to a marked accumulation of triose phosphates and a depletion of ribulose-1,5-bisphosphate (RuBP) and PGA. In high light and especially in high light and elevated carbon dioxide, however, the triose phosphates remained very low, RuBP remained high, and PGA was higher in the transformants than in wild-type plants. There was a dramatic inhibition of starch synthesis, an accumulation of glucose-6-phosphate which in leaves is preferentially located in the cytosol, and an increase of sucrose and reducing sugars in the leaves of the transformants. These results strongly indicate that decreased expression of aldolase inhibits photosynthesis for different reasons in low and high light: in low light photosynthesis is inhibited due to a restriction in the regeneration of RuBP as expected because aldolase catalyses two reactions in the regenerative part of the Calvin cycle, whereas in high light photosynthesis is inhibited due to an inhibition of starch synthesis, the resulting accumulation of phosphorylated intermediates and the depletion of free inorganic phosphate. This is the first example we are aware of where an enzyme controls a given flux via two different mechanisms, depending on the conditions.

7.2 *Small changes in the expression of transketolase lead to an inhibition of photosynthesis*

Tobacco transformants were also produced with decreased expression of plastid transketolase. There was a strong inhibition of ambient photosynthesis when transketolase activity was decreased below 40% of wild-type activity, and there was a near-linear relation between transketolase activity and the maximum rate of photosynthesis in saturating light and carbon dioxide (V. Henkes, U. Sonnewald and M. Stitt, unpublished data). Thus, at least two of the 'non-regulated' enzymes in the Calvin cycle are expressed at levels that are critical for the rate of photosynthesis.

7.3 *Relaxation of 'fine' feedback regulation can compensate for large changes in gene expression of highly regulated enzymes*

The studies on the Calvin cycle provide further striking examples of how metabolic changes can compensate for a large inhibition of the expression of a highly regulated enzyme. The activity of PRK assayed in optimal conditions *in vitro* is 20-fold higher than the maximum rate of photosynthesis (Gardeman *et al.*, 1983). However, PRK is subject to feedback inhibition by several metabolites including ADP, PGA and RuBP. When PRK was assayed in an assay in which typical stromal concentrations of these effectors are reconstituted, the activity resembled the rate of photosynthesis (Gardeman *et al.*, 1983). Small changes of these effectors would suffice to compensate for decreased expression of PRK in antisense transformants, and such changes are found in transformants with decreased expression of PRK (see Paul *et al.*, 1995).

7.4 *Changes in the levels of pathway metabolites due to decreased expression of a non-regulated enzyme can have dramatic effects on the fluxes through other pathways in the plant*

In addition to the decreased rate of photosynthesis, decreased expression of plastid aldolase and transketolase had further dramatic effects on other metabolic pathways. Decreased expression of aldolase consistently led to a marked decrease in NR activity. The most likely explanation for this decrease is that the decreased levels of starch lead to a transient shortage of sugars at the end of the night, which in turn leads to decreased expression of NR, as recently shown for wild-type tobacco plants growing in short days (Matt *et al.*, 1997).

Even more dramatic changes were seen in transformants with decreased expression of transketolase, which contained lower levels of the aromatic amino acids, the major phenylpropanoids, the precursors for lignin, and a decreased lignin content (S. Henke and M. Stitt, unpublished data). These changes were already apparent in lines with a 30% decrease in transketolase expression. The simplest explanation for these results is that erythrose-4-phosphate, which is one of the products of transketolase, has been decreased and that this limits the metabolic flux into the shikimate pathway. These results imply that transketolase activity is not only barely adequate to catalyse the fluxes around the Calvin cycle, but also restricts the flux into the shikimate pathway, which is essential for lignin production and plant defence.

7.5 *Acclimation of photosynthesis includes and requires co-ordinate changes in the expression of all the enzymes in the Calvin cycle including the 'non-regulated' ones*

Clearly, 'non-regulated' enzymes that catalyse reversible reactions are *not* necessarily expressed in large excess, whereas enzymes that are highly regulated are often present in a considerable excess compared to the amount that is needed to maintain pathway flux in stable ambient conditions. This empirical result contradicts earlier ideas about regulation (see Section 1) but it can be rationalized, if we consider how a decrease in the expression of one enzyme will interact with the operation of the other enzymes in the pathway *in vivo*. Enzymes that are subject to 'fine' regulation by feedback loops originating from within the pathway will often be able to compensate for decreased expression, because the residual enzyme can be stimulated by changes in the concentrations of a wide range of metabolites (substrates, products, inhibitors, activators). In contrast, enzymes that lack 'fine' regulatory properties can only compensate for decreased expression via an alteration in the concentrations of their substrate and product and, once these changes start to affect the operation of other enzymes in the pathway, pathway flux will be inhibited.

This conclusion has interesting consequences with respect to the 'coarse' regulation of enzymes that lack 'fine' regulatory properties. It is well documented that leaves acclimate to the environment by changing the levels of the proteins involved in photosynthesis. Most studies of acclimation, whether at the level of enzyme activities or investigation of transcription, have concentrated on regulated enzymes that catalyse irreversible reactions, in particular, Rubisco. If non-regulated enzymes like aldolase are not present in large excess, then either (i) they will need to be constitutively expressed at a high level to avoid such limitations, which our results already show not to be the case, or (ii) they will become limiting when the expression of other 'regu-lated' enzymes in the pathway is increased above a critical level or (iii) their expression or turnover will need to be regulated to allow co-ordinate changes in the activities of regulated *and* the non-regulated enzymes.

We therefore grew wild-type potato plants in low light, in high light, and in high light plus elevated carbon dioxide, and measured the activity of each Calvin cycle enzyme and total leaf protein in comparable leaves in each condition (Haake *et al.*, 1999). The enzyme activities increased by 20–80% in response to high light, and decreased by 5–30% in response to elevated carbon dioxide. The largest changes in enzyme activity were found, not for Rubisco or the other regulated enzymes, but for aldolase and transketolase. As already discussed (Section 7.2), experiments with anti-sense transformants have demonstrated that aldolase is slightly limiting for photosyn-thesis across a wide range of growth light intensities, and strongly limiting in high light and elevated carbon dioxide. When these two sets of results are combined, they demonstrate that acclimation includes marked changes in the expression of suppos-edly 'non-regulated' enzymes like aldolase, and that these changes are an essential component of acclimation. In the case of aldolase, the 'coarse' regulation is still not large enough to prevent a significant increase in the extent to which the rate of pho-tosynthesis is controlled by this 'non-regulated' enzyme in high light plus elevated carbon dioxide.

8. General conclusions about the nature of control site pathways and regulation networks

Drawing on the case studies of enzymes involved in primary carbon and nitrogen metabolism presented in this chapter, the following conclusions can be drawn.

(i)   The control of flux is usually shared between several enzymes. This can include 'regulatory' enzymes that catalyse irreversible reactions and 'non-regulatory' reactions that catalyse reversible reactions, whereby the latter often make a sur-prisingly large contribution.

(ii) The contribution of a given enzyme to the control of flux depends on the short-term conditions in which flux is measured.

(iii) The contribution of a given enzyme to the control of flux depends on the long-term conditions in which the plant has been growing. By implication, it may also depend on the genetic background.

(iv) In general, adjustments in 'fine' regulation are usually able to compensate for small changes in expression. In many cases, they can even compensate for large changes in expression. The extent to which this happens will depend on the abil-ity of the metabolic network to accommodate the requisite changes in the con-centrations of the substrates and effectors of the targeted enzyme.

(v)   Contrary to expectations, 'non-regulated' enzymes that catalyse readily reversible reactions are not present in a large excess. In the examples given here, these enzymes were, at the most, expressed in a *2–3-fold* excess over that needed to avoid a severe limitation of flux in ambient conditions, and often co-limited the maximum flux through the pathway. In some cases, they exerted a significant degree of control over fluxes in ambient growth conditions. This unexpected result can be understood by considering how these enzymes operate in the context of the pathway: when the expression of a non-regulated enzyme is decreased, metabolic compensation will be restricted to a shift of the substrate/product ratio, and pathway flux will be inhibited once these changes start to impair the operation of other enzymes in the pathway.

(vi) 'Regulated' enzymes are usually expressed in excess, the extent depending on the enzyme and the conditions. Since highly regulated enzymes are regulated by a large number of feedback loops, more options are open for compensation than for non-regulated enzymes, which will decrease the risk that deleterious side effects develop.

(vii) 'Coarse' regulation via changes in expression or protein turnover is not only found for 'regulatory' enzymes that catalyse irreversible reactions, but also for 'non-regulated' enzymes that catalyse reversible reactions.

(viii) 'Coarse' regulation may be especially important for enzymes that are not susceptible to 'fine' control, because there is no other way to increase their activity other than by altering the substrate/product ratio, which (see above) often has deleterious effects on the operation of other enzymes in the pathway.

(ix) In many cases, the 'regulatory' properties found for enzymes that catalyse an irreversible reaction may not, seen from an evolutionary standpoint, have been developed to allow the regulation of flux. Rather, they may have been a response to the problems that the highly exogonic nature of the reaction they catalyse would create in the cellular *milieu*, if it were to run to equilibrium. This would result in a massive accumulation of the product and/or depletion of the substrate, and would disrupt the operation of other enzymes and seriously disturb cell function. These reactions therefore have to be maintained away from equilibrium, which will require either that the enzyme is expressed at a very low level, or, if the enzyme is to be expressed at a higher level, that it is susceptible to 'fine' regulation. The latter, in many cases, may have the advantage that pathway flux can respond more rapidly to changes in the circumstances, but implies that 'fine' regulation of such enzymes will often link their activity to the resource availability and growth processes, rather than acting as a switch to drive these processes. In individual cases, the 'fine' regulation may take on further functions linked to co-ordinating the activity of that enzyme with the activity of other enzymes in the pathway, and sometimes also a role in controlling the flux through the pathway, but there is no *a priori* reason why a given 'regulated' enzyme should make a major contribution to the control of pathway flux, nor why a reaction that catalyses a readily reversible reaction should be excluded from such a function.

These conclusions have some implications for the structure of regulation networks, and for the design of strategies to manipulate pathways by reversed genetics. The effect of over-expression of a single enzyme is often constrained by feedback loops or other regulation mechanisms which act as a corset to limit the impact of the genetic alteration on pathways flux, and is damped because control is distributed between several enzymes in the pathway and will tend to re-distribute towards other enzymes when one enzyme is over-expressed. For this reason, it may be helpful to use genes that have been modified to weaken or remove their regulatory properties. This approach, of course, carries the risk that deleterious effects could appear as a result of constitutive over-activity of the enzyme. An alternative and complementary strategy is to alter the activity of several enzymes in the pathway, including those that catalyse the reversible and 'non-regulated' steps. Many features of plant metabolism and physiology allow us to predict that endogenous mechanisms exist to allow co-ordinate regulation at multiple sites, either due to a global change in transcription, and/or co-ordinate regulation by post-translational mechanisms. Since pathways are regulated at several sites and many enzymes are not present in large excess, a major change in flux will often require an increase in activity of several or all of the enzymes involved.

Acknowledgements

I thank the Deutsche Forschungsgeneinschaft, the Bundesministerium für Bildung und Forschung and BASF AG for support. The experiments discussed in this chapter were carried out by many industrious and intelligent doctoral and postdoctoral researchers and students in the last 10 years, including Ekkehard Neuhaus, Paul Quick, Uli Schurr, Klaus Fichtner, Marianne Lauerer, Wolf-Rüdiger Scheible, Anne Krapp, Rita Zrenner, Volke Haake and Stefen Henkes. I am also indebted to Leslie Gottlieb, Lawrence Bogororad and Steve Rodermel, Chris Somerville, Michel Caboche and Uwe Sonnewald for collaborations without which the experiments would not have been possible.

References

ap Rees, T. and Hill, S.A. (1994) Metabolic control analysis of plant metabolism. *Plant Cell Environ.* 17: 587–599.
Crawford, N.M. (1995) Nitrate: nutrient and signal for plant growth. *Plant Cell* 7: 859–868.
Gardeman, A., Stitt, M. and Heldt, H.W. (1983) Control of CO_2 fixation: regulation of spinach ribulose-5-phosphate kinase by stromal metabolite levels. *Biochim. Biophys. Acta* 722: 51–60.
Gojon, A., Dapoigny, L., Lejay, L., Tillard, P. and Rufty, T.W. (1998) Effects of genetic modification of nitrate reductase expression on $^{15}NO_3^-$ uptake and reduction in *Nicotiana* plants. *Plant Cell Environ.* 21: 43–53.
Haake, V., Zrenner, R., Sonnewald, U. and Stitt, M. (1998) A moderate decrease of plastid aldolase activity inhibits photosynthesis, alters the levels of sugars and starch, and inhibits growth of potato plants. *Plant J.* 142: 147–157.
Haake, V., Geiger, M. and Stitt, M. (1999) Changes in aldolase activity in wildtype potato plants are important for acclimation to growth irradiance and carbon dioxide concentration because plastid aldolase exerts control over ambient photosynthesis across a range of growth conditions. *Plant J.* (in press).
Harrison, E.P., Willingham, N.M., Lloyd, J.C. and Raines, C.A. (1998) Reduced sedoheptulose-1,7-bisphosphatase levels in transgenic tobacco lead to decreased photosynthetic capacity and altered carbohydrate accumulation. *Planta* 204: 27–36.
Hoff, T., Truong, H.-N. and Caboche, M. (1994). The use of mutants and transgenic plants to study nitrate assimilation. *Plant Cell Environ.* 17: 489–506.
Huber, S.C., Bachmann, M. and Huber, J.L. (1996) Post-translational control of nitrate reductase: a role for calcium and 14-3-3 proteins. *Trends Plant Sci.* 1: 432–438.
Kaiser, W.M. and Huber, S.C. (1994) Posttranslational regulation of nitrate reductase in higher plants. *Plant Physiology* 106: 817–821.
Kaiser, W.M. and Huber, S.C. (1997) Correlation between apparent activation state of nitrate reductase (NR), NR hysteresis and degradation of NR protein. *J. Exp. Bot.* 48: 1367–1374.

Koch, K.E. (1996) Carbohydrate-modulated gene expression in plants. *Annu. Rev. Plant Physiol. Plant Mol. Biol.* **47**: 509–540.

Kossmann, J., Sonnewald, U. and Willmitzer, L. (1995) Reduction of the chloroplast fructose-1,6-bisphosphatase in transgenic potato plants impairs photosynthesis and plant growth. *Plant J.* **6**: 637–650.

Kruckeberg, A., Neuhaus, H.E., Feil, R., Gottlieb, L. and Stitt, M. (1989) Reduced-activity mutants of phosphoglucose isomerase in the cytosol and chloroplast of *Clarkia xantiana*. I. Impact on mass action ratios and fluxes to sucrose and starch. *Biochem. J.* **261**: 457–467.

Lauerer, M., Saftic, D., Quick, W.P., Fichtner, K., Schulze, E.-D., Rodermel, S., Bogorad, L. and Stitt, M. (1993) Decreased ribulose-1,5-bisphosphate carboxylase/oxygenase in transgenic tobacco transformed with anitsense rbcS. VI Effects on photosynthesis in plants grown at different irradiance. *Planta* **190**: 332–445.

Matt, P., Schurr, U., Krapp, A. and Stitt, M. (1997) Growth of tobacco in short day conditions leads to high starch, low sugars, altered diurnal changes of the nia transcript and low nitrate reductase activity, and an inhibition of amino acid synthesis. *Planta* **207**: 27–41.

Moorhead, G., Douglas, P., Morrice, N., Scarabel, M., Aitken, A. and MacKintosh, C. (1996) Phosphorylated nitrate reductase from spinach leaves is inhibited by 14-3-3 proteins and activated by fusicoccin. *Curr. Biol.* **6**: 1104–1113.

Neuhaus, H.E. and Stitt, M. (1990) Control analysis of photosynthate partitioning: impact of reduced activity of ADP-glucose pyrophosphorylase or plastid phosphoglucomutase on the fluxes to starch and sucrose in *Arabidopsis thaliana* L. Heynh. *Planta* **182**: 445–454.

Neuhaus, H.E., Kruckeberg, A.L., Feil, R., Gottlieb, L. and Stitt, M. (1989) Reduced-activity mutants of phosphoglucose isomerase in the cytosol and chloroplast of *Clarkia xantiana*. II. Study of the mechanisms which regulate photosynthate partitioning. *Planta* **178**: 110–122.

Newsholme, E.A. and Start, C. (1973) *Regulation in Metabolism*. Wiley and Sons, London.

Paul, M.J., Knight, J.S., Habash, D., Parry, M.A.J., Lawlor, D.W., Barnes, S.A., Loynes, A. and Gray, J.C. (1995) Reduction in phosphoribulokinase activity by antisense RNA in transgenic tobacco: effect on CO_2 assimilation and growth in low irradiance. *Plant J.* **7**: 535–542.

Preiss, J. (1988) Biosynthesis of starch and its regulation. In: *The Biochemistry of Plants*, Vol. 14 (ed. J. Preiss). Academic Press, New York, pp. 181–254.

Price, G.D., Evans, J.R., von Caemmerer, S., Yu, J.-W. and Badger, M.R. (1995) Specific reduction of chloroplast glyceraldehyde-3-phosphate dehydrogenase activity by antisense RNA reduces CO_2 assimilation via a reduction in ribulose bisphosphate regeneration in transgenic tobacco plants. *Planta* **195**: 369–378.

Quick, W.-P., Schurr, U., Scheibe, R., Schulze, E.-D., Rodermel, S.R., Bogorad, L. and Stitt, M. (1991) Decreased ribulose-1,5-bisphosphate carboxylase/oxygenase in transgenic tobacco transformed with 'antisense' rbcS: I. Impact on photosynthesis in ambient growth conditions. *Planta* **183**: 542–554.

Quick, W.-P., Fichtner, K., Schulze, E.-D., Wendler, R., Leegood, R.C., Mooney, H., Rodermel, S.R., Bogorad, L. and Stitt, M. (1992) Decreased ribulose-1,5-bisphosphate carboxylase/oxygenase in transgenic tobacco transformed with 'antisense' rbcS: IV. Impact on photosynthesis in conditions of altered nitrogen supply. *Planta* **188**: 522–531.

Rolleston, F.S. (1972) A theoretical background to the use of measured intermediates in the study of the control of intermediary metabolism. *Curr. Topics Cell. Reg.* **5**: 47–75.

Scheible, W.-R., Lauerer, M., Schulze, E.-D., Caboche, M. and Stitt, M. (1997a) Accumulation of nitrate in the shoot acts as a signal to regulate shoot–root allocation in tobacco. *Plant J.* **11**: 671–691.

Scheible, W.-R., Gonzales-Fontes, A., Lauerer, M., Müller-Röber B., Caboche M. and Stitt, M. (1997b) Nitrate acts as a signal to induce organic acid metabolism and repress starch metabolism in tobacco. *Plant Cell* **9**: 783–789.

Scheible, W.-R., Gonzalez-Fozes, A., Morcuende, R., Lauerer, M., Geiger, M., Glaab, J., Gojon, Schulze, E.-D. and Stitt, M. (1997c) Tobacco mutants with a decreased number of functional nia genes compensate by modifying the diurnal regulation of transcription, post-translational modification and turnover of nitrate reductase. *Planta* **203**: 304–319.

Stitt, M. (1989) Control analysis of photosynthetic sucrose synthesis: assignment of elasticity coefficients to the cytosolic fructose-1,6-bisphosphatase and sucrose phosphate synthase. *Proc. Trans. Phil. Soc. Lond.* **323**: 327–338.

Stitt, M. (1994) Flux control at the level of the pathway: studies with mutants and transgenic plants having decreased activity of enzymes involved in photosynthate partitioning. In: *Flux Control in Biological Systems* (ed. E.-D. Schulze). Academic Press, London, pp. 13–36.

Stitt, M. (1995) The use of transgenic plants to study the regulation of plant carbohydrate metabolism. *Aust. J. Plant Physiol.* **22**: 635–646.

Stitt, M. (1996) Metabolic regulation of photosynthesis. In: *Advances in Photosynthesis*, Vol. 3, *Environmental Stress and Photosynthesis* (ed. N. Baker). Academic Press, London.

Stitt, M. and Schulze, E.-D. (1994) Does Rubisco control the rate of photosynthesis and plant growth? An exercise in molecular ecophysiology. *Plant Cell Environ.* **17**: 465–487.

Stitt, M. and Sonnewald, U. (1995) Regulation of metabolism in transgenic plants. *Annu. Rev. Plant Physiol. Plant Mol. Biol.* **46**: 341–368.

Stitt, M., Quick, W.P., Schurr, U., Schulze, E.-D., Rodermel, S.R. and Bogorad, L. (1991) Decreased ribulose-1,5-bisphosphate carboxylase/oxygenase in transgenic tobacco transformed with 'antisense' rbcS: II. Flux control coefficients for photosynthesis in varying light, CO_2 and air humidity. *Planta* **183**: 555–565.

2

Modelling metabolic pathways and analysing control

David A. Fell, Simon Thomas and Mark G. Poolman

1. Introduction

Why is the control of metabolism so hard to understand? Apart from the experimental difficulties, which are particularly severe in the case of plants, there are significant conceptual difficulties. Metabolism is a network of multiply connected reactions (e.g. *Figure 1*), each of which has complex non-linear kinetics. It is difficult for the human mind to cope intuitively with either the system properties of a network or the non-linearity. It is for this reason that many of the traditional concepts of metabolic control

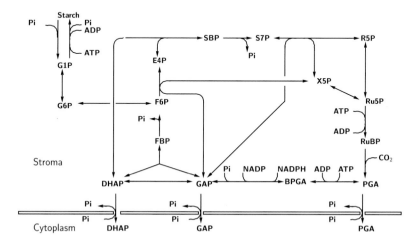

Figure 1. *The Calvin cycle model. The Calvin cycle as an example of a metabolic pathway. Each of the steps shown has an individual rate expression in the kinetic model described in Section 4; full details in Poolman (1998).*

Plant Carbohydrate Biochemistry, edited by J.A. Bryant, M.M. Burrell and N.J. Kruger.
© 1999 BIOS Scientific Publishers Ltd, Oxford.

are either misleading or just wrong (Fell, 1997). As other contributors emphasize, Tom ap Rees was particularly influential (ap Rees and Hill, 1994) in promoting the application of newer concepts of metabolic control, such as Metabolic Control Analysis (Fell, 1997; Kacser and Burns, 1973) to plant biochemistry.

Given these difficulties, the construction of quantitative models of metabolism can assist in a number of ways.

(i)   The process of formulating a model can reveal whether the components of the system have been adequately characterized and whether the available data are complete and consistent.

(ii) The model can be used qualitatively to investigate whether or not the known components and their known interactions can give rise to the types of behaviour that are exhibited by the real system.

(iii) The model can be used quantitatively to determine whether the experimental values for the components' properties can be applied to make accurate predictions of the system's *in vivo* responses.

Some of the types of metabolic model that have potential application in plant metabolism are:

(i)   Structural models. These need no more than a reaction list and explore the consequences of the constraints imposed by the requirement that the formation of each intermediate metabolite is exactly balanced by its utilization in a metabolic steady state. With such a model, it can be determined what pathways, if any, exist through the metabolic network. Then, if routes exist, it is possible to calculate the overall stoichiometries of each metabolic conversion and to find the most efficient route (e.g. Fell and Small, 1986). Whilst this may seem simple, the calculations are far from trivial with large metabolic networks. Importantly, the kinetics and control properties of the reactions only allow selection from the possibilities allowed by these structural constraints. A simple example of the application of these principles is the pathway of starch formation in amyloplasts: none of the models proposed elsewhere in this book (see Chapters 16 and 18) can be complete because they do not allow the amyloplast phosphate to reach steady state; the operation of the pathways suggested would result in continual transfer of phosphate from cytosol to amyloplast.

(ii) Control analysis models. These apply the principles of Metabolic Control Analysis (Fell, 1997) to determine where in the system the control of steady-state metabolic fluxes and concentrations is located, and thereby to predict the response of the steady state to changes in enzyme activities or concentrations of pathway effectors. Apart from the same structural information as the previous class of model, these models require partial kinetic information in the form of elasticities of each step with respect to each metabolite that affects it (see below) in a known steady state. However, because the information is partial, the predictions soon lose validity away from the initial state.

(iii) Kinetic models. These require a full kinetic description of each enzyme or step of the structural model. This means a single equation that describes the effects of all the substrates, the products, the reverse reaction and any effectors, with all the necessary parameters such as the limiting rate and K_m values obtained under close to *in vivo* conditions for the cell type under consideration. The pay-off for the larger information requirement is the ability to predict both time courses and steady states, as well as the control distribution, for a wide range of different conditions.

There are limited examples of the use of structural models in plant metabolism at the moment, though this is likely to change since they could have application in the functional interpretation of genomic information and the analysis of the effects of knockout mutations. Therefore in the rest of this chapter we shall give examples of applications of the other two types of model.

2. Theoretical

A number of concepts of Metabolic Control Analysis (Kacser and Burns, 1973) are used in this chapter; a fuller explanation of the underlying principles can be found in Fell (1997). The following gives an abbreviated description of the terms used.

(i) The *flux control coefficient* C^J_{xase} can be thought of as the percentage change in the steady-state flux J through a metabolic pathway for a 1% change in the activity of the enzyme *xase*. Its value varies as the metabolic state changes.

(ii) Similarly the *concentration control coeffcient* C^M_{xase} is approximately the percentage change in the concentration of metabolite M for a 1% change in the activity of enzyme *xase*.

(iii) The *summation theorem* for flux control coefficients states that the sum of the flux control coefficients of all the enzymes that affect a given flux is 1 exactly.

(iv) The *elasticity* ϵ^{xase}_M of an enzyme *xase* with respect to a metabolite M, which could be a substrate,product or effector, is the percentage change in the rate of the enzyme caused by a 1% change in the concentration of M with all other substrates etc. kept at constant concentrations. The value varies with the concentration of metabolite M, but the value of particular interest in the control analysis of a working pathway is that with the concentrations of all the metabolites equal to their steady-state values in the *in vivo* environment. It represents a measure of the enzyme's kinetic response.

(v) The *connectivity theorem* is a relationship that applies at steady state for each of the metabolites in the pathway. For a given metabolite M, it links the elasticities and control coefficients of all the enzymes that respond to the metabolite. Its importance here is that the set of connectivity theorem relationships for a pathway, plus the summation theorem, allow the values of the control coefficients to be determined from the elasticities (Fell, 1997; Fell and Sauro, 1985; Sauro *et al.*, 1987). That is, the control properties of the whole pathway can be related to the kinetic characteristics of the enzymes.

3. A control analysis model of potato tuber glycolysis

An example of a control analysis model is given by our study of glycolysis in potato tubers (Thomas *et al.*, 1997a,b), in collaboration with Burrell and Mooney of Advanced Technologies (Cambridge) Ltd. This work had its origins in studies by ap Rees and his group relating to the cold sweetening of potatoes on storage. They had noted that one factor in this process could be that the glycolytic enzyme phosphofructokinase was cold-labile (Dixon *et al.*, 1981); there was therefore the possibility that hexoses derived from starch breakdown on storage could be diverted into sucrose synthesis because of a reduction in their consumption by glycolysis. Burrell *et al.* (1994) therefore genetically manipulated potatoes to over-express phosphofructoki- nase (PFK) up to 30 times wild-type levels in the tubers in order to reduce sucrose lev- els. This gave a series of plants from which it was also possible to determine the flux control coefficient of PFK on glycolysis, but since the glycolytic flux remained unchanged, it appeared the control coefficient was close to zero. (Note that this is not necessarily inconsistent with a lower rate of glycolysis on loss of the enzyme, since the effects of an increase and a decrease in activity are not symmetrical.) The issue we addressed with our model was whether this finding of a near-zero flux control coeffi- cient for PFK was consistent with the known properties of the glycolytic enzymes.

3.1 *Model construction*

In order to build our model (Thomas *et al.*, 1997b) of glycolysis from the hexose phosphates to pyruvate [with our 'pyruvate kinase' representing the catabolism of phosphoenolpyruvate (PEP)], we needed expressions for the elasticities of each of the potato tuber enzymes with respect to the glycolytic intermediates. For several of the enzymes, we derived expressions for the elasticities from the enzyme rate laws (cf. Fell, 1997) and then inserted kinetic parameters reported in the literature and the metabolite concentrations measured in the wild-type tubers by Burrell *et al.* (1994). For example, for the elasticity of PFK with respect to PEP, we obtained:

$$
\epsilon^{PFK}_{PEP} = \frac{-\dfrac{nL[PEP]}{K_{PEP}}\left(1 + \dfrac{[PEP]}{K_{PEP}}\right)^{n-1}}{L\left(1 + \dfrac{[PEP]}{K_{PEP}}\right)^{n} + \left(1 + \dfrac{[F6P]}{K_{F6P}}\right)^{n}}
$$

where the parameters were obtained by non-linear fitting of experiments reported by Sasaki *et al.* (1973). We lacked reliable expressions for the elasticities for some of the near-equilibrium enzymes, but were able to make approximate estimates from the dis- placement of the reactions from equilibrium. We then used the program METACON (Thomas and Fell, 1993) to calculate the control coefficients from the elasticities, essentially by using the summation and connectivity theorem equations as described above (Section 2).

3.2 *Control coefficients*

Calculation with the model (*Figure 2*) showed that the flux control coefficient of PFK was expected to be small. However, it was important to check whether this conclusion relied strongly on some of those elasticities of whose values we were less certain. The program METACON can also calculate the dependence of each of the control coeffi- cients on each of the elasticities. For the two key results, the values of the flux control coefficients of PFK and pyruvate kinase, there are two elasticities that together have by far the largest impact: the elasticities of the two enzymes with respect to PEP ϵ^{PFK}_{PEP} and ϵ^{PK}_{PEP}). *Figure 3* shows how the calculated control coefficients depend on the value

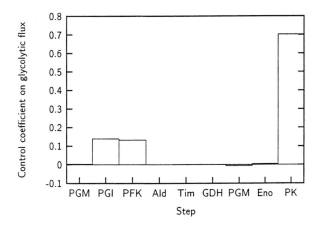

Figure 2. *Predicted flux control coefficients of glycolytic enzymes in potato tubers. Abbreviations: PGM, phosphoglucomutase; PGI, glucose-6-phosphate isomerase; PFK, phosphofructokinase; Ald, aldolase; Tim, triose phosphate isomerase; GDH, glyceraldehyde-3-phosphate dehydrogenase + phosphoglycerate kinase; PGM, phosphoglycerate mutase; Eno, enolase; PK, pyruvate kinase and subsequent oxidation of pyruvate.*

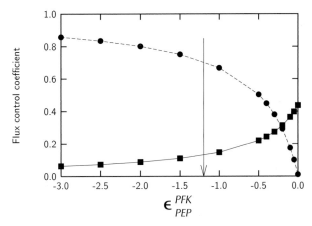

Figure 3. *Sensitivity of flux control coefficients to ϵ_{PEP}^{PFK}. ■: phosphofructokinase; ●: pyruvate kinase. The vertical arrow shows the value estimated for* in vivo *conditions.*

assumed for the elasticity of PFK, ϵ_{PEP}^{PFK}. It is clear that the value would have to be very different from our calculated estimate before there would be a significant effect on the control distribution. The same conclusion also applies for the other elasticity, ϵ_{PEP}^{PK}. Another implication of this result is that it does not matter very much that we did not know all of the other elasticities very accurately; approximate values are good enough for determining the flux control coefficients.

The result was different for the concentrations of the glycolytic intermediates. Even though changing the level of PFK in the tubers did not affect the glycolytic flux, it did have a noticeable effect on the metabolite concentrations of metabolites downstream of PFK that increased progressively with the increase in enzyme activity (Burrell *et al.*,

1994). This in itself is interesting because it is a typical instance of the consequences of changing a single enzyme activity in a pathway: the effects on the metabolites are more pronounced than the effects on fluxes. Since physiologically induced flux changes typically involve the reverse, larger changes in flux than metabolites, it is clear that *in vivo* flux changes are unlikely to be brought about by the control of single enzymes (Thomas and Fell, 1996). For the purposes of the control analysis model, the importance is that the same model also predicts the values of the concentration control coefficients of the enzymes. The concentration control coefficients of PFK indicate the relative size and direction of the changes in metabolite concentrations expected to occur when PFK is increased. Combined with the 'finite change analysis' of Small and Kacser (1993), these values enable the actual concentration changes to be predicted for particular transgenic plants. When we calculated these expected effects in the transgenic potatoes (Thomas *et al.*, 1997b), there was general qualitative agreement with the observations (Thomas *et al.*, 1997a), which lends support to the validity of the model's results for the flux control distribution. The predicted concentration control coefficents do however show more sensitivity to some of the elasticities of which we were less certain, in particular the product feedback elasticity of fructose-1,6-bisphosphate on PFK.

3.3 *Conclusion*

In conclusion, what does the model show? Firstly, it explains the experimental finding that increasing the activity of PFK has no effect on glycolytic flux because its flux control coefficient is low. The reason for this is that the enzyme is subject to strong feedback inhibition by PEP, which has the effect of transferring control of flux to the steps that utilize this metabolite (pyruvate kinase and subsequent steps in this model) as can be seen in *Figure 3*. This is the characteristic effect of feedback inhibition, as originally pointed out by Kacser and Burns (1973). Secondly, it shows how control analysis models can produce valid conclusions without exact numerical values for every molecular interaction in the system.

4. A kinetic model of the Calvin cycle

There have been several kinetic models (e.g. Giersch *et al.*, 1990; Hahn, 1991; Laisk *et al.*, 1989; Pettersson and Ryde-Pettersson, 1988; Woodrow, 1986) of various aspects of photosynthesis, implemented in a variety of ways, but we shall illustrate the simulation of the Calvin cycle using a generic biochemical simulation package, SCAMP (Sauro, 1993). Whereas most of the previous models have simplified the topology of the cycle in order to simplify the subsequent analysis, we have individually represented (Poolman, 1998) each of the reactions shown in *Figure 1*. The starting point for this model was that of Pettersson and Ryde-Pettersson (1988), which was the most complete of those cited. However, these authors had assumed that certain of the reactions, and even blocks of reactions, could be assumed to be at equilibrium so that the system was reduced to a set of six reactions from a total of 20, with the other interconversions effectively being infinitely fast. We did not wish to simplify the system in this way unless we could show that it did not affect the model behaviour; in addition, for most of the simulations we included the starch phosphorylase reaction.

For a kinetic model there needs to be a kinetic equation for each reaction represented. For example, for Rubisco this was:

$$v = \frac{V_{\text{Rbco}}\text{RuBP}}{\text{RuBP} + K_{\text{RuBP}}\left(1 + \dfrac{\text{PGA}}{K_{\text{iPGA}}} + \dfrac{\text{FBP}}{K_{\text{iFBP}}} + \dfrac{\text{SBP}}{K_{\text{iSBP}}} + \dfrac{P_i}{K_{\text{iP}_i}} + \dfrac{\text{NADPH}}{K_{\text{iNADPH}}}\right)}.$$

Note that this equation contains a product inhibition term for PGA as well as several other intermediates. There is no term for CO_2 since this is not a variable of the model and it is included in the calculation of the value of the apparent limiting rate, V_{Rbco}. For incorporation into the kinetic simulation, this is entered into a command file for the simulation package SCAMP. The first few lines of this file, including the entry for Rubisco are:

```
Title Calvin Cycle ;

Simulate ;

dec ATP_ch, ADP_ch, RuBP_ch, PGA_ch, BPGA_ch, . . . . . ;

reactions

[Rubisco]
$CO2 + RuBP_ch = 2PGA_ch ;
CO2 * Rbco_vm * RuBP_ch/((Rbco_kmCO2 + CO2) *
(RuBP_ch + Rbco_km * (1 +
PGA_ch/Rbco_KiPGA +
FBP_ch/Rbco_KiFBP +
SBP_ch/Rbco_KiSBP +
Pi_ch /Rbco_KiPi +
NADPH_ch/Rbco_KiNADPH ))) ;
```

A kinetic model actually consists of a set of differential equations for the net rate of change of each metabolite, composed of the rates of production or consumption by each reaction in which it is involved. However, the mechanical procedure of relating the metabolic model to the set of differential equations is performed automatically by SCAMP, which also detects and takes into account the mass conservation of the entities that are not synthesized or degraded by the reactions in the model (such as total phosphate and the adenosine moiety of the adenine nucleotides). In this case, the completed model contains 21 reactions (including the exchange reactions between the chloroplast and the cytosol), 16 independent metabolite concentrations, two dependent metabolite concentrations that are derived from the others via the equations for mass conservation and over 90 parameters. Even so, models such as this can be rapidly simulated without any difficulty with an average modern desktop computer using appropriate numerical integration routines such as those included in biochemical simulators such as SCAMP (Sauro, 1993) and Gepasi (Mendes, 1993).

This model was used to calculate the two different types of solution: time courses and steady states.

4.1 *Time course simulations*

A kinetic model can be used to simulate the time courses of events by introducing a perturbation to one of the external conditions at a specified time (one of the operations easily specified in the simulation package SCAMP) and recording the concentrations of the metabolites and the reaction rates at successive time intervals after this. One of the most interesting parameters to perturb is the light intensity (represented in the model by the rate of ATP generation). In certain ranges, the model can settle into either a low or a high rate of carbon fixation; which it adopts depends on the state it is currently in. For example, if the light intensity is increased slowly from low levels, the carbon assimilation flux changes little until a critical value when it switches suddenly to a higher flux. Once in the higher flux state, it remains in that state as the light intensity is reduced until at another critical value (at a lower light intensity than the upward transition) it returns to the low flux state. In the vicinity of the region where these alternate flux states coexist, changes in light intensity initiate oscillations in the model with a period of about 10 s (*Figure 4*). The period is somewhat faster than the observed oscillations in plants *in vivo*, but no adjustments of the model have been made in order to generate these oscillations.

4.2 *Steady-state simulations*

A special case of a kinetic model is the steady-state solution, where the rates of change of all the metabolites have become zero. Instead of the problem of integrating a set of differential equations, the mathematical problem becomes one of finding a solution for a set of simultaneous non-linear equations. The SCAMP package offers this as an alternative to integration, so the steady-state solutions can be obtained for the same model used for the time courses. It is true that a kinetic model can be integrated until no further changes are apparent, but this has dangers; if there are processes with very different intrinsic rates, the user might erroneously think steady state has been reached by failing to notice a very slow change. In any case another advantage of computing steady-state solutions is that it is easier to automate accurate numerical calculation of the control coefficients, a facility offered by SCAMP, and certain other simulators such as Gepasi (Mendes, 1993).

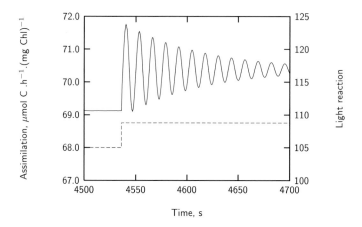

Figure 4. *Dynamics of a response to a change in illumination. Part of a simulation in which light intensity was decreased then increased in discrete steps. Solid line: rate of carbon assimilation. Dashed line: light intensity, arbitrary units.*

An important determinant of whether chloroplasts store carbon or export it to the cytoplasm is the concentration of cytosolic phosphate (Heldt *et al.*, 1977). This was investigated by determining the steady state of the model at a range of phosphate concentrations (*Figure 5*). As can be seen, the carbon assimilation flux hardly responds at all to a change in phosphate, but the partitioning between starch and export to the cytosol is much more sensitive. At low phosphate concentrations, most of the carbon is retained in the chloroplast as shown by the flux into starch being larger than the export flux. As the phosphate level increases, more of the carbon is exported at the expense of starch synthesis. Eventually, at a concentration that is very sensitive to the model parameters, the starch synthesis flux reaches zero. If starch phosphorylase is not included in the model, it breaks down at this point because all the Calvin cycle intermediates are exported and assimilation ceases. With starch breakdown possible, export to the cytosol starts to exceed the assimilation flux, supported by breakdown of stored starch, as shown by the negative value of the starch synthesis flux.

As an illustration of the use of kinetic models to determine the distribution of control in a metabolic system, we will use our results on the calculation of the flux control coefficients of the Calvin cycle enzymes. For the carbon assimilation flux, there are only two enzymes with significant amounts of control: Rubisco and SBPase. The prediction of the exact distribution of control between these enzymes depends, amongst other things, on their relative activities, for which we do not have sufficiently precise data. Therefore in *Figure 6* we have plotted the two control coefficients as a function of Rubisco activity. There are reasons to think that in tobacco at least, the activity of Rubisco will be close to the point where the two control coefficients cross over. Thus when Stitt *et al.* (1991) estimated the flux control coefficient of Rubisco in tobacco by making transgenic plants with decreased enzyme content, they obtained flux control coefficients between 0.1 and 0.8 depending on conditions. Similarly, Harrison *et al.* (1998) have found that transgenic tobacco with slightly reduced contents of SBPase showed reduced carbon assimilation, consistent with the enzyme having a significant flux control coefficient.

The model also illustrates how the distribution of control is very different depending

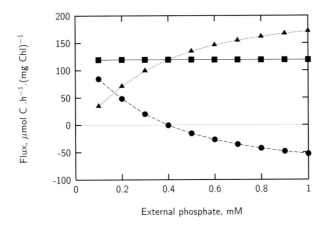

Figure 5. Steady-state flux responses to external phosphate. ■: assimilation flux; ●: starch synthesis flux (degradation when negative); ▲: export flux.

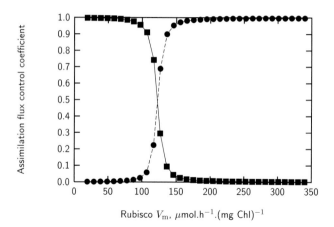

Figure 6. *Control of assimilation flux.* ■ *: Rubisco;* ●*: SBPase.*

on which flux in photosynthesis is considered. In *Figure 7* we show the corresponding result for the distribution of control when the flux under consideration is the export of carbon to the cytosol. Now the triose phosphate translocator, and to a lesser extent GAPDH also have significant control coefficients. Since the translocator has no control over the assimilation flux, it can only affect the export flux by altering the partitioning of carbon between starch synthesis and export to the cytosol. Again, this is consistent with the evidence from transgenic plants with reduced translocator activity (in this case potatoes, Reismeier *et al.*, 1993; see also Chapter 17).

5. Conclusion

These case studies show that in spite of the lack of detailed knowledge of plant bio-chemistry, there is already sufficient information available for generating models that

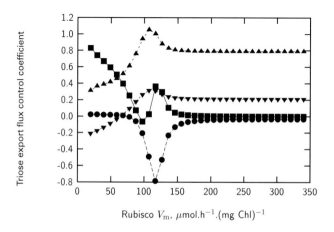

Figure 7. *Control of export flux.* ■ *: Rubisco;* ●*: SBPase;* ▲ *: triose phosphate translocator;* ▼ *: chloroplast GAPDH.*

exhibit physiologically realistic properties. Such models can be used to investigate, in advance, the potential metabolic consequences of manipulating the expression of a plant's genes. There is therefore hope that despite the complexity of metabolic networks and the kinetics of the enzymes, it will in future prove possible to design alterations to plant metabolism rationally.

References

ap Rees, T. and Hill, S.A. (1994) Metabolic control analysis of plant metabolism. *Plant Cell Environ.* **17**: 587–599.

Burrell, M.M., Mooney, P.J., Blundy, M., Carter, D., Wilson, F., Green, J., Blundy, K.S. and ap Rees, T. (1994) Genetic manipulation of 6-phosphofructokinase in potato tubers. *Planta* **194**: 95–101.

Dixon, W.L., Franks, F. and ap Rees, T. (1981) Cold-lability of phosphofructokinase from potato tubers. *Phytochemistry* **20**: 969–972.

Fell, D.A. (1997) *Understanding the Control of Metabolism.* Portland Press, London.

Fell, D.A. and Sauro, H.M. (1985) Metabolic Control Analysis: additional relationships between elasticities and control coefficients. *Eur. J. Biochem.* **148**: 555–561.

Fell, D.A. and Small, J.R. (1986) Fat synthesis in adipose tissue: an examination of stoichiometric constraints. *Biochem. J.* **238**: 781–786.

Giersch, C., Lammel, D. and Farquhar, G. (1990) Control analysis of photosynthetic CO_2 fixation. *Photosynthesis Res.* **24**: 151–165.

Hahn, B. (1991) Photosynthesis and photorespiration: modelling the essentials. *J. Theor. Biol.* **151**: 123–139.

Harrison, E.P., Willingham, N.M., Lloyd, J.C. and Raines, C.A. (1998) Reduced sedoheptulose-1,7-bisphosphatase levels in transgenic tobacco lead to decreased photosynthetic capacity and altered carbohydrate accumulation. *Planta* **204**: 27–36.

Heldt, H., Chon, J., Maronde, D., Herold, A., Stankovic, Z., Walker, D., Kraminer, A., Kirk, M. and Heber, U. (1977) Role of orthophosphate and other factors in the regulation of starch formation in leaves and isolated chloroplasts. *Plant. Physiol.* **59**: 1146–1155.

Kacser, H. and Burns, J.A. (1973) The control of flux. *Symp. Soc. Exp. Biol.* **27**: 65–104. Reprinted in *Biochem. Soc. Trans.* **23**: 341–366, 1995.

Laisk, A., Eichelmann, H.V.O., Eatherall, A. and Walker, D. (1989) A mathematical model of the carbon metabolism in photosynthesis – difficulties in explaining oscillations by fructose 2,6-bisphosphate regulation. *Proc. R. Soc. Lond. Ser. B-Biol. Sci* **237**: 389–415.

Mendes, P. (1993) Gepasi – a software package for modeling the dynamics, steady-states and control of biochemical and other systems. *Comput. Applic. Biosci.* **9**: 563–571.

Pettersson, G. and Ryde-Pettersson, U. (1988) A mathematical model of the Calvin photosynthesis cycle. *Eur. J. Biochem.* **175**: 661–672.

Poolman, M. G. (1998) Ph.D. thesis, Oxford Brookes University.

Reismeier, J.W., Flugge, U., Schulz, B., Heineke, D., Heldt, H.W., Willmetzer, L. and Fromner, W.B. (1993) Antisense repression of the chloroplast triose phosphate translocator affects carbon partitioning in transgenic potato plants. *Proc. Natl Acad. Sci. USA* **90**: 6160–6164.

Sasaki, T., Tadokoro, K. and Suzuki, S. (1973) Phosphofructokinase of *Solanum tuberosum* tuber. *Phytochemistry* **12**: 2843–2849.

Sauro, H.M. (1993) SCAMP: a general-purpose simulator and metabolic control analysis program. *Comput. Applic. Biosci.* **9**: 441–450.

Sauro, H.M., Small, J.R. and Fell, D.A. (1987) Metabolic control and its analysis: extensions to the theory and matrix method. *Eur. J. Biochem.* **165**: 215–221.

Small, J.R. and Kacser, H. (1993) Responses of metabolic systems to large changes in enzyme activities and effectors. 2. The linear treatment of branched pathways and metabolite concentrations. *Eur. J. Biochem*. **213**: 625–640.

Stitt, M., Quick, W.P., Schurr, U., Schulze, E.-D., Rodermel, S.R. and Bogorad, L. (1991) Decreased ribulose-1,5-bisphosphate carboxylase-oxygenase in transgenic tobacco transformed with 'antisense' *rbcs*: II flux control coefficients for photosynthesis in varying light, CO_2 and air humidity. *Planta* **183**: 555–566.

Thomas, S. and Fell, D.A. (1993) A computer program for the algebraic determination of control coefficients in Metabolic Control Analysis. *Biochem. J*. **292**: 351–360.

Thomas, S. and Fell, D.A. (1996) Design of metabolic control for large flux changes. *J. Theor. Biol*. **182**: 285–298.

Thomas, S., Mooney, P.J.F., Burrell, M.M. and Fell, D.A. (1997a) Finite change analysis of lines of transgenic potato (*Solanum tuberosum*) overexpressing phosphofructokinase. *Biochem. J*. **322**: 111–117.

Thomas, S., Mooney, P.J.F., Burrell, M.M. and Fell, D.A. (1997b) Metabolic control analysis of glycolysis in tuber tissue of potato (*Solanum tuberosum*). Explanation for the low control coefficient of phosphofructokinase over respiratory flux. *Biochem. J*. **322**: 119–127.

Woodrow, I. (1986) Control of the rate of photosynthetic carbon dioxide fixation. *Biochim. Biophys. Acta* 851: 181–192.

3

Carbohydrate: where does it come from, where does it go?

John Farrar

1. Introduction

Tom ap Rees' work on carbohydrate metabolism and enzymology was characterized by rigour of both method and thought. He did not simply consider each enzyme solely as a discrete entity, the experimenter's task done when it is isolated and characterized. Two of his interests show that he was curious about the flux of carbon catalysed by an enzyme or sequence of enzymes. One was his early attempt to use the C1: C6 ratio to compare fluxes through glycolysis and the pentose phosphate pathway (ap Rees, 1980), and the other his much more recent involvement with metabolic control analysis (ap Rees and Hill, 1994).

In this chapter I wish to place the study of non-structural carbohydrates in a whole-plant context. First, I will show that they are usually turned over rapidly, since the fluxes through even relatively large pools is high. Then I will ask: what controls the flux? Is it demand for the products of carbohydrate metabolism, or is it the amount of available carbohydrate? I will argue that respiration is an ideal tool for helping answer such questions. Finally, I will take one example of current interest where carbohydrate fluxes are altered – plants growing in elevated atmospheric CO_2 – and use this example to illustrate some of the questions raised.

Although Tom ap Rees did consider fluxes in a whole plant context – for example, his early assay into the unloading of phloem in roots (Dick and ap Rees, 1975) – he always seemed happier working with relatively well-defined pathways such as glycolysis, where at least there was a reasonable hope that rigour could be applied and would be rewarded with an understanding rooted in certainty. In consciously sacrificing this rigour by moving into the complexities of the whole plant, there are both dangers and rewards. The dangers are obvious: over-simplification, ignorance and simple error. The rewards are the building of bridges between the metabolic and molecular on the one hand, and the whole-plant, crop and ecological levels on the other.

I also wish to address implicitly the importance of post-photosynthetic processes. Not only does respiration cause the loss of half the carbon gained in photosynthesis, but partitioning determines the amount of photosynthetic surface, and the rate of carbon

Plant Carbohydrate Biochemistry, edited by J.A. Bryant, M.M. Burrell and N.J. Kruger.
© 1999 BIOS Scientific Publishers Ltd, Oxford.

fixation in photosynthesis is partly controlled by post-photosynthetic processes (Pollock and Farrar, 1996); the rate of growth is determined by each of these and by the amount of assimilate which is stored.

2. Production of carbohydrate

The primary source of non-structural carbohydrate is photosynthesis; indeed, the main product of photosynthesis is carbohydrate. The main products of photosynthesis in source leaves are sucrose, starch and fructan; starch and fructan tend to dominate in different species. A wide variety of different types of non-structural carbohydrate including sugar alcohols are produced by distinct species (Lewis, 1984). In general, the pathways of synthesis are well known and, with the notable exception of fructan, the enzymes are well-characterized. In the biochemical sense, the regulation of carbohydrate production is therefore well understood.

In a whole-plant context, our understanding is poorer. For example, it is clear that the rate of photosynthesis is dependent on internal factors such as the amount and activity of sink tissue. There is a clear hypothesis of how photosynthetic rate may be regulated by metabolism in sinks (Jang and Sheen, 1997; Koch, 1997; Pollock and Farrar, 1996) but the mechanism is understood only partially and provisionally. Further, this hypothesis concerns the regulation of rate of photosynthesis per unit of leaf material. The amount of leaf material relative to total amount of plant is also regulated, in response to both environmental and internal factors (Farrar and Gunn, 1998; Wilson, 1987), and will obviously have a major influence on whole-plant photosynthesis. A complete theory of photosynthetic regulation will therefore integrate the partitioning that underlies leaf area per plant with the mechanisms which control density of photosynthetic machinery per unit area of leaf. We do not have such a theory currently (Farrar, 1999).

3. Turnover of non-structural carbohydrate

Some non-structural carbohydrate, although ultimately produced from photosynthesis, is formed indirectly after translocation and metabolism of assimilate. Examples include synthesis of starch and fructan in storage organs, and the formation of starch in fibrous roots. Such pools, along with non-structural carbohydrates in source leaves, are subject to turnover. The bulk of what we know about turnover has come from the use of isotopes, often accompanied by some form of compartmental analysis (Farrar, 1989, 1992). The main outcome is an appreciation of the rapidity of flux through, and turnover of, major pools of non-structural carbohydrate. In many cases, flux is high relative to pool size, and so for sucrose in source leaves and growing sinks, relatively small changes of fluxes into or out of a pool as a result of internal or environmental changes will rapidly alter the size of the pool. Rates of turnover can be high: in source leaves and fibrous roots of barley, cytosolic sucrose turns over every 1–3 h (Farrar, 1989). Even starch turns over relatively quickly: the bulk of starch produced in source leaves during one photoperiod is remobilized during the following night. The traditional methods of mass balance and isotope techniques have recently been supplemented by magnetic resonance imaging, which has been used to produce direct estimates of flux in sieve tubes. Clearly, the study of carbohydrates is a study of compounds in flux.

Perhaps the main gap in our knowledge is of events at the level of the single cell. We now appreciate the diversity of metabolism at the single cell level, including the diversity of carbohydrates (Koroleva *et al.*, 1997, 1998). Techniques for measuring fluxes at this level are not yet available, but may preface exciting increases in understanding of the regulation of flux. Indeed, many advances in understanding of control may await the resolution at the cell level of a wider range of processes.

A single species often stores three different types of carbohydrate in a single organ (e.g. many cereals store sucrose, fructan and starch; Farrar, 1989). A possible explanation for such profligacy lies in turnover since different turnover times for multiple types of total non-structural carbohydrate (TNC) would provide a flexible means of regulating carbon flux. In barley roots starved of carbohydrate, soluble sugars are depleted before fructan or starch (Bingham and Farrar, 1988). Similarly in leaves of barley in extended darkness, the half times of sucrose, fructan and starch are 9, 16 and 19 h, respectively, providing a rapid initial loss of TNC followed by a much slower one. Since fructan is more readily mobilized than starch at low temperatures (Farrar, 1988), the plant storing both can mobilize TNC over a wider a range of temperatures than if it stored just one.

4. Flux control: storage versus export in a source leaf

If carbohydrate is mainly in flux, understanding carbohydrate metabolism becomes largely a question of understanding how flux is controlled. I will illustrate the nature of the problem by taking an apparently simple example, and ask what determines whether sucrose is stored in or exported from a source leaf. There are a number of hypotheses in the literature about what controls export: I will deal with them in turn.

The simplest hypothesis is that export is directly related to photosynthesis. This is not true: Ho (1978) showed elegantly that in tomato leaves the rate of export is less variable than that of photosynthesis, being buffered by the storage or mobilization of carbohydrates, both starch and sucrose. A related hypothesis is that sucrose export is a function of sucrose synthesis, itself determined by the activity of the enzyme sucrose phosphate synthase (SPS), since activities of SPS and rates of export correlate well (Huber, 1981); further SPS is allosterically regulated (G6P +, P_i −) (G6P, glucose-6-phosphate; P_i, inorganic phosphate) in a way that makes it a good potential regulator. This hypothesis does not explain the independence of export from rate of sucrose synthesis demonstrated by Ho (1978), and also has been discredited by the use of genetic variants. Maize lines have a positive correlation between genetic difference in SPS activity and growth, but do not demonstrate a simple relationship between SPS and export (Rocher *et al.*, 1989). Transgenic plants over-expressing SPS provide more specific evidence: tomato over-expressing SPS up to six-fold partitions relatively more assimilate to sucrose than to starch, and shoot:root ratio is changed to favour the shoot – but export is not affected and overall growth is unaltered (Galtier *et al.*, 1995). Similarly export is unaffected in potato under-expressing SPS: partitioning favours starch rather than sucrose, and shoot:root ratio and growth are unaltered (Ferrario-Mery *et al.*, 1997; Geigenberger *et al.*, 1995). Although we know a great deal about the transporters which load sucrose into phloem in plants that load phloem apoplastically (Buckhout and Tubbe, 1996; Rentsch and Frommer, 1996) we do not know if the number or activity of these transporters regulates phloem loading and thus transport *in vivo*. Plants which are antisense for a sucrose transporter export less sucrose from source leaves (Riesmeier *et al.*, 1994). However, this only demonstrates the use of that particular transporter and not that it has a role in control. In sum, hypotheses based solely on the amount of metabolic machinery in the leaf do not seem to be valid.

Another suite of hypotheses is based on the idea that the sugar status of the source leaf determines export: effectively, these hypotheses assume the machinery is adequate and that rate-control is exerted by first-order kinetics. The earliest hypotheses of this type suggested that the total non-structural carbohydrate content drove export (Ho and Thornley, 1978; Swanson and Christy, 1976). Although correlations were found, their slopes were different in light and dark (Ho and Thornley, 1978), suggesting they were not solely causal. Most importantly, these hypotheses ignored compartmentation of sugars, weighting equally starch in a mesophyll cell remote from the phloem, and sucrose in phloem parenchyma. Better are attempts to define just that pool of sucrose which is readily available for transport, using compartmental analysis (Farrar, 1989; Grantz and Farrar, 1999; *Figure 1*). There is a hyperbolic relationship between export rate and the size of a pool of sucrose which is readily transported (this pool is about 20% of the total sucrose in a barley leaf) (Farrar, 1989; *Figure 1*). A K_m for sucrose entry to phloem of 7.5 mM can be derived from this relationship; this compares with the K_m for one transporter of 0.5–2 mM (Buckhout and Tubbe, 1996).

Whether based on machinery or substrate, the hypotheses considered so far share a major flaw: they aim to explain export without considering the mechanism by which phloem operates. Currently this is believed to be pressure flow. A key feature of pressure flow is that it is driven by gradients of turgor pressure between source and sink. The turgor pressure within phloem in a source leaf is the difference between the concentration of solutes within the phloem and in the apoplast surrounding it. Whilst sucrose is a major solute within phloem, it is not the only one, and may account for about 52% of the osmotic pressure in *Ricinus* (Milburn and Baker, 1989). In sum, the consequence of phloem operating by pressure flow is that neither machinery for

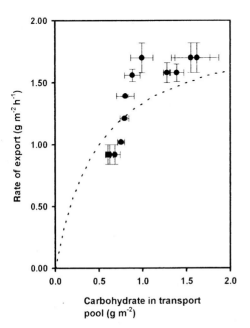

Figure 1. Export of sucrose from source leaves of barley as a function of leaf sucrose status. Export was estimated as the difference between photosynthesis and carbohydrate accumulation; the size of the transport pool of sucrose was estimated by compartmental analysis after feeding $^{14}CO_2$. Second leaves of barley; different rates of export were the result of varying atmospheric CO_2 concentration. Unpublished data of B.E. Collis, C.J. Pollock and J.F. Farrar.

sucrose production, nor the amount of sugar in the leaf in any pool, would be expected to determine the rate of export. Rather, the rate must be a function of several factors in source leaf and in sink which jointly determine turgor gradients in the phloem. Export from the source leaf becomes a whole-plant property, not defined by the source leaf alone. Sadly, we do not yet know the relationships between sucrose synthesis, its concentration, its transport across the leaf, and its loading into the phloem, which along with the loading of other solutes into the phloem and the concentration of solutes in the phloem apoplast together determine phloem turgor in the source leaf.

The conclusion to the question, what determines flux of sucrose from a source leaf, is that it is a whole-plant property in which the metabolism within the source leaf is a major, but not the only, controller. The reciprocal question – what determines flux to storage within a source leaf? – must therefore have a formally similar answer. It cannot only be the activities of enzymes and transporters within the source leaf which regulate storage. Storage is one branch of the pathway describing the fate of assimilate, and export the other; both simple considerations of substrate availability, and of metabolic control analysis (see below) mean that the flux down one branch must depend on the properties of the other. Even such apparently independent processes as loading of sucrose into the vacuole or synthesis of fructan must proceed at a rate which is in part dependent on processes occurring in remote parts of the plant. The converse – that export is only of sugars left over after the processes of storage have occurred – is rendered improbable by the demonstration of the buffering of export rate by synthesis and mobilization of stored compounds by Ho (1978).

5. The fate of newly synthesized and of remobilized carbohydrate

Catabolism of carbohydrate results in two main fates for the carbon skeletons: incorporation into newly synthesized molecules, or loss as respiratory CO_2. These two fates are precisely related, since the respiratory cost of synthesis is met by simultaneous catabolism. Together, they account for the bulk of carbohydrate which enters growing sinks (in source leaves and storage sinks, storage of TNC can be a major fate of assimilate). Typically, as a sucrose molecule is degraded, about half of its carbon ends up in new metabolites and half is respired. Considerable effort by whole-plant physiologists has established the quantitative relationships between the acquisition of carbon in photosynthesis and its use in growth and respiration (Amthor, 1989). We are thus provided with a powerful tool to probe the processes and control of plant growth: if we can understand the regulation of respiration, which can be easily measured non-destructively, we might be closer to understanding the control of the metabolism which underlies it and of growth itself. I next ask if our understanding of respiration is yet good enough.

6. Respiration: a window on metabolism

There are two ways of looking at respiration: as a series of enzyme-catalysed reactions, or as a major component of the carbon flux of a whole plant. These two views are not mutually exclusive: indeed there is a real opportunity in the way the one illuminates the other. I argue, first that control of respiration is distributed amongst its component processes, then that in the short term the bulk of control is exercised through adenylate turnover; then I will explore the role of supply of substrate in long-term regulation.

6.1 *Flux control of respiration: described by metabolic control analysis*

As described in Chapters 1 and 2, metabolic control analysis is a theory of control of a sequence of reactions in which it is implicit that control is distributed between all steps on a pathway, and suggests the type of experiments which ascribe to individual reactions in a pathway the degree of control exercised by each (ap Rees and Hill, 1994; Cornish-Bowden and Cardenas, 1990; Fell, 1997). The values for the degree of control of each reaction are only true for the set of conditions under which the experiments were performed. With the insight that control analysis provides, questions such as 'is respiration controlled by adenylate turnover or by supply of substrate?' are seen to be of little use. Rather, since both will always contribute to control, we should ask 'what is the degree of control due to each?' and it is clear that any experiment which seeks to claim control for one component of respiration alone must be very carefully interpreted. For example, if adenylate turnover was regulating respiration in part, an experiment which directly relieved that limitation would increase respiration – but would demonstrate merely that adenylate was exercising some, but not necessarily all, control over respiration rate. For glycolysis itself, control in animal systems may be mainly centred on hexokinase and phosphofructokinase (PFK) (Fell, 1997). Gratifyingly, for whole hepatocytes respiratory control is largely with adenylate turnover (Fell, 1997).

Metabolic control analysis can therefore provide a description of how control is distributed. As useful as such a description is, it remains just a description. There are key questions which metabolic control analysis cannot address, such as: how does control differ between the long and the short term? What happens when supply of substrate is substantially less than the capacity of the pathway? With the caveat that it may be a convenient but not precise approximation to speak of control as residing in just one part of the respiratory network, I will turn to some of these questions.

6.2 *Evidence for control by turnover of adenylates*

Respiration energizes biosynthesis, maintenance of cellular machinery and membrane transport. In the short term (minutes to hours) the rate of these processes controls respiration rate. The mechanism is simple: respiration is limited by the availability of ADP and $NAD(P)^+$ which are released as ATP and $NAD(P)H$ are used by energy-consuming processes. This mechanism is elegant; its corollary is that substrate supply *in vivo* is adequate and does not exercise appreciable control, and indeed when sugars are added to tissues there is a lack of, or very small, increase in respiration rate (*Table 1*; Bingham and Farrar, 1988; Day and Lambers, 1983). Direct experimental evidence does indeed suggest that the bulk of control of respiration in tissues such as leaves and root tips resides in adenylate turnover (*Table 1*; Farrar and Williams, 1991).

Adenylate control is ideal for a plant growing in relatively constant conditions, but once conditions change it will have severe disadvantages. If the demand for respiratory energy rises, the amount of respiratory machinery may not be sufficient. If either the demand or the supply of substrate falls, the respiratory flux may be well below the capacity of the machinery, which will then incur unnecessarily high maintenance costs. These disadvantages could be minimized if the amount of respiratory machinery was regulated to match the flux through it. Accordingly, I will now turn to the question of how plants respond when the supply of sugar comes to be greater, or much less, than the capacity of the respiratory machinery to metabolize it.

Table 1. *Regulation of respiration in roots of barley. Freshly excised roots were assayed in oxygen electrodes with additions specified in the table. Rates in nmol O$_2$ g fresh weight^{-1} s^{-1}. Adapted from Bingham and Farrar (1988)*

Addition	Respiration rate
None	4.54 ± 0.19
Sucrose (25 mM)	4.79 ± 0.14
FCCP (0.5 μM)	6.13 ± 0.24
FCCP + sucrose	6.31 ± 0.27

6.3 *The long term and the role of sugars in control*

Earlier I argued that the rapid turnover of pools of TNC such as sucrose means that a small change in flux can result in a large change in pool size. The sugar status of a tissue thus has the potential to be a time-integrated sensor of the balance between production and consumption of sugars under varying environmental conditions. Rather than simply be a substrate, sugars may be a signal with profound metabolic consequences (Pollock and Farrar, 1996). An efficient respiratory control system might ensure that the ability to metabolize carbohydrate just exceeds the rate of supply; however, supply is altered by environmental and internal perturbations. I now ask: is there evidence to support this idea?

6.4 *What happens when supply of substrate is lower than the capacity of the respiratory network to use it*

It is easy to design experiments where the supply of sucrose or other sugar to a tissue is greatly reduced. Root tips can be excised and kept sugar-free; plant cell cultures can be transferred to sugar-free media. The consequences of sugar starvation are clear and rapid: the endogenous sugar is reduced to a very low concentration, the rate of respiration falls, and alternative substrates such as lipids and proteins are used to support this low rate (Aubert et al., 1996; Dieuaide-Noubhani et al., 1997; James et al., 1993). There is clear evidence from cereal roots that the amount of machinery falls and the expression of key genes is reduced, as a result of sugars acting as a signal (Koch and McCarty, 1988; McDonnell and Farrar, 1992; Williams et al., 1992; B. E. Collis and J. F. Farrar, unpublished), maize root tips (Dieuaide-Noubhani et al., 1997; James et al., 1993; Koch et al., 1992; Williams and Farrar, 1992) and cultured cells (Aubert et al., 1996; Godf et al., 1995; Graham et al., 1994; Journet et al., 1986; Krapp and Stitt, 1994). The idea is appealing because of its neatness and its symmetry with photosynthetic regulation: plentiful carbohydrate regulates whole-plant metabolism to both reduce its production and to increase its consumption; a shortage does the reverse (Koch, 1997). In Koch's terminology, there is a 'famine' response to sugars: key genes are down-regulated when carbohydrate supply is low (Koch, 1997).

Respiration also declines in darkness (*Table 2*). We have kept barley, *Arabidopsis* (both wild-type and a mutant poor at starch degradation), and pea in the dark for up

Table 2. Respiration in leaves and roots of barley in extended darkness. Values
are % of the value before darkening. At the time of darkening, roots contain
enough carbohydrate to sustain respiration for 6 h at control rates (P. Dwivedi and
J.F. Farrar, unpublished data)

	Leaf			Root		
Days in dark	2	4	8	2	4	8
Rate of respiration	58	—	—	100	92	46
Carbohydrate content	11	5	9	53	53	65
Protein content	61	23	5	67	72	32

to 6 days. What we saw was remarkable. In the potentially vulnerable root tissue,
which has rather low contents of non-structural carbohydrate and is remote from the
relatively carbohydrate-rich source leaves, root respiration and growth rate declined,
but only after several days during which sugar, starch and protein contents, and activities of fumarase and PFK(ATP) were relatively little affected (although that of β-
hydroxyacyl CoA dehydrogenase declined). Roots thus only lost respiratory capacity
after prolonged (*ca* 6 days) darkening. Source leaves were quite different: carbohydrate content, protein content and respiration rate all fell within 1–2 days of darkening (*Table 2*), and there was loss of Rubisco, cytochrome *c* oxidase II and acid
invertase. At first sight this is remarkable, since the initial content of carbohydrate in
these roots is only sufficient to sustain respiration for about 6 h at the control rate.
The unexpected reason is that the shoot continues to export carbohydrate to the root
for many days, and as it is both more massive than the root and has a high carbohydrate content, root respiration can be supported by carbohydrate from the shoot for
many days. Surprisingly, it is the shoot, far richer than the root in carbohydrate when
the plant is first darkened, which first shows reduced respiration and breakdown of
protein. The protein is presumably used as a respiratory substrate within the shoot
itself.

Overall, a clear picture emerges: a shortage of sugars results in a down-regulation of
respiratory machinery. Is the reverse true? Is there a 'feast' response when sugars are
in abundance, which results in an up-regulation of respiratory machinery?

6.5 *What happens when supply of sugar is greater than respiratory capacity?*

If a piece of tissue has been excised until its respiration rate has fallen, sugars can be supplied to it to examine how respiration rate is affected by them. Shortly after excision, the
restoration is rapid, complete, and entirely due to replacing depleted substrate (Williams
and Farrar, 1990). Longer after excision, when respiratory machinery has been down-regulated, adding sugar results in the supply of substrate exceeding the capacity of the respiratory machinery to metabolize it. Under these circumstances, respiration does indeed
rise, but only after a lag consistent with inducing coarse control. The amount of transcript
for a sucrose synthase and an acid invertase (Koch *et al.*, 1992; Xu *et al.*, 1996), the amount
of cytochrome *c* oxidase (B.E. Collis, C.J. Pollock and J.F. Farrar, unpublished data) and
the activity of fumarase (B.E. Collis, C.J. Pollock and J.F. Farrar, unpublished data), and
acid and neutral invertases and the amounts of several unidentified proteins (Williams *et*

al., 1992) all increase in such experiments. Feeding glucose to spinach leaves via their transpiration stream increases respiration rate, and the activity of sucrose synthase, but activities of alkaline invertase, PFK, pyruvate kinase and NAD-malate dehydrogenase are unaltered; the ATP/ADP ratio increases three-fold (Krapp *et al.*, 1991).

A second example is the effect of cooling, where supply of carbohydrate from photosynthesis is reduced less than the ability of sinks to metabolize it (Farrar, 1989). In the short term, temperature alters respiration with a Q_{10} of about 2, but in the long term has rather little effect (Atkin and Lambers, 1998). Does respiration acclimate (i.e. adjust the amount of machinery) to temperature in response to sugar status? Acclimation can be demonstrated directly by measuring the amount of one or more proteins, or indirectly by showing that flux depends on the conditions under which the plant has been grown rather than those under which it is measured, as long as it is also shown that the flux is not just substrate-limited. When several species are grown at temperatures differing by 4°C, their root respiration acclimates (*Table 3*) since if respiration did not acclimate, then rates when measured at 16° or 20°C should be independent of growth temperature. However, rates following growth at 20° are less than after growth at 16°C. Further, growth rate was also lower at 16°, and respiration was not substrate-limited as sucrose could not increase it in the short term. Whilst temperature can cause changes in amount of respiratory enzymes (Burke, 1995), and low temperature results in increased expression of the gene for the alternative oxidase (Vanlerberghe and McIntosh, 1992), we do not know if the acclimation typified by *Table 3* is mediated at the level of protein abundance and gene expression.

6.6 *What happens in the real world?*

The changes listed above happen in conditions that are far from natural, and so the question arises: does sugar starvation occur routinely in the life of a typical plant? Until critical experiments are performed under field conditions, we are restricted to making deductions from laboratory experiments. It would be surprising if sugar starvation occurred in a well-tended crop; even at the end of the night, respiration rate and sugar content of barley roots is high (Farrar, 1985). We know too little of naturally growing plants to be definite, and one of Tom ap Rees' last projects was to find out. His student, Padmanabh Dwivedi, moved to my laboratory with the problem and the

Table 3. *Respiration rate of roots acclimates to temperature. Values are rate of oxygen consumption relative to that in plants grown and measured at 16°C*

	Bellis perennis[ab]		Dactylis glomerata[ab]		Holcus lanatus		Poa annua[b]		Trifolium repens	
Growth temperature (°C)	16	20	16	20	16	20	16	20	16	20
Measured at 16°C	1	0.56	1	1.00	1	1.05	1	0.77	1	0.90
Measured at 20°C	1.27	0.71	1.38	1.17	1.43	0.92	1.33	0.97	1.00	0.83

[a]Roots do not respire faster when sucrose is supplied to them, after growth at 16 or 20°C.
[b]Root growth is slower at 16 than at 20°C (Gunn and Farrar, 1999).

start of an answer to it. His experiments on prolonged darkening are mentioned above (Section 6.4). However, he has also looked at respiration of roots of *Fraxinus oregona* from the Cambridge botanical garden after prolonged drought; the idea was that stomatal closure during drought might lead to a reduced supply of carbohydrates to the roots, with a consequent fall in respiration rate. There was no evidence of a systematic reduction in respiration mediated in this way.

In the laboratory, we have shown that the shoot supports the metabolism of the root at its own expense: if this finding can be reproduced in the field it would suggest that roots do not normally endure carbohydrate-limited constraints on respiration. There may be a satisfactory ultimate explanation based in selection and fitness if the phenomenon of substrate-limited sink respiration is of selective importance. A proximate explanation based on phloem transport is simpler: perhaps active phloem loading in the leaf continues while there is sugar present, maintaining high turgor in the source, and meanwhile respiration and associated metabolism continue in the root, keeping the phloem unloaded and its turgor there low. While there is a turgor gradient in the phloem, transport from leaf to root will continue. Again we see that the response of the whole plant could hardly have been predicted from the behaviour of its component parts.

6.7 *Problems*

More information is needed before this system can be considered established. Whilst the use of hexose analogues supports the idea that the signal molecule is a hexose phosphate or metabolically related compound (Jang and Sheen, 1997; B. E. Collis and J. F. Farrar, unpublished data), the identity of the signal molecule and the nature of the signal transduction pathway are unknown. Sugars are compartmented between and within cells: in which cell types and compartments do they act as gene regulators? The relative roles of the cytochrome and alternative pathways has not been determined (nor will this be an easy task as the inhibitor technique once used to separate them is now known to be invalid). There are other unknowns: 'do sugars mediate in the response of respiratory machinery to temperature?', 'in what way is the sugar abundance integrated over time?', 'under what range of circumstances do sugars act as signals to regulate respiration?', 'is respiration regulated independently of those processes which it energizes, or is it matched to the capacity for growth?'

6.8 *Consequences in the whole plant*

Prolonged darkening illustrates the whole-plant nature of the problem of respiratory control: the carbohydrates which provide substrate and possibly control for root respiration are supplied by the shoot, and this supply continues to some degree independently of photosynthesis in the shoot.

It is likely to be not only respiration which is responding and acclimating to internal and external changes, but also many of the processes for which it is providing energy and carbon skeletons – particularly growth. For example, temperature history can modify subsequent growth rate of potato tubers (Engels and Marschner, 1986) and supplying sugar increases growth rate as well as respiration of barley roots (Williams and Farrar, 1990). However, one view in general currency seems to be that temperature and light limit growth rate directly via metabolic rate and substrate supply, respectively. I would argue that if plants acclimate, altering the amounts of metabolic machinery, then this more complex response will be a significant component of yield and merits attention. Here respiration may give us a deep insight into the acclimatory process, which is no less than the integral of how a tissue adjusts to a new internal or external environment. If the mechanism is regulation by sugars, there is a unifying and simplifying theory to explain the way in which a number of discrete variables regulate a major and central component of plant metabolism.

7. Control of carbohydrate flux: where it is localized?

Clearly, the carbohydrate metabolism of shoot and root can be closely linked, as the experiments with prolonged darkening have shown. Obviously, the root depends on the shoot for a supply of carbon skeletons for its growth; less obviously, its metabolism is in part regulated by the sugars which the shoot provides (Farrar, 1996; Koch, 1997), at the level of the expression of key genes. Does the root act passively as a sink for the assimilate which the shoot provides once its requirements have been met, or is there more precise control? Can distant sinks such as the root regulate the production, export and partitioning of sugars? Where is regulation exercised?

There is considerable evidence that remote parts of the plant do interact to regulate carbon flux (Farrar and Gunn, 1996). Some responses are seen very quickly, for example C-11 tracer studies indicate that carbon export from a source leaf is altered within minutes of warming, or of adding galactose to, the environment of a root (Farrar et al., 1994; Minchin et al., 1994). Other responses take longer, and the variety in speed of response indicates that at least two types of mechanism must be at work. Simple mass-action, where sugars are products and substrates, is one; sugars as regulators of gene expression is another. The interaction between these two types of control in an intact plant could produce a sophisticated and complex outcome.

We have recently proposed a scheme showing a series of 'decisions' which are taken about the fate of carbon assimilated in leaves (Farrar and Gunn, 1998). Here I elaborate on that scheme, which was mostly confined to mass-action effects. Whilst the decisions are of course not conscious, they have presumably been exposed to natural selection. Many are branch points in the flow of carbon (*Figure 2*). The size of the pools that accumulate during carbon flux are a consequence of the fluxes to and from them, and are held to be the regulators of gene expression. The scheme in *Figure 2* is clearly a gross over-simplification of the real plant (intracellular compartmentation is ignored, many tissue types are missing, and the multiplicity of organs which the phloem connects are not represented) but it is still complex enough to convey a key message: whole plant growth is an integral of many component processes, the control of each of which is scarcely understood. It is hardly surprising that it remains difficult to relate whole-plant flux control to the biochemistry which underlies it.

8. The plant at elevated atmospheric CO_2: the role of carbohydrate

A rather general hypothesis about the control of carbohydrate metabolism has emerged from considering fluxes in a whole-plant context; it is summarized in *Figure 2*. Sugars have a central role: in excess, they down-regulate photosynthesis and up-regulate respiration, and do the reverse when in short supply. An independent test of this general hypothesis has recently become available with the appearance of data on the effects of growing of plants at elevated atmospheric concentrations of CO_2.

A number of generalizations can be made about the growth of plants at elevated

Mesophyll cell in mature source leaf
Store starch or synthesize sucrose?
Store or export sucrose?
Mobilize stored starch, sucrose, fructan or not?
Regulate expression of photosynthetic and carbohydrate storage genes with sucrose?

Vascular bundle in mature source leaf (including parenchymatous bundle sheath)
Store sucrose or load it into phloem?
Regulate phloem turgor with osmolytes other than sucrose?
Regulate expression of sucrose transporter genes?

Phloem network
Sucrose exported from source leaf?
Sucrose transported to shoot or to root?
Regulate relative amounts of leaf and root?

Phloem in sink
Sucrose unloaded or not?

Receiver cells in sink
Store sucrose or metabolize it for growth and respiration?
Mobilize or not carbohydrate stores?
Grow at existing meristem or start new meristem?
Regulate expression of key genes controlling growth and respiration?

Figure 2. Some decision points in the control of carbon flux. Regulation at the level of gene expression is shown in italics.

CO_2 (Drake *et al.*, 1997). They grow faster, but although absolute growth rates may stay higher, relative growth rate usually increases only transiently. Photosynthesis per unit of leaf frequently declines with prolonged exposure. Content of non-structural carbohydrate in source leaves rises, whilst the content of nutrients such as nitrogen falls (Drake *et al.*, 1997). Partitioning of dry weight between organs is little affected (Farrar and Gunn, 1996; Gunn and Farrar, 1999). Unfortunately a further generalization can be added: there is enormous variability in the effect of elevated CO_2, depending upon species, nutrition and environmental conditions. This variability is perhaps most extreme in the reports of effects on the rate of dark respiration (Drake, *et al.*, 1999; Poorter, 1994). The challenge is to test the general model for regulation of carbohydrate fluxes against the data from work on elevated CO_2.

Most of the work I will describe is from our laboratory, on barley and *Dactylis glomerata*. It is distinct from much of the literature on elevated CO_2 because both source leaves, overall growth, and roots as representative sinks, have been examined. Plants were grown hydroponically, in atmospheres containing either 350 or 700 ppm CO_2. They grew faster at high CO_2, and both shoots and roots were 30% larger at 14 days old. Photosynthesis of leaf 2 of barley was higher (Plum *et al.*, 1996) because of the greater CO_2 and only when it was older was there down-regulation of photosynthesis, which was not due to a loss of Rubisco protein (Hibberd *et al.*, 1996). When sinks are constrained by low temperature, the TNC content of source leaves rises and their photosynthetic rate falls in parallel (Plum *et al.*, 1996). The increased photosynthesis was clearly directly responsible for the increased growth. It is not quite so clear how the two are linked, since there are other potential fates for increased assimilate other than increasing the growth of sinks, such as storage in source leaves. I will now discuss the likely links.

Some of the increased photosynthesis of leaf 2 results in a greater accumulation of non-structural carbohydrate (Hibberd et al., 1996b); this increases specific leaf weight (SLW). Sucrose, fructan and starch are all increased in a manner closely dependent on the age of the leaf (Hibberd et al., 1996b). Although the increase in TNC is enough to increase SLW, it is small when expressed as a function of photosynthesis – that is, the extra TNC is equivalent to about 2 h of photosynthesis. For a leaf which has been fully expanded for 8 days, this modest increase means that most of the extra assimilate fixed in photosynthesis must have been exported, and indeed there is direct evidence that it is (Collis et al., 1996; Hibberd et al., 1996b; Table 4). But why? Why is a proportion retained in the leaf while most is exported? Why is all the extra not stored within the leaf, or all exported? It emphasizes that neither carbon storage nor export is controlled simply by either photosynthetic flux, or by a measurement of sugar status such as the amount of sucrose. (At least, on a whole-leaf basis; but compartmentation between transport and vacuolar pools in barley is unaltered by elevated CO_2; Hibberd et al., 1996). One possibility is that export is driven by sugar status and the rate of export rises only when sugar has accumulated; a new steady-state rate of export depends on the prior accumulation of sugar. Alternatively, of course, this relationship could be fortuitous and export controlled by quite different factors.

That there is at least one such factor is shown by the data in Table 4. Just 24 h after transferring plants from 350 to 700 ppm CO_2 and vice versa, both the rate of photosynthesis and the sucrose content of the second leaf reflect only the current, but not the previous, CO_2 concentration. If either photosynthesis or sugar status of the source leaf control export, then the rate of export too should match the current CO_2 concentration and not the historical one. But it does not; on the contrary, all the plants which are transferred, whether from 350 to 700 ppm or from 700 to 350 ppm, have rates of export intermediate between those of plants maintained at 350 ppm and those maintained at 700 ppm. Whilst there may be some feature of source leaf metabolism which we have not measured that can explain this effect, by far the simplest explanation resides in sinks. The roots and shoots of plants grown at 350 ppm are growing more slowly than those of plants grown at 700 ppm; if this difference is due to a different capacity to grow, and not solely due to instantaneous differences in rate of supply of assimilate, then the results can

Table 4. Export from second leaf of barley at elevated CO_2: effect of sinks. Barley plants were grown at either 350 or 700 ppm CO_2 and some switched (>) to the other CO_2 concentration 2 days after expansion of the second leaf. Carbohydrate content, and rates of photosynthesis and export, were measured 24 h later. Unpublished data of B.E. Collis, C.J. Pollock and J.F. Farrar

CO_2 concentration	350	700	350>700	700>350
Net photosynthesis (μmol m^{-2} s^{-1})	11.2	19.0	18.2	11.8
Carbohydrate (g m^{-2})	6.1	10.4	9.6	5.3
Export of sucrose (g m^{-2} h^{-1})	0.93	1.73	1.58	1.22

be explained. Plants historically at 350 ppm have sinks which cannot grow fast enough to use the extra assimilate which the source leaves can produce after transfer to 700 ppm; accordingly export to them is lower than to sinks of plants which have remained at 700 ppm. Plants historically at 700 ppm have sinks with the capacity to grow faster than the supply of assimilate will allow when their source leaves are in 350 ppm, and so they regulate export to exceed the rate it would be to sinks with a lower capacity to grow. Whilst this model needs rigorous testing, it is at least consistent with the known effects of sugars on root metabolism, regulating its capacity rather than just acting as substrates.

The idea that the capacity of root metabolism can be set by supply of sugars suggests that it will be higher in plants grown at 700 than in plants grown at 350 ppm. At once a problem is raised, since the rate of elongation of seminal axes is extremely rapid at 350 ppm. Although the axes of seminal roots elongate faster at 700 than 350 ppm CO_2 (*Table 5*), most of the increase in weight at elevated CO_2 is due to the earlier and greater production of nodal roots and the greater rate of production of laterals on seminal roots (*Table 5*). Do the roots differ in metabolism? Single seminal roots respire faster but do not have a proportionately higher sugar content (*Table 5*); however, in seminal root tips the sugar content as well as respiration rate is increased (*Table 5*). We can ask: does the increased growth rate depend on a permanent increase in sugar content? Are respiration, growth rate and sugar content inextricably linked? To attempt to answer the latter question, we have added a severe perturbation to treatment with elevated CO_2.

We subjected barley plants to a severe defoliation to reduce greatly the supply of sugars to roots. The result was striking: growth of, and import of carbon into, roots was completely stopped in the 7 days after defoliation at 350 ppm CO_2, but roots did grow at 700 ppm. In spite of this qualitative difference in growth and carbon transport, the concentration of carbohydrate was little affected: the only difference was in a one-third reduction in the roots of plants defoliated at 350 ppm (*Table 6*). It is hard to see sugar content as the sole regulator of root metabolism in this experiment, and certainly neither respiration, nor growth rate, nor metabolism of incoming [^{14}C]-sucrose to insoluble material, correlate with it. (Note that here the root suffers: it is not protected by export from the shoot as in the response to prolonged darkening described above.)

Table 5. *The growth and root metabolism of barley at elevated CO_2. Barley was grown hydroponically and harvested when 14 days old. Unpublished data of B.E. Collis, C.J. Pollock and J.F. Farrar*

CO_2 concentration (ppm)	350	700
Dry weight (mg)	154	222
Root weight ratio	0.31	0.31
Rate of elongation of seminal axis (cm h^{-1})	0.12	0.18
Number of nodal roots	3.4	5.8
Rate of lateral root production (h^{-1})	1.1	1.8
Carbohydrate content of seminal axes (mg gfw^{-1})	3.2	4.0
Respiration of seminal axes (nmol g^{-1} s^{-1})	2.7	3.9
Carbohydrate content of root tips (mg gfw^{-1})	17.3	37.9
Respiration of root tips (pmol tip^{-1} s^{-1})	22	33

gfw, gram fresh weight.

Table 6. *Barley at elevated CO_2: effect of severe defoliation on growth and metabolism of, and carbon import into, roots. Barley was grown hydroponically until 14 days old at either 350 or 700 ppm CO_2 and then all but half of the second leaf blade was removed from half of the plants. The table shows data for the status of the roots 7 days after the partial defoliation. Unpublished data of D.Jones and J.F. Farrar*

CO_2 concentration (ppm), treatment	350	350, defoliated	700	700, defoliated
Total growth (mg)	279	17	342	48
Net partitioning fraction to root	0.24	0	0.20	0.30
^{14}C Partitioning fraction to root	0.09	0.03	0.13	0.02
Gross import to root (mg day^{-1})	23	0	23	4
% Insoluble ^{14}C in root	55	25	63	40
Carbohydrate in root (mg gdw^{-1})	27	18	27	27
Respiration (nmol g^{-1} s^{-1})	3.4	1.3	3.1	1.7

gdw, gram dry weight.

The simplest conclusion is that sugar is a necessary but not a sufficient signal for regulation of root metabolism. We do not know what the other signals are, but the effects of nitrate, amino acids and phosphate on metabolic and physiological regulation at the gene level suggest that a variety of resource compounds (metabolites representing the amounts of resources acquired from the environment) may interact to regulate the metabolism underlying growth.

I noted above that roots grow faster after plants are transferred to 700 ppm CO_2, mainly due to more rapid production of lateral and nodal roots: what features of metabolism have led to the greater production of nodal roots and to the more rapid production of laterals? The formation of nodal roots is linked to the production of tillers, and so we can ask two more precise questions: why are tillers and their nodal roots produced sooner? And why are more laterals produced on seminal roots? Both are questions of development – why is a new meristem initiated in preference to expanding growth at an existing meristem? We may hypothesize that sugars have a role in triggering such developmental switches.

What of respiration at high CO_2? It would be expected that if growth is faster, respiration would be too, to provide the carbon skeletons and energy for it. Surprisingly, the existing data do not unequivocally support this view; there is some evidence for short-term inhibition of respiration by CO_2, the so-called 'direct effect'. The long-term effects of CO_2 may involve acclimation, and be related to reduced nitrogen content of tissues (Amthor, 1997; Drake et al., 1999). The fascination of the literature on high CO_2 and respiration is precisely that it suggests that it may not be possible to explain it by the phenomena considered in this review, and that novel explanations may be needed.

References

Amthor, J.S. (1989) *Respiration and Crop Productivity.* Springer, Berlin.

Amthor, J.S. (1997) Plant respiratory responses to elevated carbon dioxide. In: *Advances in Carbon Dioxide Effects Research*. ASA special publication 61, pp. 35–77.

ap Rees, T. (1980) Integration of pathways of synthesis and degradation of hexose phosphates. In: *The Biochemistry of Plants*, Vol. 3 (ed. D.D. Davies). Academic Press, New York, pp.1–42.

ap Rees, T. and Hill, S.A. (1994) Metabolic control analysis of plant metabolism. *Plant Cell Environ.* **17**: 587–599.

Atkin, O.K. and Lambers, H. (1998) Slow growing alpine and fast-growing lowland species. In: *Inherent Variation in Plant Growth* (eds H. Lambers, H. Poorter and M.M.I. van Vuuren). Backhuys, Leiden, pp. 259–287.

Aubert, S., Gout, E., Bligny, R., Martymazars, D., Barrieu, F., Alabouvette, J., Marty, F. and Douce, R. (1996) Ultrastructural and biochemical characterization of autophagy in higher plant cells subjected to carbon deprivation – control by the supply of mitochondria with respiratory substrates. *J. Cell Biol.* **133**: 1251–1263.

Bingham, I.J. and Farrar, J.F. (1988) Regulation of respiration in roots of barley. *Physiol. Plant.* **70**: 491–498.

Buckhout, T.J. and Tubbe, A. (1996) Structure mechanisms of catalysis and regulation of sugar transporters in plants. In: *Photoassimilate Distribution in Plants and Crops* (eds E. Zamski and A.A. Schaffer). Dekker, New York, pp. 229–260.

Burke, J.J. (1995) Enzyme adaptation to temperature. In: *Environment and Plant Metabolism.* (ed. N. Smirnoff). BIOS, Oxford, pp. 63–78.

Collis, B.E., Plum, S.A., Farrar, J.F. and Pollock, C.J. (1996) Root growth of barley at elevated CO_2. *Aspects Appl. Biol.* **45**: 181–185.

Cornish-Bowden, A. and Cardenas, M.L. (1990) *Control of Metabolic Processes.* Plenum, New York.

Day, D.A and Lambers, H. (1983) The regulation of glycolysis and electron transport in roots. *Physiol. Plant.* **58**: 155–160.

Dick, P.S. and ap Rees, T. (1975) The pathway of sugar transport in roots of *Pisum sativum. J. Exp. Bot.* **26**: 305–314.

Dieuaide-Noubhani, M., Canioni, P. and Raymond, P. (1997) Sugar-starvation-induced changes of carbon metabolism in excised maize root tips. *Plant Physiol.* **115**: 1505–1513.

Drake, B.G., Gonzalez-Meler, M.A. and Long, S.P. (1997) More efficient plants: a consequence of rising atmospheric CO_2? *Annu. Rev. Plant Physiol. Plant Mol. Biol.* **48**: 609–639.

Drake, B.G., Berry, J., Bunce, J., Farrar, J.F., Lambers, H., Azcon-Bieto, K., Gonzalez-Meler, M. and Wullshleger, S. (1999) Does elevated atmospheric CO_2 inhibit mitochondrial respiration in green plants? *Plant Cell Environ.* In press.

Engels, C. and Marschner, H. (1986) Allocation of photosynthate to individual tubers of *Solanum tuberosum* L. *J. Exp. Bot.* **37**: 1795–1803.

Farrar, J.F. (1985) Fluxes of carbon in roots of barley plants. *New Phytol.* **99**: 57–69.

Farrar, J.F. (1998) Temperature and the partitioning and translocation of carbon. In: *Plants and Temperature* (eds S.P. Long and F.I. Woodward). Company of Biologists, Cambridge, pp. 203–235.

Farrar, J.F. (1989) Fluxes and turnover of sucrose and fructans in healthy and diseased plants. *J. Plant Physiol.* **134**: 137–140.

Farrar, J.F. (1992) The whole plant: carbon partitioning during development. In: *Carbon Partitioning Within and Between Organisms* (eds C.J. Pollock, J.F. Farrar and A.J. Gordon) BIOS, Oxford, pp. 163–179.

Farrar, J.F. (1996) Sinks – integral parts of a whole plant. *J. Exp. Bot.* **47**: 1273–1279.

Farrar, J.F. (1999) The acquisition, partitioning – and loss – of carbon. A whole plant problem. In: *Advances in Physiological Plant Ecology* (eds M.C. Press, J.D. Scholes and M.G. Barker). Blackwell, Oxford. In press.

Farrar, J.F. and Gunn, S. (1996) Effects of temperature and atmospheric carbon dioxide concentration on source sink relations in the context of climate change. In: *Photoassimilate Distribution in Plants and Crops* (eds E. Zamski and A.A. Sheffer). Dekker, New York, pp. 389–406.

Farrar, J.F. and Gunn, S. (1998) Allocation: allometry, acclimation – and alchemy? In: *Inherent Variation in Plant Growth* (eds H. Lambers, H. Poorter and M.M.I. van Vuuren). Backhuys, Leiden, pp. 183–198.

Farrar, J.F. and Williams, J.H.H. (1991) Control of the rate of respiration in roots: compartmentation, demand and the supply of substrate. In: *Compartmentation of Plant Metabolism in Non-photosynthetic Tissues* (ed. M. Emes). Cambridge University Press, Cambridge, pp. 167–188.

Farrar, J.F., Minchin, P.E.H. and Thorpe, M.R. (1994) Carbon import into barley roots: stimulation by galactose. *J. Exp. Bot.* **45**: 17–22.

Fell, D. (1997) *Understanding the Control of Metabolism.* Portland Press, London.

Ferrario-Mery, S., Murchie, B., Hirel, B., Galtier, N., Quick, W.P. and Foyer, C.H. (1997) Manipulation of the pathways of sucrose biosynthesis and nitrogen assimilation. In: *A Molecular Approach to Primary Metabolism in Higher Plants* (eds C.H. Foyer and W.P. Quick). Taylor and Francis, London, pp. 125–153.

Galtier, N., Foyer, C.H., Murchie, E., Alred, R., Quick, P., Voelker, T.A., Thepenier, C., Lasceve, G. and Betsche, T. (1995) Effects of light and carbon dioxide on photosynthesis and carbon partitioning in leaves of tomato overexpressing sucrose phosphate synthase. *J. Exp. Bot.* **46**: 1335–1344.

Geigenberger, P., Krause, K.P., Hill, L.M., Reimholz, R., MacRae, E., Quick, P., Sonnewald, U. and Stitt, M. (1995) The regulation of sucrose synthesis in leaves and tubers of potato plants. In: *Sucrose Metabolism, Biochemistry, Physiology and Molecular Biology* (ed. H.G. Pontis). American Society of Plant Physiologists, Maryland, pp.14–24.

Godf, D.E., Reigel, A. and Roitsch, T. (1995) Regulation of sucrose synthase expression in *Chenopodium rubrum*: characterization of sugar induced expression in photoautotrophic suspension cultures and sink tissue specific expression in plants. *J. Plant Physiol.* **146**: 231–238.

Graham, I.A., Denby, K.J. and Leaver, C.J. (1994) Carbon catabolite repression regulates glyoxylate cycle gene expression in cucumber. *Plant Cell* **6**: 761–772.

Grantz, D.A. and Farrar, J.F. (1999) Acute exposure to ozone inhibits rapid carbon translocation from source leaves of Pima cotton. *J. Exp. Bot.* In press.

Gunn, S., Bailey, S.J. and Farrar, J.F. (1999) Partitioning of dry weight and leaf area within plants of three species grown at elevated CO_2. *Funct. Ecol.* (in press).

Hibberd, J.M., Richardson, P., Whitbread, R. and Farrar, J.F. (1996a) Effects of leaf age, basal meristem, and infection with powdery mildew on photosynthesis in barley grown in 700 μmol mol^{-1} CO_2. *New Phytol.* **134**: 317–325.

Hibberd, J.M., Whitbread, R. and Farrar, J.F. (1996b) Carbohydrate metabolism in source leaves of barley grown in 700 μl l^{-1} CO_2 and infected with powdery mildew. *New Phytol.* **133**: 659–671.

Ho, L.C. (1978) The regulation of carbon transport and the carbon balance of mature tomato leaves. *Ann. Bot.* **42**: 155–164.

Ho, L.C. and Thornley, J.H.M. (1978) Energy requirements for translocation from mature tomato leaves. *Ann. of Bot.* **42**: 481–483.

Huber, S.C. (1981) Inter- and intra-specific variation in photosynthetic formation of starch and sucrose. *Z. Pflanzenphysiol.* **101**: 49–54.

James, F., Brouquisse, R., Pradet, A. and Raymond, P. (1993) Changes in proteolytic activities in glucose-starved maize root-tips – regulation by sugars. *Plant Physiol. Biochem.* **31**: 845–856.

Jang, J.C. and Sheen, J. (1997) Sugar sensing in higher plants. *Trends Plant Sci.* **2**: 208–214.

Journet, E.P., Bligny, R. and Douce, R. (1986) Biochemical changes during sucrose deprivation in higher plant cells. *J. Biol. Chem.* **261**: 3193–3199.

Koch, K.E. (1997) Molecular cross-talk and the regulation of C- and N- responsive genes. In: *A Molecular Approach to Primary Metabolism in Higher Plants.* (eds C.A. Foyer and W.P. Quick). Taylor and Francis, London, pp. 105–124.

Koch, K.E. and McCarty, D.R. (1988) Induction of sucrose synthase by sucrose depletion in maize root tips. *Plant Physiol.* **86**: 35.

Koch, K.E., Nolte, K.D., Duke, E.R., McCarty, D.R. and Avigne, W.T. (1992) Sugar levels modulate differential expression of maize sucrose synthase genes. *Plant Cell* **4**: 59–69.

Koroleva, O.A., Farrar, J.F., Tomos, A.D. and Pollock, C.J. (1997) Patterns of solute in individual mesophyll, bundle sheath and epidermal cells of barley leaves induced to accumulate carbohydrate. *New Phytol.* **136:** 97–104.

Koroleva, O.A., Farrar, J.F., Tomos, A.D. and Pollock, C.J. (1998) Carbohydrates in individual cells of epidermis, mesophyll and bundle sheath in barley leaves with changed export or photosynthetic rate. *Plant Physiol.* **118:** 1525–1532.

Krapp, A. and Stitt, M. (1994) Influence of high – carbohydrate content on the activity of plastidic and cytosolic isoenzyme pairs in photosynthetic tissues. *Plant Cell Environ.* **17:** 861–866.

Krapp, A., Quick, P.D. and Stitt, M. (1991) Ribulose-1,5-bisphosphate carboxylase and chlorophyll decrease when glucose is supplied to mature spinach leaves via the transpiration stream. *Planta* **186:** 58–69.

Lewis, D.H. (1984) Occurrence and distribution of storage carbohydrates in vascular plants. In: *Storage Carbohydrates in Vascular Plants* (ed. D.H. Lewis). Cambridge University Press, Cambridge, pp. 1–52.

McDonnell, E. and Farrar, J.F. (1992) Substrate supply and its effect on mitochondrial and whole tissue respiration in barley roots. In: *Molecular, Biochemical and Physiological Aspects of Plant Respiration* (eds H. Lambers and L.H.W. van der Plas). Academic Press, The Hague, pp. 455–462.

Milburn, J.A. and Baker, D.A. (1989) Physico-chemical aspects of phloem sap. In: *Transport of Photoassimilates* (eds D.A. Baker and J.A. Milburn). Longman, Harlow, pp. 345–359.

Minchin, P.E.H., Farrar, J.F. and Thorpe, M.R. (1994) Partitioning in split root systems of barley: effect of temperature of the root. *J. Exp. Bot.* **45:** 1103–1109.

Plum, S.A., Farrar, J.F. and Stirling, C. (1996) Carbon partitioning in barley following manipulation of source and sink. *Aspects Appl. Biol.* **45:** 177–180.

Pollock, C.J. and Farrar, J.F. (1996) Source-sink relations: the role of sucrose. In: *Photosynthesis and the Environment* (ed. N. Baker). Kluwer, Dordrecht, pp. 261–279.

Poorter, H. (1994) Interspecific variation in the growth response of plants to an elevated ambient CO_2 concentration. *Vegetatio* **104:** 77–97.

Rentsch, D. and Frommer, W.B. (1996) Molecular approaches towards an understanding of loading and unloading of assimilates in higher plants. *J. Exp. Bot.* **47:** 1199–1204.

Riesmeier, J.W., Frommer, W.B. and Willmitzer, L. (1994) Evidence for an essential role of the sucrose transporter in phloem loading and assimilate partitioning. *EMBO J.* **13:** 1–7.

Rocher, J.P., Prioul, J.L., Leehamy, A., Reyss, A. and Joussaume, M. (1989) Genetic variability in carbon fixation, sucrose-P synthase and ADP-Glc pyrophosphorylase in maize plants of differing growth rate. *Plant Physiol* **89:** 416–420.

Swanson, C.A. and Christy, A.L. (1976) Control of translocation by photosynthesis and carbohydrate concentration of the source leaf. In: *Transport and Transfer Processes in Plants* (eds I.F. Warlaw and J.B. Passioura). Academic Press, New York, pp. 329–338.

Vanlerberghe, G.C. and McIntosh, L. (1992) Lower growth temperature increases alternative oxidase protein in tobacco. *Plant Physiol.* **100:** 115–119.

Williams, J.H.H. and Farrar, J.F. (1990) Control of barley root respiration. *Physiol. Plant.* **79:** 259–266.

Williams, J.H.H. and Farrar, J.F. (1992) Substrate supply and respiratory control. In: *Molecular, Biochemical and Physiological Aspects of Plant Respiration* (eds H. Lambers and L.H.W. van der Plas). Academic Press, The Hague, pp. 471–475.

Williams, J.H.H., Winters, A.L. and Farrar, J.F. (1992) Sucrose: a novel plant growth regulator. In: *Molecular, Biochemical and Physiological Aspects of Plant Respiration* (eds H. Lambers and L.H.W. van der Plas). Academic Press, The Hague, pp. 463–469.

Wilson, J.B. (1987) A review of evidence on the control of shoot: root ratio, in relation to models. *Ann. Bot.* **61:** 433–449.

Xu, J., Avigne, W.T., McCarty, D.R. and Koch, K.E. (1996) A similar dichotomy of sugar modulation and developmental expression affects both paths of sucrose metabolism: evidence from a maize invertase gene family. *Plant Cell* **8:** 1209–1220.

4

The metabolism of sugars based upon sucrose

Chris Pollock, Andy Cairns, Joe Gallagher, Maria A. Machado de Carvalho and Anuradha Kochhar

1. Introduction

Plant metabolism is complex, variable and poorly understood. Competition for resources, a sedentary habit and herbivory all increase the selective advantage of a distinctive range of metabolic products matched to the constraints of specific environments. Corner (1964) cogently argues that land plants are not generally limited by the availability of fixed carbon and that they build their metabolic diversity on an excess of the primary products of photosynthesis. Although plant secondary metabolism is commonly held to encompass the synthesis and breakdown of compounds such as phenolics, tannins, pigments or alkaloids, this variety of metabolism is also exhibited within the central processes of carbohydrate storage and transport. Sucrose is the central element in carbohydrate metabolism (see Chapter 3), but there are significant variations on the theme, some of which are illustrated in *Table 1*. These variations are the subject of this chapter. We concentrate on sugars based upon sucrose, rather than alternatives such as trehalose or the sugar alcohols, and we emphasize storage rather than transport. There are a number of recent reviews of the subject and, wherever possible, we have cited these reviews. However, some current studies are discussed by reference to the primary literature.

2. Sucrose – the parent

Sucrose is a non-reducing, neutral disaccharide which is freely soluble in water and which forms relatively non-viscous solutions at high concentration. Its synthesis and subsequent catabolism are mediated by enzymes whose activities are tightly regulated in both time and space (see Chapter 3) and its transport through the phloem makes it a pivotal metabolite which has been implicated in the 'global' regulation of processes within plants (Pollock and Farrar, 1996). Arnold (1968) has argued that the protected nature of the glycosidic bond in sucrose is a necessary precondition for its role in both transport and storage, given that subsequent metabolism cannot proceed without

Plant Carbohydrate Biochemistry, edited by J.A. Bryant, M.M. Burrell and N.J. Kruger.
© 1999 BIOS Scientific Publishers Ltd, Oxford.

Table 1. *The occurrence of soluble storage and transport sugars in vascular plants (from Corner, 1964; Kardler and Hopf, 1984; Keller, 1995; Lewis, 1984)*

Compound	Occurrence	Role	Notes
Sucrose	Ubiquitous	Transport and storage in leaves and storage organs	Storage organs usually vegetative, e.g. beet taproot and cane internodes
Starch	Ubiquitous but present in very low amounts in Alliaceae	Storage in leaves and storage organs	Chloroplast starch turns over rapidly. Storage organs, both vegetative and reproductive, have much slower turnover rates
Monosaccharides	Ubiquitous, generally in low amounts	Mainly as intermediates or breakdown products	Glucose, fructose and galactose predominate. Also found in glycosides
Sugar alcohols – cyclic and acyclic	Inositol probably ubiquitous, others widespread but discontinuous	Inositol mainly as an intermediate; others transport and storage; possible role in stress tolerance	Wide range of chemical forms, each with restricted distribution. Particularly common in parasitic plants
Oligosaccharides not based upon sucrose (i) trehalose and selaginose	Very specific and restricted	Storage transport in fungi; stress tolerance	Interactions between sucrose and trehalose important in quiescence transition in some resurrection plants
(ii) maltose	Widespread but usually at low concentrations	Breakdown product of starch but also synthezised *de novo*, possibly as a primer for starch synthesis	–
Oligosaccharides based on sucrose (series based on raffinose, lychnose, isolychnose, planteose, and umbelliferose. Also gentianose)	Raffinose and stachyose widespread but often in very low amounts. Remainder specific and restricted	Transport and storage	Apart from gentianose, all based on addition of galactose to the sucrose motif
Polysaccharides based on sucrose (fructans)	Confined to a few groups but very abundant within them	Storage; limited evidence for transport of short-chain fructans	Based on addition of fructose residues to sucrose. Occur with starch in most cases but often in different organs. Wide structural and size variation

cleavage of this linkage. There is, however, a further property of this linkage which facilitates the role of sucrose as a substrate for polymer synthesis. *Table 2* demonstrates the high free energy of hydrolysis of this linkage, and illustrates why the donation of fructose residues from sucrose to an acceptor sugar can become a significant process in certain species.

3. Soluble oligosaccharides – the children

Figure 1 indicates the range of structures found in higher plants for those carbohydrates which include the sucrose motif (Lewis, 1984). Some of these sugars are of extremely restricted distribution, and their physiology and biochemistry have not been studied extensively. For example, the trisaccharide gentianose [glu (1,6) glu (1,2) fru*] is only found in the Gentianaceae (*Table 1*; Kandler and Hopf, 1984). In other cases, such as raffinose [gal (1,6) glu (1,2) fru], they are found in small quantities in many plants, but are accumulated in large quantities only in genera such as the Labiatiae. All of these sugars retain certain of the properties of sucrose. They are all non-reducing and freely soluble in water. There is excellent evidence that both raffinose and stachyose [gal (1,6) gal (1,6) glu (1,2) fru] are transported in the phloem (Haritatos and Turgeon, 1995; Keller, 1995). It is possible that other short-chain oligosaccharides based on sucrose may serve a similar function (Wang and Nobel, 1998).

The role of raffinose and stachyose as transport sugars is intimately concerned with a distinctive method of loading sugars into the phloem through symplastic connections, rather than the more widespread apoplastic route. Synthesis of these larger sugars from sucrose inside the phloem elements makes them too large to diffuse easily through the plasmodesmata into the mesophyll cells and allows the internal concentrations within the phloem to increase to the values necessary to generate pressure-driven mass flow (Haritatos and Turgeon, 1995).

Within species, the pattern of galactosyl-sucrose oligosaccharides is usually fairly simple, with a fairly small range of sizes based upon only one or two series. In contrast,

Table 2. *Free energy of hydrolysis (ΔG°) of glycosidic linkages (Lewis, 1984)*

Compound	$\Delta G°$ (kJ mol^{-1})
UDPglucose (pH 7.4)	–31.8
Glucose-1-phosphate (pH 8.5)	–20.1
Glucose-6-phosphate (pH 7.0)	–13.8
Fructose-6-phosphate (pH 7.0)	–15.9
Sucrose	–27.6
Maltose	–19.3
Trehalose	–18.4
Glycogen (1,4-glucosyl linkage)	–18.0
Dextran (1,6-glucosyl linkage)	–8.4
Levan (2,6-fructosyl linkage)	–16.8

* In all the linear structures described in this chapter, the sucrose motif is underlined for clarity.

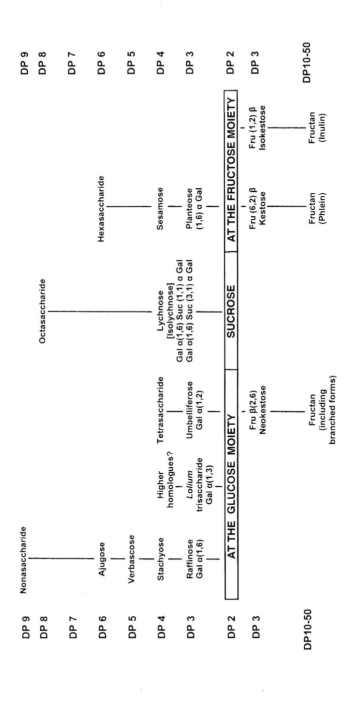

Figure 1. Structural relationships within sucrosyl oligo- and polysaccharides. Sugars in the upper part of the figure relate to galactosyl sucroses. In the lower part both glucosyl and fructosyl sucroses are presented. DP, degree of polymerization. Redrawn from Lewis (ed.), Storage Carbohydrates in Vascular Plants, 1984, with permission from Cambridge University Press.

fructans show a large degree of structural heterogeneity. Mean size can vary from 5 to almost 50 kDa, and different species have different mixtures of isomeric oligosaccharides. The three different ways by which fructose varieties can be attached to sucrose lead to a very large number of possible structures including branched forms (Pollock *et al.*, 1996). Interestingly, within species, the 'fingerprint' of fructan oligosaccharides is very consistent, implying that the synthetic mechanism must be very tightly regulated. We have no idea why some species (such as Jerusalem artichoke) accumulate almost exclusively a single homologous series of linear molecules with a terminal sucrose motif, whereas others (particularly in the monocots) can accumulate extremely complex mixtures with as many as 12 different isomeric pentasaccharides, all of which appear to be metabolic precursors of high molecular mass fructan (Pollock *et al.*, 1996).

4. Building on sucrose; a range of options

The synthesis of sucrosyl oligo- and polysaccharides in the various species and organs where they accumulate has been studied both by the use of isotopic tracers and by the demonstration of synthesis *in vitro* using cell-free extracts and partially purified enzymes. There are four main ways in which additional sugar residues may become attached to sucrose.

4.1 *Galactinol as an intermediate*

Raffinose and stachyose biosynthesis appears to be mediated via the involvement of the intermediate galactinol (galactosyl *myo*-inositol). This is formed from UDPgalactose and *myo*-inositol by a specific galactosyl transferase catalysing Reaction 1:

$$\text{UDPgalactose} + \textit{myo}\text{-inositol} \rightleftharpoons \text{galactinol} + \text{UDP.} \tag{1}$$

Raffinose is then formed by a second galactosyl transfer:

$$\text{galactinol} + \text{sucrose} \rightleftharpoons \text{raffinose} + \textit{myo}\text{-inositol.} \tag{2}$$

Higher homologues such as the tetrasaccharide stachyose are synthesized by further galactose transfer from galactinol:

$$\text{galactinol} + \text{raffinose} \rightleftharpoons \text{stachyose} + \textit{myo}\text{-inositol.} \tag{3}$$

Crude or partially purified enzyme preparations which can catalyse these reactions *in vitro* have been prepared from leaves of a number of species (Kandler and Hopf, 1984; Keller and Pharr, 1996). In leaves of *Lamium maculatum* fed $^{14}CO_2$, galactinol shows the accumulation and subsequent loss of radioactivity which is typical of a precursor, with a steady increase in the radioactivity in the end-products raffinose and stachyose (Kandler and Hopf, 1984). There seems little doubt that this represents the general pattern of synthesis in such species.

4.2 *UDP galactose as an intermediate*

It is believed that UDPgalactose can function in the direct donation of a glycosyl residue to generate neutral oligosaccharides containing the sucrose motif (Reaction 4):

$$\text{UDPgalactose} + \text{sucrose} \rightarrow \text{UDP} + \text{sucrosyl galactose.} \tag{4}$$

Enzyme preparations from a range of species have been isolated which catalyse this reaction to give either umbelliferose [gal (1,2) glu (1,2) fru] or planteose [glu (1,2) fru 6–1 gal] (Kandler and Hopf, 1984).

4.3 *Raffinose as a galactose donor*

Recent studies by Keller and co-workers (Bachmann *et al.*, 1994) have demonstrated that leaves of *Ajuga reptans* also accumulate higher homologues of the raffinose series (up to a degree of polymerization of *ca* 15) as storage carbohydrates within the vacuole. As well as Reactions 1–3, these leaves also exhibit non-specific galactosyl transfer between donor and acceptor oligosaccharides (Reaction 5):

$$gal_{(n)} \text{-glu-fru} + gal_{(m)} \text{-glu-fru} \rightleftharpoons gal_{(n+1)} \text{-glu-fru} + gal_{(m-1)} \text{-glu-fru}. \qquad (5)$$

Thus, the synthesis, via galactinol, of raffinose and stachyose will provide substrates for chain elongation, releasing sucrose and raffinose, respectively, in direct proportion to the number of moles of galactose transferred to growing chains.

Vegetative organs from members of the Caryophyllaceae accumulate mixtures of raffinose and the lychnose/isolychnose series. In these sugars, galactosyl residues are attached to the parent raffinose through the fructosyl moiety of sucrose, to yield oligosaccharides where the sucrose motif is included within the molecule [gal (1,6) glu (1,2) fru - (1,1/1,6) gal$_{(n)}$]. In these plants, radiochemical evidence suggests that raffinose acts as both the galactosyl donor and the primary acceptor, with the enzyme system presumably favouring donation to the fructosyl moiety rather than to the glucosyl moiety as in the raffinose series.

4.4 *Sucrose as a fructosyl donor*

The fructans are a complex series of oligo- and polysaccharides based on the addition of fructose residues to sucrose (*Figure 1*). In this case sucrose appears to act as both fructosyl donor and acceptor (Pollock *et al.*, 1996) according to the overall pattern shown in Reaction 6:

$$n \text{ glu-fru} \rightarrow \text{glu-fru fru}_{(n-1)} + (n-1) \text{ glu}. \qquad (6)$$

Feeding $^{14}CO_2$ to leaves of fructan-accumulating species leads to progressive transfer of radioactivity from sucrose into trisaccharides and then into larger fructans (*Figure 2*), suggesting that chain elongation is progressive. Interestingly, relatively little radioactivity accumulates in free glucose, suggesting that it is resynthesized into sucrose which can then act further as a fructosyl donor.

With the exception of the transfer of glucosyl residues to sucrose to generate gentianose (for which no mechanism has yet been elucidated), it would appear that three major reactions occur in higher plants to produce the wide range of structures shown in *Figure 1*. Firstly, there is the direct involvement of sugar nucleotides in the synthesis of sucrose itself, galactinol, umbelliferose and planteose. Secondly, there is the participation of galactinol as the galactosyl donor in the synthesis of raffinose and stachyose, and finally, there is the direct transfer of galactosyl or fructosyl residues from a donor to an acceptor sugar. In the final section, we will consider some ways in which these reactions may have evolved.

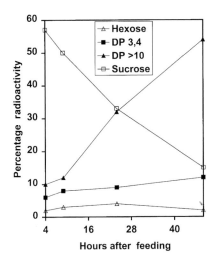

Figure 2. Changes in the radioactivity of individual components of extracts from leaf bases of Dactylis glomerata *harvested at different times after exposure to* $^{14}CO_2$. *DP, degree of polymerization. Reprinted from* Phytochemistry, *vol. 18, Pollock, C.J. Pathway of fructan synthesis in leaf bases of* Dactylis glomerata, *pp. 777–779, 1979, with permission from Elsevier Science. (See also Pollock* et al., *1996.)*

5. Occurrence and selective advantage

The occurrence of many of the sucrosyl oligo- and polysaccharides is highly discontinuous. Fructans, for example, are abundant in the Asterales and Campanulales among the dicots and in the Poales, the Liliales and Asparagales among the monocots (*Table 1*). In an analysis of distribution based upon the phylogeny of flowering plants, it appears that the majority of the groups accumulating such sugars are evolutionarily advanced (Lewis, 1984), and there seems little doubt that the ability to divert significant amounts of carbon into such compounds has arisen more than once during evolution. We assume, therefore, that this trait imposes some selective advantage within the flora where it occurs. By contrast, raffinose, and to some extent stachyose, are distributed widely in higher plants, although they are often found in very low amounts. It would appear in this case that up-regulation of an existing biosynthetic pathway to provide a major sink for fixed carbon has occurred in only a few cases.

There have been a number of suggestions made concerning the selective advantage conferred by such sugars. Apart from the role of raffinose and stachyose in transport, where the advantage has been clearly identified (Haritatos and Turgeon, 1995), none of these suggestions have received general acceptance. A number of authors have suggested that accumulation of high concentrations of soluble sugars may contribute to increased osmotic pressure and thus promote tolerance to environmental stresses such as drought and freeze-dehydration (Pontis and del Campillo, 1985). Unfortunately, the correlation between soluble sugar accumulation and stress tolerance can be spurious, since species or cultivars that acclimate well to stress often show low rates of growth during acclimation. The resulting excess of photosynthesis over consumption leads to an accumulation of stored material, much of which will be consumed when the stress is relieved but which need not contribute directly to tolerance (Pollock *et al.*, 1988). Undoubtedly, the osmotic pressure of cell sap will be affected by the accumulation of soluble carbohydrates, but, in cereals at least, leaf tissues show a significant ability to adjust (J. F. Farrar, A. D. Tomos and J. H. H. Williams, unpublished observations). In an analysis of the distribution of fructan-containing plants, Hendry and Wallace (1993) concluded that such compounds tended to be associated with species where growth is concentrated into narrow seasons within the annual cycle, particularly where rainfall may have been the limiting factor during evolution. The ability to store large amounts of easily mobilizable sugars in perennating organs could, therefore, be of selective advantage and this advantage would be increased where, for example, herbivory caused abrupt changes in the balance between supply of, and demand for, fixed carbon.

6. Physiological issues – where, when and why?

Apart from the transport function discussed above, sucrosyl oligo- and polysaccharides function predominantly as storage carbohydrates. In many cases their storage is seasonal, in seeds or in perennating organs such as bulbs, corms and rhizomes (Kandler and Hopf, 1984). In such cases, storage is initiated during the development of the relevant organ and mobilization occurs during growth or germination (Edelman and Jefford, 1968). Although the mechanisms and structures are different, the physiology effectively mimics that of starch metabolism in grains and tubers.

Of more biochemical interest is the short-term storage which occurs predominantly in leaves. The dynamics of this process have been studied most extensively for fructans (Pollock *et al.*, 1996), although sucrosyl galactans of the raffinose series serve a similar function in species such as *Ajuga reptans* (Bachmann *et al.*, 1994). In mature leaves of temperate grasses any treatment which causes the supply of fixed carbon to exceed the rate of export will trigger fructan accumulation. As discussed above, the first indication of this is an increase in sucrose content, followed in a few hours by the appearance of progressively larger amounts of fructan (Pollock *et al.*, 1996). The most extreme expression of this imbalance between carbon fixation and export can be seen if excised grass or cereal leaves are stood with their cut ends in water and illuminated. Under such conditions the rate of photosynthesis does not decline for at least 2 days and well over 90% of the carbon which is fixed is accumulated as soluble neutral sugars (*Figure 3*). If the leaves are supplied only with water, then about 60% of this material is as fructan. If inhibitors of gene expression such as cycloheximide are supplied for a short period immediately after excision, the net accumulation of carbohydrate is not affected, but the conversion of sucrose into fructan is completely prevented, suggesting very strongly that development of the ability to synthesize fructans requires the expression of previously quiescent genes. These genes are presumed to code for the enzymes of fructan biosynthesis (Pollock *et al.*, 1996).

In all cases where it has been studied in detail, the vacuole is the main site of storage for these sugars. Both the carbohydrates themselves and the enzymes believed to be responsible for their synthesis have been isolated from vacuoles prepared via the enzymatic production of protoplasts (Pollock and Kingston-Smith, 1997). There is, however, a considerable discrepancy between the substrate and enzyme contents of isolated vacuoles and those required *in vitro* to obtain significant rates of fructan synthesis (Pollock and Kingston-Smith, 1997) (see below). Suggestions have been made that the site of synthesis may be small vesicles within the cytoplasm, and that these fuse with the vacuole when synthesis is complete (Kaeser, 1983).

7. The enzymes of synthesis and breakdown

As indicated above, flow of carbon into sucrosyl oligo- and polysaccharides can be directly via UDPgalactose, via galactinol or via reversible transfer of glycosyl residues

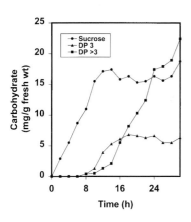

Figure 3. *Progressive accumulation of sucrose, trisaccharides and oligosaccharides following blockage of sucrose export in leaves of* Lolium temulentum. *DP, degree of polymerization. Reprinted from* New Phytologist, *vol. 109, Cairns, A.J. and Pollock, C.J. Fructan biosynthesis in excised leaves of* Lolium temulentum L. I: Chromatographic characterisation of oligofructans and their labelling patterns following CO₂ feeding, pp. 399–405, 1988, with permission from New Phytologist. (See also Pollock and Farrar, 1996.)*

between donor and acceptor sugars. All of these activities have been demonstrated *in vitro* using relatively crude preparations, but few of the proteins involved have been purified. We are not aware of any detailed examination of the enzymology of storage oligosaccharide biosynthesis involving UDPgalactose as the galactosyl donor.

In the case of the raffinose series, the enzymes from *A. reptans* responsible for the synthesis of galactinol (Reaction 1), raffinose (Reaction 2) and stachyose (Reaction 3) are found in the cytosol, with the galactosyl transferase activity (Reaction 5) within the vacuole. The cytosolic enzymes have neutral pH optima and relatively high affinities for their substrates. By contrast, the vacuolar galactosyl transferase has an acidic pH optimum and appears to have a lower affinity for its substrates (Bachmann *et al.*, 1994; Keller and Pharr, 1996) (see below).

Enzymological studies on fructan biosynthesis in plants has been dominated by the model proposed by Edelman and Jefford (1968) to explain synthesis of the 2,1-linked fructans in tubers of Jerusalem artichoke (*Helianthus tuberosus*). This model proposed the concerted action of two monofunctional enzymes: (i) sucrose: sucrose fructosyl transferase (SST) which forms a trisaccharide from two molecules of sucrose (Reaction 7) and (ii) fructan: fructan fructosyl transferase (FFT) which reversibly elongates acceptor fructans by transfer of a single fructosyl residue from a donor in a manner analogous to Reaction 5 (Reaction 8).

$$\underline{\text{glu (1,2) fru}} + \underline{\text{glu (1,2) fru}} \rightleftharpoons \underline{\text{glu (1,2) fru}}\text{ (1,2) fru} + \text{glu} \qquad (7)$$

$$\underline{\text{glu (1,2) fru}}\text{ (1,2) fru}_{(n)} + \underline{\text{glu (1,2) fru}}\text{ (1,2) fru}_{(m)}$$
$$\rightleftharpoons \underline{\text{glu (1,2) fru}}\text{ (1,2) fru}_{(n+1)} + \underline{\text{glu (1,2) fru}}\text{ (1,2) fru}_{(m-1)} \qquad (8)$$

There is growing evidence that this model is not adequate to explain the complexities of fructan synthesis in other species, and there are pronounced inconsistencies between the properties of enzyme preparations *in vitro* and the physiology of fructan metabolism *in vivo*. These inconsistencies can be grouped into four main areas.

(i) *The synthesis of other linkages.* The Edelman model (Edelman and Jefford, 1968) deals only with the synthesis of the homologous fructan series found in *H. tuberosus*. In this series, adjacent fructose residues are linked 2,1 and all the members of the series are observed. Other species with more complex oligosaccharide distributions have been shown to have extractable activities which can lead to the formation *in vitro* of 2,6-linked fructans and complex mixed linkage and branched forms (Shiomi, 1989; Sprenger *et al.*, 1995). Some of these enzymes have been extensively purified and appear to be catalytically distinct from the SST/FFT activities described by Edelman and Jefford (1968).

(ii) *Substrate effects.* The affinity of most preparations for sucrose is very low; K_m estimates are in the range of 100–300 mM and it is often difficult to saturate the activity. These values are as high or much higher than most estimates of vacuolar sucrose concentration. Whilst this would make the operation of fructan synthesis extremely sensitive to sucrose concentrations (which it appears to be *in vivo*), matching *in vivo* rates of synthesis with *in vitro* activity is only possible at apparently non-physiological substrate contents.

(iii) *Promiscuity.* Most, if not all, the enzyme preparations made from plants that can synthesize fructans can also catalyse hydrolysis of sucrose or fructans. It may be significant that purified invertases from species that do not accumulate fructan are able to catalyse the synthesis of small fructans when incubated at high sucrose concentrations (Pollock *et al.*, 1996). Multiple activities mean that a reaction catalysed *in vitro* by a particular preparation may not be of physiological relevance *in vivo*. Indirect evidence to support a role *in vivo* is provided when variation in the amount of extractable activity correlates with changes in the capacity of the tissue to synthesize fructan, but it seems possible that net synthesis *in vivo* may occur against a background of continued turnover or hydrolytic modification of fructans.

(iv) *Concentration effects.* Few of the studies carried out on purified or partially purified preparations have attempted to reproduce *in vitro* the full range of fructan structures observed *in vivo*. Where this has been attempted, the nature of the products has been shown to be strongly affected by the concentration of protein used in the assay mixture (Cairns *et al.*, 1999). Increasing enzyme concentration increases the size of the fructan products formed within a given time, such that we have been able to generate *in vitro* linear 2,6-linked fructan polymers with degree of polymerization values around 50 and which are chemically identical to the material accumulated by the species (*Phleum pratense*) from which the extracts were made (Cairns *et al.*, 1999). The enzymological consequences of the observation are significant. Researchers studying the catalytic properties of a particular preparation at low enzyme concentration could assign an entirely different activity to that which would be recognized by assay at higher concentrations. Thus it is difficult to extrapolate data from one study to another and to obtain an unequivocal understanding of the enzymology of synthesis. It may be that resolution of this issue will require *in vivo* function of specific proteins to be identified via antisense technology, but there is no doubt that the difficulty in obtaining large amounts of pure, catalytically active protein has mitigated against the type of detailed *in vitro* characterization which is absolutely necessary if we are to understand the complexities of fructan synthesis.

Fructan breakdown, by contrast, appears to be a relatively simple process. Physiological and radiochemical data show conclusively that hydrolysis proceeds by the iterative removal of the terminal fructose residue, leaving the chain one unit smaller. Activities which catalyse these reactions have been isolated from a range of species, and in some cases have been shown to increase in abundance when fructan breakdown is stimulated (Cairns *et al.*, 1999; Simpson and Bonnett, 1993). Once again, there is some confusion where preparations can both degrade and synthesize fructan *in vitro*, but the broad nature of the process seems clear.

8. Genes, genetics and evolution

In recent years, genes have been isolated which code for proteins that are strong candidates for involvement in fructan metabolism. Sprenger *et al.* (1995) cloned a gene for a fructosyl transferase from barley leaves via enzyme purification, protein sequencing and the screening of a cDNA library using a synthetic oligonucleolide. This gene itself has been used as a heterologous probe to screen other libraries. Other workers have used differential screening or heterologous screening to identify genes which are up-regulated when sucrose concentrations rise.

The majority of the relevant genes which have been isolated show strong homology with acid invertase, that is the enzyme which hydrolyses sucrose and which has an acidic pH optimum consistent with a vacuolar or apoplastic location (Sprenger *et al.*, 1995) (*Figure 4*). This is consistent with the reaction chemistry involved in both hydrolysis and glycosyl transfer where either water or another sugar can be considered as the acceptor (Pollock *et al.*, 1996) (*Table 3*).

The similarity in DNA sequence and the catalytic similarities of the reactions both raise the possibility that the fructan syndrome may have arisen via changes in the amino acid sequence of specific invertases which would affect the balance between hydrolysis and fructosyl transfer. Acid invertase is ubiquitous, and if the changes involved are relatively minor (as suggested by the strong sequence homology observed to date) then a polyphyletic origin is possible. Support for this hypothesis is provided by the protein engineering of the bacterial fructan-synthesizing enzyme levan sucrase. Here, single amino acid changes at position 331 close to the active site were shown to alter the relative activity of the enzyme when measured as a fructosyl transferase, a fructan hydrolase, a SST or an invertase (Chambert and Petit-Glatron, 1991).

There is much less known about the evolutionary basis of the synthesis of raffinose-series or other galactosyl sucrose oligosaccharides. By analogy, the galactosyl transferases could also have evolved from an invertase and there are a range of cell wall glycosyl transferases that have neutral sugars as acceptors and that use UDPgalactose as a donor. The evolutionary origin of galactinol synthesis is not known. However, the reaction is widespread, and it may be that low levels of accumulation are a by-product of some other ubiquitous reaction. In this case, evolution would have been based upon the selective advantage gained by increasing the flux through the reaction. However, until genes are isolated and homologies assessed, such considerations must remain speculative.

Table 3. Similarities between the group transfer reactions relevant to fructan metabolism (*Pollock* et al., 1996)

1. G-O-F + H-OH sucrose + water	→ G-O-H + F-O-H glucose + fructose	Sucrose hydrolysis (cleavage of G–F bond)
2. G-F + G-F sucrose + sucrose	→ G-F-F + G trisaccharide + glucose	Fructosyl transfer from sucrose (synthesis of F–F bond)
3. G-F-F + G-F-F trisaccharide + trisaccharide	→ G-F-F-F + G tetrasaccharide + glucose	Fructosyl transfer from fructan (no net synthesis of F–F bond)
4. G-F-F-O-F + H-OH fructan + water	→ G-F-F-O-H + F-O-H fructan + fructose	Fructan hydrolysis (cleavage of F–F bond)

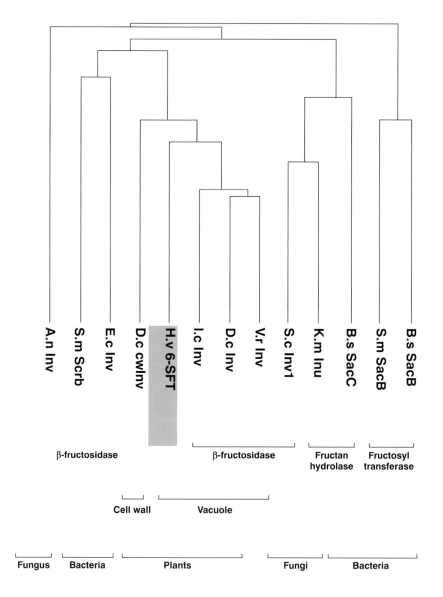

Figure 4. Comparison of deduced amino acid sequences for proteins which catalyse group transfer reactions involving fructose. The dendrogram shows the relatedness of sequences with the order of branching indicating statistical similarity. 6-SFT is a fructosyl transferase from barley which is implicated in fructan biosynthesis. The other enzymes catalyse sucrose or fructan hydrolysis or fructan synthesis from a range of plants, fungi and bacteria. Reprinted from Proceedings of the National Academy of Sciences, USA, vol. 92, Sprenger et al., Purification, cloning and functional expression of sucrose-fructan 6-transferase, a key enzyme of fructan synthesis in barley, pp. 11652–11656. Copyright (1995) National Academy of Sciences, USA.

9. Concluding remarks

Study of alternative storage and transport sugars in plants has suffered from their relative lack of economic importance or dietary significance. They do, however, provide interesting information on the mechanisms controlling enzyme specificity, on the role of assimilate abundance in regulating plant metabolism at the level of gene expression, and on the ways in which the diversity of plant metabolism may have evolved. Their broad, yet discontinuous distribution implies a significant physiological role and their occurrence in temperate cereals indicates their relevance to crop science. As the techniques of molecular biology become applied more widely to the study of diversity, we will gain a greater understanding of the functions of these sugars.

Acknowledgements

We acknowledge financial support for our own studies in this area from BBSRC (UK) and FAPESP (Brazil). We are grateful to Jo Spikes for her valiant attempts to decipher the senior author's handwriting.

References

Arnold, W.N. (1968) The selection of sucrose as the translocate of higher plants. *J. Theor. Biol.* **21:** 13–20.

Bachmann, M., Matile, P. and Keller, F. (1994) Metabolism of the raffinose family oligosaccharides in leaves of *Ajuga reptans* L. Cold acclimation, translocation, and sink to source transition: discovery of chain elongation enzyme. *Plant Physiol.* **105:** 1335–1345.

Cairns, A.J., Pollock, C.J., Gallagher, J. and Harrison, J. (1999) Fructans: synthesis and regulation in plants. In: *Photosynthesis Physiology and Metabolism* (eds R.C. Leegood, T.D. Sharkey and S. von Caemmerer). Kluwer Academic Publishers, The Netherlands (in press).

Chambert, R. and Petit-Glatron, M.F. (1991) Polymerase and hydrolase activities of *Bacillus subtilis* levansucrase can be separately modulated by site-directed mutagenesis. *Biochem. J.* **279:** 35–39.

Corner, E.J.H. (1964) *The Life of Plants.* Weidenfeld and Nicholson, London, 315 pp.

Edelman, J. and Jefford, T.G. (1968) The mechanism of fructosan metabolism in higher plants as exemplified in *Helianthus tuberosus. New Phytol.* **67:** 517–531.

Haritatos, E. and Turgeon, R. (1995) Symplastic phloem loading by polymer trapping. In: *Sucrose Metabolism, Biochemistry, Physiology and Molecular Biology* (eds H.G. Pontis, G.L. Salerno and E.J. Echeverria). American Society of Plant Physiologists, Rockville, USA, pp. 216–244.

Hendry, G.A.F. and Wallace, R.K. (1993) The origin distribution and evolutionary significance of fructans. In: *Science and Technology of Fructans* (eds M. Suzuki and N.J. Chatterton). CRC Press, Boca Raton, FL, pp. 119–140.

Kaeser, W. (1983) Ultrastructure of storage cells in Jerusalem artichoke tubers *(Helianthus tuberosus* L.). Vesicle formation during inulin synthesis. *Z. Pflanzenphysiol.* **111:** 253–260.

Kandler, O. and Hopf, H. (1984) Biosynthesis of oligosaccharides in vascular plants. In: *Storage Carbohydrates in Vascular Plants* (ed. D.H. Lewis). Cambridge University Press, Cambridge, pp. 115–131.

Keller, F. (1995) Role of the vacuole in raffinose oligosaccharide storage. In: *Sucrose Metabolism, Biochemistry, Physiology and Molecular Biology* (eds H.G. Pontis, G.L. Salerno and E.J. Echeverria). American Society of Plant Physiologists, Rockville, USA, pp. 156–166.

Keller, F. and Pharr, D.M. (1996) Metabolism of carbohydrate in sinks and sources: galactosyl-sucrose oligosaccharides. In: *Photoassimilate Distribution in Plants and Crops* (eds L.E. Zamski and A.A. Shaffer). Marcel Dekker, New York, pp. 157–183.

Lewis, D.H. (1984) Occurrence and distribution of storage carbohydrates in vascular plants. In: *Storage Carbohydrates in Vascular Plants* (ed. D.H. Lewis). Cambridge University Press, Cambridge, pp. 1–52.

Pollock, C.J. and Farrar, J.F. (1996) Source–sink relations—the role of sucrose. In: *Photosynthesis and the Environment* (ed. N.R. Baker). Kluwer Academic Publishers, The Netherlands, pp. 261–279.

Pollock, C.J. and Kingston-Smith, A.K. (1997) The vacuole and carbohydrate metabolism. *Adv. Bot. Res.* **25:** 195–215.

Pollock, C.J., Eagles, C.F. and Sims, I.M. (1988) Effect of photoperiod and irradiance changes upon development of freezing tolerance and accumulation of soluble carbohydrate in seedlings of *Lolium perenne* grown at 2°C. *Ann. Bot.* **62:** 95–100.

Pollock, C.J., Cairns, A.J., Sims, I.M. and Housley, T.L. (1996) Fructans as reserve carbohydrates in crop plants. In: *Photoassimilate Distribution in Plants and Crops: Source–Sink Relationships, Chapter 5* (eds L.E. Zamski and A.A. Shaffer). Marcel Dekker, New York, pp. 97–113.

Pontis, H.G. and del Campillo, E. (1985) Fructans. In: *Biochemistry of Storage Carbohydrates in Green Plants* (eds P.M. Dey and R.A. Dixon). Academic Press, San Diego, pp. 205–226.

Shiomi, N. (1989) Properties of the fructosyltransferases involved in the synthesis of fructan in liliaceous plants. *J. Plant Physiol.* **134:** 151–155.

Simpson, R.J. and Bonnett, G.D. (1993) Fructan exhydrolase from grasses. *New Phytol.* **123:** 453–464.

Sprenger, N., Bortlik, K., Brandt, A., Boller, T. and Wiemken, A. (1995) Purification, cloning and functional expression of sucrose-fructan 6-transferase, a key enzyme of fructan synthesis in barley. *Proc. Natl Acad. Sci. USA* **92:** 11652–11656.

Wang, N. and Nobel, P.S. (1998) Phloem transport of fructans in the crassulacean acid metabolism species *Agave deserti. Plant Physiol.* **116:** 709–714.

5

Regulation of sucrose metabolism by protein phosphorylation: stimulation of sucrose synthesis by osmotic stress and 5-aminoimidazole-4-carboxamide riboside

Steven C. Huber, Werner M. Kaiser, Dikran Toroser, Gurdeep S. Athwal, Heike Winter and Joan L. Huber

1. Introduction

Phosphorylation can affect an enzyme in a number of different ways, including effects on: (i) catalytic activity (activation or inhibition); (ii) intracellular localization (e.g. reversible membrane association); and (iii) protein:protein interactions. In plants, enzymes of primary carbon and nitrogen metabolism represent examples of each type of interaction. For example, the activity of sucrose phosphate synthase (SPS; EC 2.4.1.14) is known to be affected by phosphorylation of Ser-158 (McMichael *et al.*, 1993) and Ser-424 (Toroser and Huber, 1997). These regulatory sites function in light/dark modulation and osmotic-stress activation, respectively. In contrast, sucrose synthase (SuSy; EC 2.4.1.13) is phosphorylated on a single seryl residue, Ser-15 (Huber *et al.*, 1996), which has little effect on activity but appears to be one of the factors controlling association with the plasma membrane with the dephosphorylated form being membrane associated (Winter *et al.*, 1998). An example of the third type of effect is NADH:nitrate reductase (NR; EC 1.6.6.1) since inactivation of the enzyme requires phosphorylation (of Ser-543; Bachmann *et al.*, 1996b; Douglas *et al.*, 1995), followed by binding of a 14-3-3 inhibitor protein (Bachmann *et al.*, 1996a; Moorhead *et al.*, 1996).

Plant Carbohydrate Biochemistry, edited by J.A. Bryant, M.M. Burrell and N.J. Kruger.
© 1999 BIOS Scientific Publishers Ltd, Oxford.

We are interested in the environmental (external) and endogenous signals that affect sucrose metabolism as a result of impacting the phosphorylation of SPS or SuSy. We are particularly interested in the effects of osmotic stress, which has been shown to affect carbon partitioning in leaves (Dancer *et al.*, 1990; Zrenner and Stitt, 1990) and potato tubers (Geigenberger *et al.*, 1997). In both tissues, activation of SPS has been shown to occur, and was associated with increased sucrose synthesis and related changes in metabolism. Increased sucrose synthesis was elegantly shown by the use of tracers, but effects on steady-state sugar pools were not always examined. However, Stewart (1971) showed that osmotic stress markedly enhances conversion of starch to sucrose in excised bean leaves in the dark resulting in elevated sucrose levels, but it is not known whether similar effects occur in other species such as spinach. Consequently, one of the objectives of the present study was to determine whether osmotic-stress activation of SPS in spinach leaves was associated with changes in the steady-state sucrose pool.

Osmotic stress can have a multitude of effects on metabolism. Of particular relevance to protein phosphorylation, and related signal transduction pathways, is the report that osmotic stress dramatically reduced the ATP content of potato tubers (Geigenberger *et al.*, 1997). The content of ADP remained low and constant but unfortunately, information on the content of AMP was not provided. Because ATP and AMP often change inversely, a rise in AMP might be expected to occur. Thus, a second objective of the present study was to determine whether adenine nucleotides were affected by osmotic stress in spinach leaves.

As one experimental approach to examine possible effects of elevated AMP that may occur under stress conditions, we adopted a technique developed in animal systems using the compound 5-aminoimidazole-4-carboxamide riboside (AICAR), which is cell permeable and readily taken up. In the cytosol, AICAR is converted to the monophosphorylated derivative, 5-aminoimidazole-4-carboxamide ribonucleoside monophosphate (ZMP), which is an AMP analogue and accumulates in the cell (Sullivan *et al.*, 1994). The amount that accumulates is relatively small, thereby reducing the impact on metabolism in general; that is, AICAR is not a phosphate-sequestering agent similar to the sugar mannose. A rise in AMP (or ZMP) can have several consequences, and one of the most surprising is the apparent disruption of 14-3-3 protein binding to target ligands such as NR (Huber and Kaiser, 1996). Because of the possibility that AMP might be elevated during osmotic stress and because SPS may interact in a regulatory fashion with 14-3-3 proteins (Toroser *et al.*, 1998 and unpublished data), a third objective of the present study was to investigate the effects of AICAR feeding on SPS activity and net sucrose metabolism in spinach leaves. The results obtained demonstrate that AICAR feeding does activate SPS in spinach leaves, but probably by a mechanism that is distinct from that which occurs during osmotic stress. Regardless of the specific mechanism involved, activation of SPS in the dark is associated with increased sucrose pools consistent with the notion that regulation of SPS activity is a key factor controlling carbon partitioning.

2. Osmotic stress activates sucrose phosphate synthase *in situ* without changes in adenine nucleotides

Osmotic-stress activation of SPS has been well documented in several photosynthetic and non-photosynthetic tissues (Dancer *et al.*, 1990; Geigenberger *et al.*, 1997; Zrenner and Stitt, 1990). In the present study, imposition of a moderate water deficit to excised spinach leaves resulted in a two- to four-fold increase in V_{SEL} SPS activity (i.e. limiting substrates plus inorganic phosphate), whereas V_{NONSEL} (i.e. maximum activity) remained constant. The change in apparent activation state, defined as $V_{SEL}/V_{NONSEL} \times 100$, is consistent with covalent modification of existing enzyme protein as opposed to synthesis of additional SPS protein and this has been confirmed in spinach leaves by immunotitration (Toroser and Huber, 1997). Importantly, the increase in SPS activity was correlated positively with leaf sucrose content in the darkened spinach leaves used in the study (*Figure 1*). These results are consistent with earlier findings (Dancer *et al.*, 1990; Geigenberger *et al.*, 1997; Zrenner and Stitt, 1990) and suggest that SPS had been activated *in situ* and that net sucrose synthesis had probably been increased as well.

Because of the earlier report (Geigenberger *et al.*, 1997) that osmotic stress can reduce the level of ATP in tubers, it was important to determine whether adenine nucleotides were altered in stressed spinach leaves. As shown in *Table 1*, none of the adenine nucleotides were reduced in stressed spinach leaves. In fact, ATP content was increased slightly as was the total adenylate content. Thus, there is no evidence to suggest that changes in adenine nucleotides occurred that could directly affect the phosphorylation status of enzymes such as SPS. Moreover, the lack of change in adenine nucleotides is a strong indication that general phosphate metabolism has not been grossly affected by the imposed stress. Lastly, the observation that AMP does not increase under osmotic stress indicates that it is not an essential component of the signal transduction pathway leading to stimulation of the protein kinase peak IV (PK_{IV}) kinase that phosphorylates Ser-424 of SPS (Toroser and Huber, 1997).

3. AICAR inhibits photosynthesis

Even though cytosolic [AMP] may not change during osmotic stress, for the reasons outlined in the Introduction, we wanted to mimic a rise in cytosolic [AMP] using

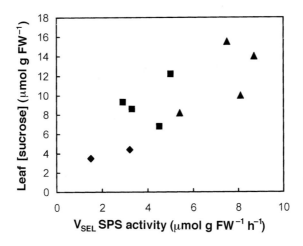

Figure 1. Positive correlation between sucrose phosphate synthase (SPS) activity and leaf sucrose content in osmotically stressed spinach leaves in the dark. Excised leaves were placed in the dark for 3 h with cut petioles in water (control; ◆), betaine (■) or sorbitol (▲). The concentration of betaine or sorbitol was 150 or 300 mM, and there was no consistent difference between the two concentrations. Each point represents a separate leaf that was sampled and assayed in duplicate. The V_{SEL} SPS activity is plotted; V_{NONSEL} activities were constant at 24.5 ± 2.0 across treatments.

Table 1. *Osmotic stress does not reduce the adenine nucleotide content of spinach leaves*

| Treatment | Nucleotide (nmol g^{-1} fresh wt) | | | |
	ATP	ADP	AMP	Total
H$_2$O	44.9 ± 3.2	33.5 ± 4.9	7.0 ± 1.6	85.5 ± 6.9
Betaine	58.4 ± 5.3	38.1 ± 4.6	5.3 ± 1.8	101.5 ± 8.7

Leaves were excised from spinach plants at mid-morning, and the cut petioles were placed either in water or 150 mM betaine. After 3 h in the dark at room temperature, leaf tissue was harvested directly into liquid nitrogen prior to adenylate analysis (Huber and Kaiser, 1996). Values are means of three determinations ± SE. Similar results to those presented were obtained when leaves were placed in an equivalent concentration of sorbitol (data not shown).

AICAR and examine the effects on metabolism. In order to look generally at metabolism, we treated spinach leaves with AICAR and examined photosynthetic capacity. As shown in *Table 2*, AICAR reduced photosynthetic capacity, measured as CO_2-dependent O_2 evolution in a leaf disc electrode, or as $^{14}CO_2$ fixation. The inhibition of light-dependent photosynthetic metabolism appears to be related to a reduction in the capacity for carbon flux into sucrose (gluconeogenesis), because in other experiments, the accumulation of sucrose in leaf discs during 3 h of CO_2 fixation was completely prevented in AICAR-treated tissue relative to controls, while starch accumulation was not reduced (data not shown). Inhibition of cytosolic fructose-1,6-bisphosphatase (FBPase) *in vivo* by ZMP, as an analogue of 5'-AMP (Herzog *et al.*, 1984), could explain the inhibition of gluconeogenesis and photosynthetic capacity. In addition, dark $^{14}CO_2$ fixation was partially inhibited by AICAR (*Table 2*), which suggests some inhibition of glycolysis in AICAR-treated leaves. Inhibition of glycolysis by AICAR has, in fact, been observed in rat hepatocytes (Vincent *et al.*, 1992). However, the basis for apparent inhibition of glycolysis in AICAR-treated plant tissue has not been elucidated.

Table 2. *Effect of AICAR on CO_2 fixation and O_2 evolution*

| Treatment | Rate (μmol CO_2 or O_2 g^{-1} FW h^{-1}) | | |
	Dark $^{14}CO_2$	Light $^{14}CO_2$	Light O_2
H$_2$O	0.71 ± 0.13	72.9 ± 2.2	156 ± 13
AICAR	0.41 ± 0.06	22.5 ± 1.7	73.5 ± 4.0

Spinach leaves were treated with 20 mM 5-aminoimidazole-4-carboxamide riboside (AICAR) and discs were then cut for measurement of CO_2 fixation in the dark (phosphoenolpyruvate carboxylation) or light (photosynthesis), measured as $^{14}CO_2$ fixation (Wuerzburg) or O_2 evolution (Raleigh, NC). Values are means of three to six determinations ± SE.

4. AICAR activates sucrose phosphate synthase and stimulates sucrose synthesis from starch

As discussed in the Introduction, recent results suggest that SPS may interact with 14-3-3 proteins. Because the inactivation of phospho-NR by 14-3-3 proteins (Bachmann et al., 1996a; Moorhead et al., 1996) can be disrupted in situ with AICAR (Huber and Kaiser, 1996), we wanted to determine whether AICAR feeding to spinach leaves would activate SPS. To do this, leaves were excised from illuminated plants and placed in the dark for 1 h to allow light-activated enzymes such as SPS to deactivate in the dark. At 1 h, AICAR was supplied to the excised leaves via the transpiration stream and at various times thereafter, leaves were harvested for extraction and assay of SPS. As shown in Figure 2, there was a rapid, partial activation of SPS as a result of AICAR feeding. Interestingly, both V_{SEL} (Figure 2a) and V_{NONSEL} (Figure 2b) activities of SPS were increased. This is in contrast to light- and osmotic-stress treatments, which only increase the V_{SEL} activity of SPS (i.e. maximum activity remains constant). Thus, AICAR activation of SPS is kinetically distinct, which is consistent with the notion that a different mechanism is involved.

One of the surprising effects of 5'-AMP (or ZMP) is the apparent interference with binding of 14-3-3s to their ligands. The basis for this effect is not entirely clear, but recent results suggest that 14-3-3s may contain an AMP binding site (Athwal et al., 1998). In the presence of millimolar concentrations of 5'-AMP, binding of 14-3-3s to peptides or native proteins is reduced. This appears to explain the AICAR activation of NR in darkened spinach leaves (Huber and Kaiser, 1996). Similarly, it may also explain AICAR activation of SPS, as 14-3-3s appear to inhibit SPS activity (Toroser et al., 1998).

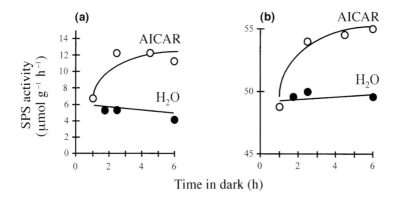

Figure 2. 5-Aminoimidazole-4-carboxamide riboside (AICAR) activation of sucrose phosphate synthase (SPS) in darkened spinach leaves. Leaves were excised at 10:00 a.m. and placed with their petioles in water for 1 h, after which half of the leaves were transferred to 10 mM AICAR. At indicated times thereafter, leaves were harvested and extracted for measurement of (a) V_{SEL} and (b) V_{NONSEL} activities of SPS. Note that both activities were increased by AICAR treatment, which is not observed in light/dark modulation (Huber et al., 1992; McMichael et al., 1993) or osmotic stress activation (see Figure 1 legend).

The activation of SPS by AICAR was partial and even though both V_{SEL} and V_{NONSEL} were increased, activation state (ratio of the two activities, expressed as a percentage) also increased. In the controls, activation state was about 10% and decreased slightly with time whereas in the AICAR-treated leaves, activation state was maintained at greater than 20%.

In order to determine whether AICAR activated SPS *in situ*, we monitored changes in leaf carbohydrates in the dark (*Table 3*). Leaves were excised from plants after several hours of photosynthesis, and placed in the dark with petioles either in water or 10 mM AICAR. At time zero, and after 3 h in the dark, leaf samples were harvested for carbohydrate analysis. In the control leaves, both the sucrose and starch pools were mobilized as indicated by a decrease in their contents. Presumably, the carbon derived from the carbohydrate reserves was used for respiration and biosynthesis of various other products such as organic acids (e.g. malate). However, in the AICAR-treated leaves, there was a net accumulation of soluble sugars and a slight increase in the rate of starch mobilization. These results suggest that SPS was activated *in situ* in the AICAR-treated leaves and that sucrose synthetic capacity was increased as a result. Furthermore, if AICAR activation does represent displacement of 14-3-3 proteins, these results would also provide at least one line of evidence for the *in situ* association of SPS with 14-3-3s that inhibits SPS activity.

5. Concluding remarks

SPS normally has a low activation state in a darkened leaf (as a result of Ser-158 phosphorylation; McMichael *et al.*, 1993), but can be activated in the dark either by osmotic-stress or AICAR feeding. In the former case, activation involves Ser-424 phosphorylation (Toroser and Huber, 1997) whereas in the latter case, displacement of inhibitory 14-3-3 proteins appears to be involved (Toroser *et al.*, 1998). In both cases, however, sucrose synthetic capacity *in situ* is increased as evidenced by changes in the steady-state leaf sucrose content (*Figure 1* and *Table 3*). It is interesting that AICAR feeding inhibits gluconeogenesis in the light (*Table 2*), but enhances the process in the dark (*Table 3*). This apparent contradiction may be resolved by a consideration of the source of intermediates for sucrose biosynthesis. In the light, sucrose is synthesized from triose phosphates produced photosynthetically and released from the chloroplast,

Table 3. AICAR feeding causes net sugar accumulation in detached spinach leaves

Treatment	Change in leaf carbohydrate (μg atoms C g^{-1} fresh weight)		
	ΔSucrose	ΔHexose	ΔStarch
H_2O	−24.5	−1.0	−8.1
AICAR	+7.2	+1.8	−11.5

Leaves were excised from spinach plants at approximately 10:00 a.m. and placed in the dark with the petioles in H_2O (control) or 10 mM 5-aminoimidazole-4-carboxamide riboside (AICAR). Immediately after excision or after 3 h in the dark, leaves were harvested for carbohydrate determinations. The change in sucrose, hexose sugars (glucose and fructose) and starch is presented. Values are means of two determinations and are representative of three separate experiments.

and thus, cytosolic FBPase is essential. It is quite likely that ZMP is a strong inhibitor of the FBPase (Herzog *et al.*, 1984). In contrast, in the dark, sucrose is probably formed primarily from glucose, derived from starch mobilization. This metabolism would not require the cytosolic FBPase and thus, inhibition by AICAR would not necessarily be expected. This postulate does not exclude other explanations for the AICAR activation of SPS (and gluconeogenesis in the dark), or that more than one factor may be involved.

Acknowledgements

Co-operative investigations of the USDA, ARS and the North Carolina ARS, Raleigh, NC 27695–7643. This work was also supported in part by grants from the US Department of Energy (Grant DE-AI05-91 ER 20031 to S.C.H.) and the Deutsche Forschungsgemeinschaft (SFB 251 to W.M.K.). Mention of a trademark or proprietary product does not constitute a guarantee or warranty of the product by the NC ARS or the USDA and does not imply its approval to the exclusion of other products that may also be suitable.

References

Athwal, G.S., Huber, J.L. and Huber, S.C. (1998) Phosphorylated nitrate reductase and 14-3-3 proteins: the site of interaction, effects of ions, and evidence for an AMP binding site on 14-3-3 proteins. *Plant Physiol.* 118: 1041–1048.

Bachmann, M., Huber, J.L., Liao, P.-C., Gage, D.A. and Huber, S.C. (1996a) The inhibitor protein of phosphorylated nitrate reductase from spinach (*Spinacia oleracea*) leaves is a 14-3-3 protein. *FEBS Lett.* 387: 127–131.

Bachmann, M., Shiraishi, N., Campbell, W.H., Yoo, B.-C., Harmon, A. and Huber, S.C. (1996b) Identification of Ser 543 as the major regulatory phosphorylation site in spinach leaf nitrate reductase. *Plant Cell* 8: 505–517.

Dancer, J., David, M. and Stitt, M.(1990) Water-stress leads to a change of partitioning in favor of sucrose in heterotrophic cell-suspension cultures of *Chenopodium rubrum*. *Plant Cell Environ.* 13: 957–963.

Douglas, P., Morrice, N. and MacKintosh, C. (1995) Identification of a regulatory phosphorylation site in the hinge 1 region of nitrate reductase from spinach (*Spinacia olereacea*) leaves. *FEBS Lett.* 377: 113–117.

Geigenberger, P., Reimholz, R., Geiger, M., Merlo, L., Canale, V. and Stitt, M. (1997) Regulation of sucrose and starch metabolism in potato tubers in response to short-term water deficit. *Planta* 201: 502–518.

Herzog, B., Stitt, M. and Heldt, H.W. (1984) Control of photosynthetic sucrose synthesis by fructose 2,6-bisphosphate. III Properties of the cytosolic fructose 1,6-bisphosphatase. *Plant Physiol.* 75: 561–565.

Huber, S.C. and Kaiser, W.M. (1996) 5-Aminoimidazole-4-carboxamide riboside activates nitrate reductase in darkened spinach and pea leaves. *Physiol. Plant.* 98: 833–837.

Huber, S.C., Huber, J.L., Campbell, W.H. and Redinbaugh, M.G. (1992) Comparative studies of the light-modulation of nitrate reductase and sucrose-phosphate synthase activities in spinach leaves. *Plant Physiol.* 100: 706–712.

Huber, S.C., Huber, J.L., Liao, P.-C., Gage, D.A., McMichael, R.W. Jr., Chourey, P.S., Hannah, L.C. and Koch, K. (1996) Phosphorylation of serine-15 of maize leaf sucrose synthase. *Plant Physiol.* 112: 793–802.

McMichael, Jr., R.W., Klein, R.R., Salvucci, M. and Huber, S.C. (1993) Identification of the major regulatory phosphorylation site in sucrose-phosphate synthase. *Arch. Biochem. Biophys.* 307: 248–252.

Moorhead, G., Douglas, P., Morrice, N., Scarabel, M., Aitken, A. and MacKintosh, C. (1996) Phosphorylated nitrate reductase from spinach leaves is inhibited by 14-3-3 proteins and activated by fusicoccin. *Curr. Biol.* **6:** 1104–1113.

Stewart, C.R. (1971) Effect of wilting on carbohydrates during incubation of excised bean leaves in the dark. *Plant Physiol.* **48:** 792–794.

Sullivan, J.E., Brocklehurst, K.J., Marley, A.E., Carey, F., Carling, D. and Beri, R.K. (1994) Inhibition of lipolysis and lipogenesis in isolated rat adipocytes with AICAR, a cell-permeable activator of AMP-activated protein-kinase. *FEBS Lett.* **353:** 33–36.

Toroser, D. and Huber, S.C. (1997) Protein phosphorylation as a mechanism for osmotic-stress activation of sucrose-phosphate synthase in spinach leaves. *Plant Physiol.* **114:** 947–955.

Toroser, D., Athwal, G.S. and Huber, S.C. (1998) Evidence for a sequence specific regulatory interaction between sucrose-phosphate synthase and 14-3-3 proteins. *FEBS Lett.* **435:** 110–114.

Vincent, M.-F., Bontemps, F. and Van den Berghe, G. (1992) Inhibition of glycolysis by 5-amino-4-imidazolecarboxamide riboside in isolated rat hepatocytes. *Biochem. J.* **281:** 267–272.

Winter, H., Huber, J.L. and Huber, S.C. (1998) Membrane association of sucrose synthase: changes during the graviresponse and possible control by protein phosphorylation. *FEBS Lett.* **420:** 151–155.

Zrenner, R. and Stitt, M. (1991) Comparison of the effect of rapidly and gradually developing water-stress on carbohydrate-metabolism in spinach leaves. *Plant Cell Environ.* **14:** 939–946.

6

Sugars, far more than just fuel for plant growth

Uwe Sonnewald and Karin Herbers

1. Introduction

Depending on their ability to produce or consume assimilates, plant organs can be divided into source and sink tissues (Ho, 1988). Source organs, such as mature leaves, produce excess carbohydrates during photosynthesis which are used to support growth and development of sink tissues unable to produce sufficient amounts of assimilates (i.e. sink leaves) or which are unable to produce assimilates at all (i.e. roots, tubers, seeds etc.). The rate of assimilate production in source leaves is controlled by environmental and developmental factors. Environmental factors include abiotic (light, pCO_2, water, minerals) and biotic (pathogens, symbionts) stimuli. In addition, source to sink relations change during development which implies that source metabolism must continuously be adapted to sink demands. As a consequence, plants are extremely responsive to external and internal perturbations in metabolism. This has direct consequences for plant biotechnology since molecular genetic approaches aiming to improve harvest index will need to consider the mechanisms underlying these acclimation processes. In this chapter we will first describe two examples demonstrating the enormous flexibility of plant metabolism towards altered photosynthetic carbon allocation, then, the concept of sugar-mediated gene regulation in higher plants will be introduced and finally strategies to unravel the underlying signal transduction pathway will be discussed.

2. Counter-regulation of plant metabolism to genetic modifications

During the last decade numerous attempts have been undertaken to improve crop productivity by altering single steps in major metabolic pathways (for a summary see Stitt and Sonnewald, 1995; Herbers and Sonnewald, 1996, 1998a, 1998b). Prerequisites for this molecular approach are efficient plant transformation systems, suitable cis-regulatory elements for the proper targeting of gene expression and the desired sub-cellular compartmentation of the gene product, and the gene of interest. Since sucrose is the major transport form of carbohydrates in most higher plants it is not surprising

Plant Carbohydrate Biochemistry, edited by J.A. Bryant, M.M. Burrell and N.J. Kruger.
© 1999 BIOS Scientific Publishers Ltd, Oxford.

that the modulation of its metabolism has extensively been studied in several plant species. With the exception of the last enzyme of the pathway, the sucrose-6-phosphate phosphatase (SPP), the genes encoding all the enzymes have been cloned allowing the *in vivo* modulation of their activities. Based on transgenic plants and available mutants (for a summary see *Table 1*) the flux control coefficient (for details see Kacser and Burns, 1995; Fell and Thomas, 1995) of most enzymes has been determined opening the theoretical possibility of modulating carbon flux through the pathway. Although important insights into the role of individual enzymes have been obtained, improved crop performance under field conditions has not been achieved. The major conclusion that can be drawn from the investigations is that plant metabolism is amazingly flexible to metabolic perturbations and any useful mathematic model must consider and interpret the extensive networking between individual routes. Adaptations to introduced alterations include the use of redundant and/or alternative enzymes, the adjustment of enzyme activities via signal metabolites (Hajirezaei *et al.*, 1993) or posttranslational protein modification (i.e. protein phosphorylation; Huber *et al.*, 1989). In addition, intracellular and extracellular partitioning can be changed to circumvent metabolic blocks set in one compartment. This has been nicely illustrated in transgenic plants characterized by a reduced activity of the triose phosphate translocator (TPT) or the cytosolic fructose-1,6-bisphosphatase (cytFBPase) (see *Figure 1*). Triose phosphates produced during photosynthesis are distributed between plastidic and cytosolic carbon metabolism via the TPT protein. Excess triose phosphates formed in the light are converted into starch which is degraded during the night to fuel metabolism in the absence of photosynthetic carbon fixation. In contrast to primary photosynthate, export of starch breakdown products (glucose and/or glucose-1-phosphate) is thought to be independent of the TPT protein but requires the activities of either the glucose- or the glucose phosphate translocator (GPT). In the anti-TPT and anti-cytFBPase transgenic plants the synthesis of sucrose during the light period is drastically reduced since either triose phosphates cannot be exported from the chloroplasts (in case of anti-TPT plants) or the exported triose phosphates cannot be metabolized to sucrose due to the inability to convert fructose-1,6-bisphosphate into F6P (in case

Table 1. Summary of plants with altered rates of photosynthetic sucrose biosynthesis

Enzyme	Origin	Species	Manipulation	Reference
TPT	Potato	Potato	Antisense	Riesmeier *et al.*, 1993
	Tobacco	Tobacco	Antisense	Hausler *et al.*, 1998
Aldolase	Potato	Potato	Antisense	Haake *et al.*, 1998
cytFBPase	Potato	Potato	Antisense	Zrenner *et al.*, 1996
	Flaveria	*Flaveria*	Mutation	Sharkey *et al.*, 1992
PGI	*Clarkia*	*Clarkia*	Mutation	Kruckeberg *et al.*, 1989
UGPase	Potato	Potato	Antisense	Zrenner *et al.*, 1993
SPS	Maize	Tomato	Over-expression	Laporte *et al.*, 1997
	Potato	Potato	Antisense	Geigenberger *et al.*, 1998
6-PF-2-K	Rat	Tobacco	Over-expression	Scott *et al.*, 1995

UGPase, UDPglucose pyrophosphorylase; SPS, sucrose-6-phosphate synthase; 6-PF-2-K, 6-phosphofructo-2-kinase.

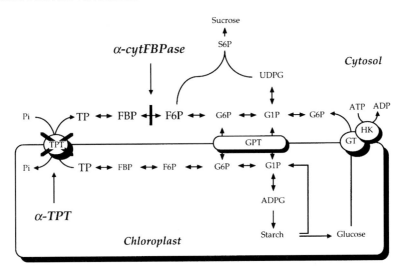

Figure 1. *Metabolic adjustments to a reduced ability to convert triose phosphates into sucrose. Perturbations have been brought about by inhibition of either the TPT or the cytFBPase activities via the antisense strategy.*

of anti-cytFBPase plants). As a consequence, starch is synthesized instead of sucrose during the light period. This starch is subsequently degraded into hexose derivatives circumventing both metabolic blocks (Hausler *et al.*, 1998; Zrenner *et al.*, 1996). Soluble sugars derived from starch are glucose (originating from amylolytic starch degradation) or glucose-1-phosphate (derived from phosphorolytic starch degradation). Both starch phosphorylases and amylases are present in leaf chloroplasts (Sonnewald *et al.*, 1995; Steup and Gerbling, 1983). The hexoses produced are either exported via the GPT or the glucose transporter (GT). In both cases neither the TPT nor the cytFBPase activity is needed for the synthesis of sucrose (*Figure 1*). cDNA clones encoding the GPT protein from several plant species have recently been cloned (Kammerer *et al.*, 1998), and the spinach GPT protein has been shown to be regulated by sugar (Quick *et al.*, 1995). *In vivo* data obtained from the analysis of a high starch *Arabidopsis* mutant suggest that the GT protein is mainly responsible for the export of starch breakdown products (Trethewey and ap Rees, 1994). A corresponding cDNA clone has recently been isolated (I. Flügge, personal communication). The exported glucose could be phosphorylated by a hexokinase localized at the outer membrane of plastids (I. Flügge, personal communication). Thus, changing the export mode from C3- to C6-intermediates allows the mesophyll cell to cope with limitations in the sucrose biosynthetic capacity. This leaves one question open: how does a leaf mesophyll cell 'know' that starch has to be accumulated during the light period for later export to sink tissues during the dark period?

3. Sugars as possible mediator between sink and source tissues

It has long been hypothesized that so-called sink signals exist adjusting the assimilate production in source leaves according to sink demand by altering the activity of photosynthetic genes (Herold, 1980). However, the nature of this signal molecule is still unknown. In recent years indirect evidence has accumulated suggesting that sucrose could play a major role in sink to source signalling. Several genes have been shown to be regulated by external sugars (for recent reviews see Koch, 1996; Smeekens, 1998). Depending on their response to sugars plant genes can be grouped into three classes: 'famine', 'feast' and non-responding genes. Expression of 'famine' genes is up-regulated by sugar depletion, whereas expression of 'feast' genes is enhanced by sugar abundance (definition by Koch, 1996). Photosynthetic genes belong to the first (sugar-repressed) class, which is in agreement with the hypothesis that sugars are involved in the sink regulation of photosynthesis.

Sugars are known to be potent regulators of gene expression in bacteria, yeast and mammals. In bacteria and yeast the expression of many genes is repressed in the presence of glucose (catabolite repression or glucose repression). Although not completely understood, hexokinases seem to be involved in the glucose-mediated signal transduction pathway in yeast (Sanz et al., 1996). Due to the inability to dissect the catalytic and regulatory functions of hexokinase, their role in signal transduction is still questionable. In contrast, the function of the GT proteins, Snf3p and Rgt2p, as the initial sensors for glucose, has recently been demonstrated by site-directed mutagenesis. Conversion of a highly conserved arginine residue in these transporters uncouples the signalling function from their glucose transport activity (Özcan et al., 1996). Both of these transporters differ from other glucose transporters by a C-terminal extension that could interact with downstream proteins transmitting the glucose signal. In plants and yeast hexose transporters are encoded by multigene families (Caspari et al., 1994). Recently, *Arabidopsis thaliana* EST sequences with characteristics similar to the yeast sugar-sensing hexose transporters have been identified (E. Neuhaus, unpublished results). Whether these transporters play a signalling function in higher plants remains to be elucidated.

Although basic regulatory mechanisms are likely to be evolutionarily conserved, in plants newly fixed carbon must be distinguished from imported carbohydrates. Theoretically, several mechanisms of sugar sensing can be imagined (*Figure 2*). Sugars could be sensed by receptor molecules, and a signal generated in the absence of further transport and/or metabolism (*Figure 2a*). Alternatively, sensing might occur by means of (limited) metabolism of the respective sugar, for instance mediated by hexokinases (*Figure 2b*). Signalling in this case could be an integrational event brought about by a

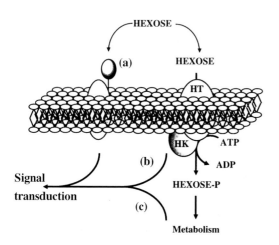

Figure 2. Theoretical model of possible routes for sugar-mediated signal transduction in plants. (a) Sugars are recognized by extracellular receptor molecules. The signal is transmitted without the necessity of sugar transport and/or metabolism. (b) Sugars are transported via specialized sugar transporters transmitting the sugar signal to downstream proteins (e.g. hexokinase). Further metabolism is not required. (c) Sugars are transported, signal transduction requires further metabolism.

sugar-sensing complex between hexokinases and transporter molecules as has been suggested by Jang and Sheen (1997). Alternatively, flux through respective metabolic pathways might be required (*Figure 2c*). In addition, changes in the carbohydrate status might give rise to metabolic signals such as altered cellular redox states which could then modulate sugar-responsive gene expression (Escoubas *et al.*, 1995).

4. Hexokinase as putative sugar sensor in higher plants

Results obtained from the application of sugar analogues and transgenic plants suggest that hexokinases are involved in sugar-mediated signal transduction in plants. 2-deoxyglucose (2-DOG), a substrate of hexokinases, was shown to repress efficiently the expression of photosynthetic genes and of genes of the glyoxylate cycle in maize protoplasts, cucumber and cell suspension cultures, respectively (Graham *et al.*, 1994; Jang and Sheen, 1994; Prata *et al.*, 1995). On the other hand, hexose analogues that cannot be phosphorylated such as 3-O-methylglucose (3-O-mG) and 6-deoxyglucose (6-DOG) as well as glucose-6-phosphate (G6P) and downstream metabolites were ineffective in these systems. Addition of mannoheptulose, a competitive inhibitor of hexokinases, confirmed that hexokinase was involved in transmitting the sugar signal.

Recently, transgenic *Arabidopsis* plants with reduced and elevated levels of hexokinase mRNAs have been created and the sugar response of seedlings has been investigated. Over-expression of hexokinase resulted in enhanced sensitivity to glucose whereas reduced levels of hexokinase led to a reduced glucose sensitivity of seedlings (Jang *et al.*, 1997).

5. Evidence for hexokinase-independent signalling pathways

In addition to a hexokinase-mediated sensing, there is good evidence for hexokinase-independent sensing involving monosaccharide transporter(s) or receptor proteins. Godt *et al.* (1995) and Roitsch *et al.* (1995) presented data showing that sugar induction of cell wall invertase and sucrose synthase could be mimicked by the addition of 3-O-mG and 6-DOG. Similarly, the sugar-responsive patatin class I promoter was shown to be equally inducible by these analogues in *Arabidopsis* plants (Martin *et al.*, 1997). More recently, sucrose-specific regulation of a sucrose transport protein has been shown (Chiou and Bush, 1998). Furthermore, translation of the ATB2 bZIP transcription factor from *Arabidopsis* has been shown to be repressed specifically by sucrose (Rook *et al.*, 1998).

Photosynthetic gene repression by sugars as a mechanism of feedback inhibition of photosynthesis has been shown in many higher plants. Transgenic tobacco plants expressing yeast invertase in the cell wall (cwInv) were found to accumulate high levels of soluble sugars and were therefore used as a model to study the feedback inhibition of photosynthesis (Stitt *et al.*, 1990; von Schaewen *et al.*, 1990). Recently, we found that transcripts encoding defence-related genes were induced in the sugar-accumulating cwInv plants (Herbers *et al.*, 1996). Moreover, comparison of source leaves of these plants with those of sugar-accumulating transgenic plants expressing yeast invertase in the vacuole (vacInv) and in the cytosol (cytInv) (Sonnewald *et al.*, 1991) revealed that the accumulation of pathogenesis-related protein transcripts was restricted to cwInv and vacInv plants. Similarly, the repression of chlorophyll *a/b* binding protein (*cab*) was found not to occur in cytInv plants but only in vacInv and cwInv plants. These data suggest that a hexokinase-independent sensing mechanism for the repression of photosynthetic genes and induction of PR-protein genes operates in source leaves. Sensing in this case appears to be associated with the secretory membrane system.

In order to evaluate further the role of hexokinases in sugar sensing of source leaves we made use of transgenic plants expressing either a 2-deoxyglucose-6-phosphate phosphatase from yeast or a heterologous glucokinase from *Zymomonas mobilis*. In addition, non-metabolizable sucrose isomers have been included in our studies.

6. 2-DOG- and glucose-mediated signalling involves different mechanisms

To study the molecular mechanism of sugar sensing, glucose- and 2-DOG-responsiveness of *AGPaseS*, *rbcS* and *PR-Q* transcription was investigated in source leaves. These genes were considered to be representatives for sugar-regulated storage function, photosynthetic and defence-related genes, respectively. Glucose and 2-DOG were equally effective in repression of *rbcS* and induction of *PR-Q* transcripts. However, accumulation of *AGPaseS* transcripts was not induced by floating leaves on 2-DOG, indicating that phosphorylation is not sufficient for the induction of AGPase expression (K. Herbers unpublished data).

To investigate whether the down-regulation of *rbcS* and the induction of *PR-Q* by 2-DOG in source leaves resulted from hexokinase-mediated signalling we decided to express yeast 2-DOGP-phosphatase in transgenic tobacco plants (DOG^{R}1) (Sanz *et al.*, 1994). In leaves floating on 2-DOG the expression of DOG^{R}1 resulted in dephosphorylation of 2-DOGP to 2-DOG which was again subjected to hexokinase action. In 2-DOG-resistant plants glucose, but not 2-DOG, was effective in modulating gene expression. A similar result has been obtained in yeast where glucose repression by 2-DOG, but not by glucose, was prevented in DOG^{R}1 over-expressing yeast (Randez-Gil *et al.*, 1995). The authors suggested that 2-DOG might trigger the process of catabolite repression through an additional pathway not shared by glucose. In view of the drastic effects on overall metabolism (i.e. altered nucleotide pools, strong inhibition of glycolysis, disorder in cell wall structures and protein glycosylation etc.) we would suggest that down-regulation of photosynthetic and other genes involved in metabolism could be a consequence of 2-DOG toxicity on plant cells. Recently, the effect of 2-DOG on *rbcS* expression and metabolism has been investigated in *Chenopodium rubrum* cell-suspension cultures (Klein and Stitt, 1998). Finding severe perturbations of metabolism led the authors to conclude that 2-DOG cannot be used to investigate the role of hexokinase in sugar sensing since the metabolic consequence overwhelms possible sugar-sensing mechanisms.

7. Hexokinase activity is not sufficient for sugar sensing

To study the involvement of hexokinase in the repression of *rbcS* and induction of *PR-Q* and *AGPaseS* transcripts, two different transgenic plant systems were analysed for sugar-modulated gene expression: (i) tobacco plants expressing a cytoplasmic glucokinase from *Zymomonas mobilis* which should continuously compete with endogenous hexokinase for substrate and (ii) sugar-accumulating transgenic tobacco plants where sucrose would either be cleaved to hexoses by yeast invertase expressed in the cytosol or in the cell wall (Sonnewald *et al.*, 1991; von Schaewen *et al.*, 1990).

Floating experiments of *glk*-expressing leaves revealed no difference in sugar-modulated gene expression of *PR-Q*, *rbcS* and *AGPaseS*. Furthermore, no phenotypic differences were observed in the transgenic plants (K. Herbers unpublished data). Both observations question the role for hexokinase in sugar sensing in leaves of adult plants. This interpretation is supported by comparing transcript levels for *rbcS* and *PR-Q* in cytInv and cwInv plants. Despite hexokinase activity only cwInv plants responded to elevated endogenous sugars by reduction of *rbcS* and induction of *PR-Q* transcripts. This strongly suggests that hexokinase action on intracellular glucose is not sufficient for sugar sensing but rather that exogenous sugars are sensed either at the outside of cells or during transport.

8. Non-metabolizable sucrose isomers are able to induce *PR-Q* expression and to repress *rbcS* transcription

Although indicative of a hexokinase-independent sugar-sensing mechanism the data discussed so far are compatible with hexokinase being associated with a hexose transport protein sensing a subpopulation of imported sugars not assessed by the heterologous glucokinase.

To obtain further evidence for the extracellular signal perception structural isomers of sucrose, that is, disaccharides composed of glucose and fructose molecules with different glucosidic linkages, were used in floating experiments. Previously it has been demonstrated that palatinose (6-O-α-D-glucopyranosyl-D-fructose) and turanose (3-O-α-D-glucopyranosyl-D-fructose) are not recognized by plant sucrose or hexose transporters (Li *et al.*, 1994; M'Batchi and Delrot, 1988) and hence are not transported into the cell. Therefore, these sucrose isomers were selected for floating experiments. Feeding palatinose or turanose to tobacco leaves led to the repression of *rbcS* and the induction of *PR-Q* transcripts, indicating a hexokinase-independent mechanism. In contrast, induction of *AGPaseS* only occurred in the presence of sucrose or glucose. Neither palatinose nor turanose were effective as inducers of *AGPaseS* indicating that extracellular sugar perception is not sufficient but rather that metabolism is required for sugar regulation of *AGPaseS* expression.

9. Conclusions

Single site manipulations of metabolic pathways will rarely lead to major changes in crop productivity. This is because of extensive cross talk between different routes and complex regulatory mechanisms. Future attempts to modulate plant metabolism should include strategies to manipulate signal transduction pathways in addition to altering the properties of selected enzymes. Sugars play a pivotal role in plant development and thus crop yield. Different transduction pathways operate to regulate different sugar-modulated genes. *AGPaseS* induction in leaves requires the metabolism of carbohydrates whereas extracellular perception of sugars is necessary and sufficient for *rbcS* and *PR-Q* regulation.

References

Caspari, T., Will, A., Opekarova, M., Sauer, N. and Tanner, W. (1994) Hexose/H+ symporters in lower and higher plants. *J. Exp. Biol.* **196**: 483–491.

Chiou, T.-J. and Bush, D.R. (1998) Sucrose is a signal molecule in assimilate partitioning. *Proc. Natl Acad. Sci. USA* **95**: 4874–4788.

Escoubas, J.M., Lomas, M., LaRoche, J. and Falkowski, P.G. (1995) Light intensity regulation of *cab* gene transcription is signaled by the redox state of the plastoquinone pool. *Proc. Natl Acad. Sci. USA* **92**: 10237–10241.

Fell, D.A. and Thomas, S. (1995) Physiological control of metabolic flux: the requirement for multisite modulation. *Biochem. J.* **311**: 35–39.

Geigenberger, P., Hajirezaei, M., Deiting, U., Sonnewald, U. and Stitt, M. (1998) Overexpression of pyrophosphatase leads to increased sucrose degradation and starch synthesis, increased activities of enzymes for sucrose–interconversion, and increased levels of nucleotides in growing potato tubers. *Planta* **205**: 428–437.

Godt, D.E., Riegel, A. and Roitsch, T. (1995) Regulation of sucrose synthase expression in *Chenopodium rubrum*: characterisation of sugar-induced expression in photoautotrophic suspension cultures and sink tissue specific expression in plants. *J. Plant Physiol.* **146**: 231–238.

Graham, I.A., Denby, K.J. and Leaver, C.J. (1994) Carbon catabolite repression regulates glyoxylate cycle gene expression in cucumber. *Plant Cell* **6**: 761–772.

Haake, V., Zrenner, R., Sonnewald, U. and Stitt, M. (1998) A moderate decrease of plastid aldolase activity inhibits photosynthesis, alters the levels of sugars and starch, and inhibits growth of potato plants. *Plant J.* **14**: 147–157.

Hajirezaei, M., Sonnewald, U., Viola, R., Carlisle, S., Dennis, D. and Stitt, M. (1993) Transgenic potato plants with strongly decreased expression of pyrophosphate: fructose-6-phosphate phosphotransferase show no visible phenotype and only minor changes in tuber metabolism. *Planta* **192**: 16–30.

Hausler, R.E., Schlieben, N.H., Schulz, B. and Flügge, U.I. (1998) Compensation of decreased triose phosphate/phosphate translocator activity by accelerated starch turnover and glucose transport in transgenic tobacco. *Planta* **204**: 366–376.

Herbers, K. and Sonnewald, U. (1996) Manipulating metabolic partitioning in transgenic plants. *Trends Biotechnol.* **14**: 198–205.

Herbers, K. and Sonnewald, U. (1998a) Transgenic plants in biochemistry and plant physiology. *Progress Botany* **59**: 534–569.

Herbers, K. and Sonnewald, U. (1998b) Molecular determinants of sink strength *Curr. Opinion Plant Biol.* **1**: 207–216.

Herbers, K., Meuwly, P., Frommer, W.B., Métraux, J.-P. and Sonnewald, U. (1996) Systemic acquired resistance mediated by the ectopic expression of invertase: possible hexose sensing in the secretory pathway. *Plant Cell* **8**: 793–803.

Herold, A. (1980) Regulation of photosynthesis by sink activity – the missing link. *New Phytol.* **86**: 131–144.

Ho, L.C. (1988) Metabolism and compartmentation of imported sugars in sink organs in relation to sink strength. *Annu. Rev. Plant Physiol. Plant Mol. Biol.* **39**: 355–378.

Huber, J.L.A., Huber, S.C. and Nielsen, T.H. (1989) Protein phosphorylation as a mechanism for regulation of spinach leaf sucrose-phosphate synthase activity. *Arch. Biochem. Biophys.* **270**: 681–690.

Jang, J.C. and Sheen, J. (1994) Sugar sensing in higher plants. *Plant Cell* **6**: 1665–1679.

Jang, J.C. and Sheen, J. (1997) Sugar sensing in higher plants. *Trends Plant Sci.* **2**: 208–213.

Jang, J.C., Leon, P., Zhou, L. and Sheen, J. (1997) Hexokinase as a sugar sensor in higher plants. *Plant Cell* **9**: 5–19.

Kacser, H. and Burns, J.A. (1995) The control of flux. *Biochem. Soc. Trans.* **23**: 341–366.

Kammerer, B., Fischer, K., Hilpert, B., Schubert, S., Gutensohn, M., Weber, A. and Flugge, U.I. (1998) Molecular characterization of a carbon transporter in plastids from heterotrophic tissues: the glucose 6-phosphate/phosphate antiporter. *Plant Cell* **10**: 105–117.

Klein, D. and Stitt, M. (1998) Effects of 2-deoxyglucose on the expression of rbcS and the metabolism of *Chenopodium rubrum* cell-suspension cultures. *Planta* **205**: 223–234.

Koch, K.E. (1996) Carbohydrate-modulated gene expression in plants. *Annu. Rev. Plant Physiol. Plant Mol. Biol.* **47**: 509–540.

Kruckeberg, A., Neuhaus, H.E., Feil, R., Gottlieb, L. and Stitt, M. (1989) Dosage mutants of phosphoglucose isomerase in cytosol and chloroplast of *Clarkia xantiana*. I. Impact on mass action ratios and fluxes to sucrose and starch. *Biochem. J.* **261**: 457–467.

Laporte, M.M., Galagan, J.A., Shapiro, J.A., Boersig, M.R., Shewmaker, C.K. and Sharkey, T.D. (1997) Sucrose-phosphate synthase activity and yield analysis of tomato plants transformed with maize sucrose-phosphate synthase. *Planta* **203**: 253–259.

Li, Z.S., Noubhani, A.M., Bourbouloux, A. and Delrot, S. (1994) Affinity purification of sucrose binding proteins from the plant plasma membrane. *Biochim. Biophys. Acta* **1219**: 389–397.

Martin, T., Hellmann, H., Schmidt, R., Willmitzer, L. and Frommer, W.B. (1997) Identification of mutants in metabolically regulated gene expression. *Plant J.* **11**: 53–62.

M'Batchi, B. and Delrot, S. (1988) Stimulation of sugar exit from leaf tissues of *Vicia faba* L. *Planta* **174**: 340–348.

Özcan, S., Dover, J., Rosenwald, A.G., Wölfl, S. and Johnston, M. (1996) Two glucose transporters in *Saccharomyces cerevisiae* are glucose sensors that generate a signal for induction of gene expression. *Proc. Natl Acad. Sci. USA* **93**: 12428–12432.

Prata, R.T.N., Williamson, J.D., Conkling, M.A. and Pharr, D.M. (1995) Sugar repression of mannitol dehydrogenase activity in celery cells. *Plant Physiol.* **114**: 307–314.

Quick, W.P., Scheibe, R. and Neuhaus, E.H. (1995) Induction of hexose phosphate translocator activity in spinach chloroplasts. *Plant Physiol.* **109**: 113–121.

Randez-Gil, F., Prieto, J.A. and Sanz, P. (1995) The expression of a specific 2-deoxyglucose-6P phosphatase prevents catabolite repression mediated by 2-deoxyglucose in yeast. *Curr. Genet.* **28**: 101–107.

Riesmeier, J., Flügge, U.I., Schulz, B., Heineke, D., Heldt, H.W., Willmitzer, L. and Frommer, W.B. (1993) Antisense repression of the chloroplast triose phosphate translocator affects carbon partitioning in transgenic potato plants. *Proc. Natl Acad. Sci. USA* **90**: 6160–6164.

Roitsch, T., Bittner, M. and Godt, D.E. (1995) Induction of apoplastic invertase by D-glucose and a glucose analog and tissue specific expression suggest a role in sink-source regulation. *Plant Physiol.* **108**: 285–294.

Rook, F., Gerrits, N., Kortstee, A., van Kampen, M., Borrias, M., Weisbeek, P. and Smeekens, S. (1998) Sucrose-specific signalling represses translation of Arabidopsis ATB2 bZIP transcription factor gene. *Plant J.* **15**: 253–263.

Sanz, P., Randez-Gil, F. and Prieto, J.A. (1994) Molecular characterization of a gene that confers 2-deoxyglucose resistance in yeast. *Yeast* **10**: 1195–1202.

Sanz, P., Nieto, A. and Prieto, J.A. (1996) Glucose repression may involve processes with different sugar kinase requirements. *J. Bacteriol.* **178**: 4721–4723.

Scott, P., Lange, A.J., Pilkis, S.J. and Kruger, N.J. (1995) Carbon metabolism in leaves of transgenic tobacco (*Nicotiana tabacum* L.) containing elevated fructose 2,6-bisphosphate. *Plant J.* **7**: 461–469.

Sharkey, T.D., Savitch, L.V., Vanderveer, P.J. and Micallef, B.J. (1992) Carbon partitioning in a *Flaveria linearis* mutant with reduced cytosolic fructose bisphosphatase. *Plant Physiol.* **100**: 210–215.

Smeekens, S. (1998) Sugar regulation of gene expression in plants. *Curr. Opinion Plant Biol.* **3**: 230–234.

Sonnewald, U., Brauer, M., von Schaewen, A., Stitt, M. and Willmitzer, L. (1991) Transgenic tobacco plants expressing yeast-derived invertase in either the cytosol, vacuole or apoplast: a powerful tool for studying sucrose metabolism and sink/source interactions. *Plant J.* **1**: 95–106.

Sonnewald, U., Basner, A., Greve, B. and Steup, M. (1995) A second L-type isozyme of potato glucan phosphorylase: cloning, antisense inhibition and expression analysis. *Plant Mol. Biol.* **27**: 567–576.

Steup, M. and Gerbling, K.P. (1983) Multiple forms of amylase in leaf extracts: electrophoretic transfer of the enzyme forms into amylose-containing polyacrylamide gels. *Anal. Biochem.* **134:** 96–100.

Stitt, M. and Sonnewald, U. (1995) Regulation of metabolism in transgenic plants (1995) *Annu. Rev. Plant Physiol. Plant Mol. Biol.* **46:** 341–368.

Stitt, M., von Schaewen, A. and Willmitzer, L. (1990) "Sink" regulation of photosynthetic metabolism in transgenic tobacco plants expressing yeast invertase in their cell wall involves a decrease of Calvin-cycle enzymes and an increase of glycolytic enzymes. *Planta* **183:** 40–50.

Trethewey, R.N. and ap Rees, T. (1994) A mutant of *Arabidopsis thaliana* lacking the ability to transport glucose across the chloroplast envelope. *Biochem. J.* **15:** 449–454.

von Schaewen, A., Stitt, M., Schmidt, R., Sonnewald, U. and Willmitzer, L. (1990) Expression of yeast-derived invertase in the cell wall of tobacco and Arabidopsis plants leads to accumulation of carbohydrate and inhibition of photosynthesis and strongly influences growth and phenotype of transgenic tobacco plants. *EMBO J.* **9:** 3033–3044.

Zrenner, R., Willmitzer, L. and Sonnewald, U. (1993) Analysis of the expression of potato uridine phosphate glucose pyrophosphorylase and its inhibition by "antisense" RNA. *Planta* **190:** 247–252.

Zrenner, R., Krause, K.-P., Apel, P. and Sonnewald, U. (1996) Reduction of cytosolic fructose-1,6-bisphosphatase in transgenic potato plants limits photosynthetic sucrose biosynthesis with no impact on plant growth and tuber yield. *Plant J.* **9:** 671–681.

7

The synthesis of the starch granule

Alison M. Smith, Kay Denyer, Samuel C. Zeeman, Anne Edwards and Cathie Martin

1. Introduction

Starch is the main form in which higher plants store carbon, and by far the major component of the harvested parts of the world's major crops. For example, starch contributes between 50 and 80% of the dry weight of mature cereal grains, potato tubers and pea seeds. This makes it the major carbohydrate of nutritional importance in the human diet. It is also of enormous value as a raw material for a wide range of industries. When cooked in the presence of water, starch swells (gelatinizes) to form gels or pastes which are used as thickeners, texturizers, stabilizers and fat-replacements in the food industry and in the manufacture of materials including paints, glues, paper, textiles, pharmaceuticals and biodegradable plastics. However, in spite of its importance, the way in which starch is synthesized in the plant is still poorly understood. In this chapter, we will describe briefly the structure of starch, then discuss recent progress in discovering how it is synthesized.

2. The structure of the granule

2.1 *The polymers and their organization*

Starch occurs in plants as semi-crystalline granules of complex structure and organization. Granules are made up of two types of glucose polymer: amylose, which consists of essentially linear chains of glucose units and contributes about 30% of storage starches, and amylopectin, a highly branched polymer which makes up the remaining 70% of storage starches. The amylopectin molecule contains regularly spaced clusters of chains of 12–20 glucose units. These clusters of shorter branches are joined by longer chains (40 or more glucose units) which span two, three or more clusters. This distinctive, polymodal distribution of chain lengths within the amylopectin molecule allows the polymer to become organized to form a semi-crystalline granule (see *Figure 1*). Within the granule, adjacent short chains within clusters form double

Plant Carbohydrate Biochemistry, edited by J.A. Bryant, M.M. Burrell and N.J. Kruger.
© 1999 BIOS Scientific Publishers Ltd, Oxford.

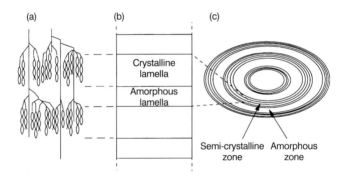

Figure 1. *Organization of the starch granule. (a) The cluster structure of the amylopectin molecule, and formation of double helices between adjacent chains within the clusters. (b) The double helices become ordered in regular arrays, giving rise to crystalline lamellae. These alternate with amorphous lamellae in which the branch points of the amylopectin molecule are primarily located. The periodicity of this repeat structure is 9 nm. (c) Semi-crystalline zones of alternating crystalline and amorphous lamellae alternate with amorphous zones, with a periodicity of hundreds of nanometres. Reprinted from* Photosynthesis: Physiology and Metabolism, *R.C. Leegood, T.D. Sharkey and S. von Caemmerer (eds), 1999, Starch synthesis in leaves, R. Trethewey and A.M. Smith, Figure 1, with kind permission from Kluwer Academic Publishers.*

helices. These helices pack together in ordered arrays to give crystalline lamellae. Crystalline lamellae are separated by amorphous lamellae containing the branch points of the polymer, giving rise to semi-crystalline zones of concentrically organized, alternating crystalline and amorphous lamellae. The semi-crystalline zones themselves alternate with a periodicity of hundreds of nanometres with amorphous zones in which the amylopectin is considerably less organized (French, 1984; Jenkins et al., 1993). Much of the amylose component of the granule is probably within the amorphous zone as single helices, interspersed with amylopectin molecules (Jane et al., 1992). A semi-crystalline-amorphous repeat is called a 'growth ring'. New microscopic and physical techniques are revealing still further levels of organization within the granule. For example, it has been proposed that the semi-crystalline regions are organized into interdigitating suprahelical structures (Oostergetel and van Bruggen, 1993), and into discrete domains or 'blocklets' (Gallant et al., 1997).

2.2 *Approaches to understanding granule synthesis*

The complexity of the starch granule potentially presents a major challenge to the biologist trying to understand its synthesis. However, there is now reason to suppose that at least part of the organization of amylopectin to form a granule may be a process of self-assembly, rather than biologically catalysed. It is proposed that amylopectin has the properties of a side-chain liquid crystal polymer. Provided that an appropriate polymodal distribution of branch lengths is generated by enzymes at the outer edge of the growing granule, the newly synthesized material will form double helices and crystallize (Waigh et al., 1996, 1997, 1998). The main tasks of the biologist are therefore to explain how a polymodal distribution of branch lengths may be generated at the outer edge of the granule, and how both a highly branched polymer, amylopectin, and an essentially unbranched polymer, amylose, may be synthesized at the same time.

A valuable approach to these problems is to examine the biochemical basis of the many mutations known to affect the structure and composition of starch in storage organs. Some such mutations apparently affect only one aspect of starch structure and composition. For example, mutations at the *waxy* loci of cereals, the *amf* locus of potato and the *lam* locus of pea all eliminate the amylose component of starch and have little effect on amylopectin or on granule organization (Denyer *et al.*, 1995; Hovenkamp-Hermelink *et al.*, 1987; Smith and Martin, 1993). However, most of these mutations affect many aspects of the starch. The mutations at the *r* locus of pea and the *sugary*1 loci of maize and rice, for example, affect radically the whole structure and organization of the granule (Colonna and Mercier, 1984; Matsuo *et al.*, 1987; Sumner and Somers, 1944). These are further discussed below.

In spite of the complex effects of many of the mutations on granule structure and composition, almost all of those thus far characterized lie in genes encoding two enzymes: starch synthase and starch-branching enzyme. These two enzymes alone can potentially synthesize both amylose and amylopectin. Starch synthase adds a glucose unit from the sugar nucleotide ADPglucose to the non-reducing end of a glucose chain. Starch-branching enzyme cleaves a length from the end of a chain and transfers it to the side of the same or an adjacent chain to form a branch. Both enzymes exist as multiple, distinct isoforms, each of which is encoded by a different gene. It seems reasonable to suggest that the presence *in vivo* of several different isoforms of the two enzymes, each with different specificities, locations on the granule and developmental timing of activity, might be very important in allowing amylopectin and amylose to be synthesized (Smith *et al.* 1997). We will examine first whether the existence of multiple isoforms of these two enzymes provides a sufficient explanation for amylopectin synthesis, then examine the role of distinct isoforms in amylose synthesis.

3. The synthesis of amylopectin

3.1 *The role of starch-branching enzymes*

The existence of multiple isoforms of starch-branching enzymes offers a possible explanation for the polymodal distribution of branch lengths of amylopectin. This type of distribution might result from the simultaneous activities of two or more isoforms with different preferences for the length of chain transferred. Support for this idea comes from examination of the isoforms of starch-branching enzyme for which primary amino acid sequences are available. These can be divided on the basis of clear differences in sequence into two classes, referred to as A and B (Burton *et al.*, 1995). Storage organs of, for example, pea, wheat, maize, rice and potato, possess members of both classes. Detailed studies of the A and B isoforms of maize endosperm *in vitro* and when expressed in *Escherichia coli* show that they do indeed differ in their preferences for the length of chain transferred (Guan and Preiss, 1993; Guan *et al.*, 1995; Takeda *et al.*, 1993), and it seems likely that the differences between the maize isoforms reflect differences between the A and B classes generally.

The existence in several species of mutations which eliminate activity of the A isoform of starch-branching enzyme has allowed us to test the idea that the differences in properties between the A and B isoforms can account for the polymodal distribution of chain lengths of amylopectin. The mutation at the *r* locus of pea lies in the gene encoding the A isoform (Bhattacharyya *et al.*, 1990). Activity of starch-branching enzyme in the embryo and leaf of the mutant pea is much lower than in the organs of the wild-type pea. All of the activity in the mutant pea is contributed by the B isoform, whereas in the wild-type pea the activity is contributed by both A and B isoforms (Smith, 1988). However, although the average chain length of amylopectin in the mutant pea is longer than that in the wild-type pea, the basic polymodal distribution of chain lengths is unaffected by the mutation (Lloyd *et al.*, 1996; Tomlinson *et al.*, 1997). The starch granules of the mutant embryo, although of unusual morphology, contain semi-crystalline zones of the same basic structure as those in starch from wild-type embryos (Jenkins and Donald, 1995). We conclude that the synthesis of amylopectin with a structure appropriate for organization to form a starch granule does not require two different isoforms of starch-branching enzyme.

3.2 *The role of starch synthases*

As with starch-branching enzyme, the isoforms of starch synthase responsible for amylopectin synthesis can be divided into several classes on the basis of distinct differences in their primary sequences. At present three such classes, known as SSI, SSII and SSIII, can be defined. Many organs possess members of more than one class, and there is good evidence that the endosperm of maize and the tubers of potato possess all three (Abel *et al.*, 1996; Gao *et al.*, 1998; Harn *et al.*, 1998; Knight *et al.*, 1998; Marshall *et al.*, 1996; Tomlinson *et al.*, 1998). Until recently, however, it was not known whether SSI, II and III played different roles in amylopectin synthesis, and what these roles might be.

   The discovery of a mutant pea specifically lacking the SSII isoform gave us the first evidence that different isoforms do indeed play different roles. A mutation at the *rug*5 locus of pea lies in the gene encoding SSII, an isoform which accounts for over half of the amylopectin-synthesizing starch synthase activity in the wild-type pea embryo (Craig *et al.*, 1998, Denyer and Smith, 1992; Dry *et al.*, 1992). In the absence of SSII, there are small increases in the activity of other isoforms of starch synthase so that total activity in the mutant is similar to that in wild-type embryos and the rate of starch synthesis is also similar throughout most of embryo development. In spite of this, the starch of the mutant is radically different from that of wild-type peas. The amylopectin is seriously deficient in chains of about 40–50 glucose units: the length required to span two clusters. It is enriched in both very short chains and in extremely long chains. These changes result in highly contorted granules, presumably because the organization of the amylopectin has been affected by the changes in its branch length distribution (Craig *et al.*, 1998).

   The phenotype of the *rug*5 mutant indicates that the SSII isoform of starch synthase plays some specific role in the synthesis of amylopectin, a role which other isoforms cannot replace. To provide further information about the extent to which individual isoforms have specific roles, and to discover whether particular classes of isoforms play particular, definable roles, we examined the roles of the two main isoforms responsible for amylopectin synthesis in the potato tuber. Isoforms of the SSIII and SSII class respectively account for about 80% and 10–15% of the soluble starch synthase activity in the tuber (Edwards *et al.*, 1995; Marshall *et al.*, 1996). Tubers of transgenic plants expressing antisense RNA for one or both of these isoforms have the same starch contents as normal tubers, but have amylopectin with very different chain length profiles. The chain length profiles of amylopectin from lines with reductions in SSII, lines with reductions in SSIII, and lines with reductions in both SSII and SSIII all differ considerably and reproducibly from each other, and this is reflected in differences between the lines in granule morphology and the gelatinization properties of the starch. The phenotype of the lines with reductions in both SSII and SSIII is very different from that which could be predicted from the phenotypes of lines in which either SSII or SSIII is reduced (Edwards *et al.*, 1999).

Two important conclusions can be drawn from these analyses. Firstly, SSII and SSIII both make distinct and specific contributions to the synthesis of amylopectin. Secondly, their contributions depend not only upon their intrinsic properties but also upon whether the other isoform is present. Thus the contribution made by SSII when SSIII is present is probably different from that which it makes when SSIII is absent, and the same is true for SSIII in the presence and absence of SSII. This complexity is not surprising. The growing surface of the starch granule, which is the substrate for an isoform of starch synthase *in vivo*, is a complex product of the actions of several different isoforms of starch synthase and starch-branching enzyme. The product of the isoform will depend both upon its intrinsic properties and upon the precise nature of the substrate with which it is presented. Its contribution will thus depend upon the complement of other isoforms of starch synthase and starch-branching enzyme present, and this in turn will depend on a host of genetic, developmental and environmental factors. Thus although apparently similar isoforms of starch synthase are present in many different starch-synthesizing organs, it is likely that their contributions to amylopectin synthesis will be different in each of these organs.

3.3 *A possible role for debranching enzymes*

Although most of the previously described mutations with radical effects on granule structure and composition lie in genes encoding starch synthase and starch-branching enzyme, there are three notable exceptions. Mutations at the *sugary*1 loci of maize and rice and the *STA7* locus of the unicellular green alga *Chlamydomonas rheinhardtii* result in the replacement of some or all of the starch with a water-soluble glucan known as phytoglycogen (Matsuo *et al.*, 1987; Mouille *et al.*, 1996; Sumner and Somers, 1944). Phytoglycogen is more highly branched than amylopectin, and does not form organized, semi-crystalline granules. All three mutations affect the activity during the period of starch synthesis of a class of enzymes known as debranching enzyme (Mouille *et al.*, 1996; Nakamura *et al.*, 1996; Pan and Nelson, 1984): in fact the mutation at the *sugary*1 locus of maize has been shown to lie in a gene encoding one such enzyme, isoamylase (James *et al.*, 1995). These enzymes are involved in starch degradation, but are also known to be present in starch-synthesizing organs. They hydrolyse the linkages which form the branch points in glucans. The phenotypes of the mutants have led to the development of a model in which debranching is an integral part of the synthesis of amylopectin (Ball *et al.*, 1996).

The model proposes that in the first step in the synthesis of a cluster of branches at the surface of the growing granule, starch synthase elongates from a bed of short chains. When chains of sufficient length have been synthesized, branching enzyme is able to act. A highly branched, unorganized glucan (known as pre-amylopectin) is synthesized. This is in turn trimmed by debranching enzyme to give a bed of short chains from which the next round of chain elongation can occur.

This 'trimming' model is appealing in that it integrates the synthesis of amylopectin with its organization to form a granule, and offers an explanation for the accumulation of phytoglycogen in mutants deficient in debranching enzyme (that is, phytoglycogen is an accumulation of pre-amylopectin). However, it is not known whether starch-branching enzyme and debranching enzymes have the rather specific properties required of them by the model. Interpretation of the phenotypes of the *sugary*1 mutants is also complicated by several major pleiotropic effects on enzymes other than debranching enzymes (Nakamura *et al.*, 1996; Singletary *et al.*, 1997). Thus although the model has been widely quoted, and has stimulated much new research into debranching enzymes, it has not yet proved possible to test it rigorously.

We have recently discovered an *Arabidopsis* mutant (*dbe*1) which sheds new light on the validity of the model. This mutant specifically lacks an isoamylase present in wild-type chloroplasts, probably because of a mutation in a gene encoding this enzyme. Effects of the mutation on other enzymes of starch synthesis and degradation are minimal. During the day the chloroplasts of the mutant accumulate both starch and an abnormal, soluble, highly branched glucan similar to the phytoglycogen of the *sugary*1 mutants. Although the rate of starch synthesis is much lower than in wild-type leaves, the starch of the mutants is indistinguishable from that of the wild type with respect to granule morphology and the chain length distribution of its amylopectin (Zeeman *et al.*, 1998).

It is not easy to reconcile the 'trimming' model with our observation that two very different branched glucans, amylopectin and phytoglycogen, accumulate in the same chloroplasts at the same time in the *dbe*1 mutant. The simple expectation from the model is that reductions in debranching enzyme activity would produce a single type of branched glucan, with a chain length distribution less like that of amylopectin and more like that of phytoglycogen as the extent of reduction of activity is increased. We propose instead that debranching enzyme is not directly involved in amylopectin synthesis, and that it plays a role in recycling soluble products of starch synthases and branching enzymes, as follows (see *Figure 2*).

Although the major substrate for starch synthases and starch-branching enzymes in the normal plastid is the surface of the granule, starch synthase can also potentially elongate small malto-oligosaccharides (e.g. maltose and maltotriose) which are likely to be present in the plastid stroma. The consequences of this activity would be deleterious for starch synthesis. Extensive elongation of the soluble glucans would allow branching by starch-branching enzyme, creating more reducing ends and allowing accelerated synthesis of branched, soluble glucan at the expense of synthesis of starch at the granule surface. We argue that the accumulation of soluble glucans is normally prevented by the presence in the stroma of a suite of enzymes capable of attacking soluble linear and branched glucans. There is good evidence for the presence of such enzymes in starch-synthesizing plastids; in addition to isoamylase many plastids are known to have high activities of, for example, phosphorylase and disproportionating enzyme. Deficiencies in this recycling mechanism, for example a lack of isoamylase, will allow soluble glucans to accumulate. The rate of starch synthesis will be reduced because the soluble glucan presents an alternative substrate for the starch synthases and starch-branching enzymes, but some normal amylopectin can continue to be synthesized.

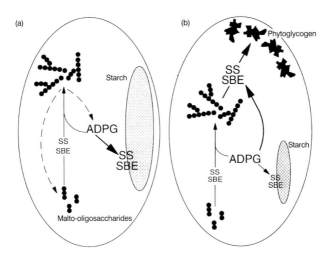

Figure 2. *A model to explain the phytoglycogen-accumulating phenotype of the* Arabidopsis *mutant* dbe1*. (a) A chloroplast from a wild-type leaf, in which starch synthase (SS) and starch-branching enzyme (SBE) are primarily involved in the synthesis of starch from ADPglucose (ADPG), at the surface of the starch granule. These enzymes may also elaborate small malto-oligosaccharides present in the stroma, but the products of this activity will be degraded (dashed lines) by stromal starch-degrading enzymes, including isoamylase. (b) A chloroplast from a* dbe1 *leaf. Loss of isoamylase activity reduces the rate at which soluble products of starch synthase and starch-branching enzyme can be degraded. Elaboration of these soluble glucans exceeds the rate of their degradation and they accumulate in the stroma as phytoglycogen. The soluble glucans provide large amounts of substrate for starch synthase and starch-branching enzyme. This reduces the amount of these enzymes available for starch synthesis, and thus reduces the rate of starch synthesis.*

We believe that this very indirect role for isoamylase provides a simpler explanation of the phenotype of the *dbe*1 mutant than does the trimming model, and that it may also offer an alternative explanation of the *sugary*1 phenotypes. However, these ideas are at present very speculative, and this remains a field in which detailed and innovative experimentation is urgently required.

4. The synthesis of amylose

The synthesis of amylose is an exclusive function of a class of isoforms of starch synthase bound to starch granules, known as granule-bound starch synthase I or GBSSI. The starch of mutants lacking GBSSI, and of transgenic plants in which GBSSI has been severely reduced by expression of antisense RNA, contains little or no amylose (Denyer *et al.*, 1995; Hovenkamp-Hermelink *et al.*, 1987; van der Leij *et al.*, 1991; Smith and Martin, 1993; Visser *et al.*, 1991). The reasons why GBSSI, and not other isoforms of starch synthase, can synthesize amylose are not yet clear, but recent studies of GBSSI are revealing unusual properties which point to an explanation.

Early speculation about the mechanism of amylose synthesis was based on the fact that GBSSI is bound to the starch granule. It was argued that the product of GBSSI, in contrast to the products of other, soluble isoforms of starch synthase, would be within the matrix of the granule, and hence inaccessible to starch-branching enzyme. The product would thus remain unbranched (Denyer *et al.*, 1993). Detailed studies of granule-bound proteins have revealed this explanation to be inadequate. Isoforms of starch synthase other than GBSSI are also found tightly bound to the granule. These isoforms are also present in the soluble fraction of the stroma, and it is likely that they become incorporated into the matrix of the granule as their amylopectin product crystallizes around them. Although they have been shown to be capable of activity within and after extraction from the granule matrix, the fact that they are present in the starch of mutants lacking GBSSI reveals that they cannot synthesize amylose (Denyer *et al.*, 1993, 1995; Edwards *et al.*, 1996; Hylton *et al.*, 1996). Thus features of GBSSI other than, or in addition to, its location on the granule must be required for amylose synthesis.

We have investigated these features using starch granules isolated from wild-type and mutant peas. Wild-type starch granules contain both the GBSSI and SSII isoforms of starch synthase. Granules of the *lam* mutant contain only SSII (Denyer *et al.*, 1995), and those of the *rug*5 mutant contain only GBSSI (Craig *et al.*, 1998). When supplied with ADPglucose, the isoforms within the isolated granules incorporate the glucose unit into the starch. The use of ADP[^{14}C]glucose, followed by fractionation of the starch into amylose and amylopectin, can thus reveal which polymers are elongated by the two isoforms. We discovered that neither SSII nor GBSSI incorporated glucose into amylose in isolated granules: both elongated chains within the amylopectin fraction. This suggested that a soluble factor, which was lost during isolation of the granules, was necessary for amylose synthesis. In agreement with this idea, addition of low concentrations of small, soluble malto-oligosaccharides such as maltose and maltotriose to the isolated granules allowed GBSSI, but not SSII, to synthesize amylose within the granule. GBSSI can use these malto-oligosaccharides as initial substrates (primers) for synthesis of considerably longer chains which are unable to diffuse out of the starch granule, whereas SSII cannot (Denyer *et al.*, 1996, and unpublished data).

We investigated whether the failure of SSII to synthesize amylose was because it cannot use malto-oligosaccharides as substrates, or because it cannot elongate them sufficiently to prevent their diffusion out of the granule. Products of about nine or more glucose units in length are likely to be trapped inside the granule. By examining the soluble products from granules containing only SSII, we found that this isoform can indeed use short malto-oligosaccharides as substrates, but that it adds only a single glucose unit before dissociating. Thus the only detectable product of elongation of maltotriose by SSII (three glucose units) was maltotetraose (four glucose units). In contrast, GBSSI synthesized a whole range of soluble products from maltotriose, ranging from maltotetraose up to about nine glucose units in length (K. Denyer *et al.*, unpublished data). This reveals that GBSSI does not dissociate from its immediate glucan product, but uses it as a substrate for the addition of further glucose units from ADPglucose. This processive mode of action allows the synthesis of long chains even at high concentrations of malto-oligosaccharide primer. Our initial studies of GBSSI in a soluble form (from an *E. coli* expression system) suggest that its mode of action and other kinetic properties may be strongly influenced by its location within the amylopectin matrix of the granule. A primary aim of further work will be to define the influence of amylopectin on the enzyme, and to identify the regions of the protein responsible for its association with the granule matrix.

The fundamental difference in reaction mechanism between SSII and GBSSI is probably central to the unique ability of GBSSI to synthesize amylose. We suggest that, *in vivo*, GBSSI within the granule processively elongates malto-oligosaccharides which diffuse in from the stroma. This produces chains too long to diffuse out of the granule, and these are subjected to further elongation to produce amylose. The factors which determine the amount of amylose synthesized and the size of the molecules produced remain to be determined.

References

Abel, G.J.W., Springer, F., Willmitzer, L. and Kossmann, J. (1996) Cloning and functional analysis of a cDNA encoding a novel 139 kDa starch synthase from potato (*Solanum tuberosum* L.). *Plant J.* 10: 981–991.

Ball, S., Guan, H.P., James, M., Myers, A., Keeling, P., Mouille, G., Buléon, A., Colonna, P. and Preiss, J. (1996) From glycogen to amylopectin: a model for the biogenesis of the starch granule. *Cell* 86: 349–352.

Bhattacharyya, M.K., Smith, A.M., Ellis, T.H.N., Hedley, C. and Martin, C. (1990) The wrinkled-seed character of pea described by Mendel is caused by a transposon-like insertion in a gene encoding starch-branching enzyme. *Cell* 60: 115–121.

Burton, R.A., Bewley, J.D., Smith, A.M., Bhattacharyya, M.K., Tatge, H., Ring, S., Bull, V., Hamilton, W.D.O. and Martin, C. (1995) Starch branching enzymes belonging to distinct enzyme families are differentially expressed during pea embryo development. *Plant J.* 7: 3–17.

Colonna, P. and Mercier, C. (1984) Macromolecular structure of wrinkled- and smooth-pea starch components. *Carbohydr. Res.* 126: 233–247.

Craig, J., Lloyd, J.R., Tomlinson, K., Barber, L., Edwards, A., Wang, T.L., Martin, C., Hedley, C.L. and Smith, A.M. (1998) Mutations in the gene encoding starch synthase II profoundly alter amylopectin structure in pea embryos. *Plant Cell* 10: 413–426.

Denyer, K. and Smith, A.M. (1992) The purification and characterisation of two forms of soluble starch synthase from developing pea embryos. *Planta* 186: 609–617.

Denyer, K., Sidebottom, C., Hylton, C.M. and Smith, A.M. (1993) Soluble isoforms of starch synthase and starch-branching enzyme also occur within starch granules in developing pea embryos. *Plant J.* 4: 191–198.

Denyer, K., Barber, L.M., Burton, R., Hedley, C.L., Hylton, C.M., Johnson, S., Jones, D.A., Marshall, J., Smith, A.M., Tatge, H., Tomlinson, K. and Wang, T.L. (1995) The isolation and characterisation of novel low-amylose mutants of *Pisum sativum* L. *Plant Cell Environ.* 18: 1019–1026.

Denyer, K., Clarke, B., Hylton, C., Tatge, H. and Smith, A.M. (1996) The elongation of amylose and amylopectin chains in isolated starch granules. *Plant J.* 10: 1135–1143.

Dry, I., Smith, A.M., Edwards, E.A., Bhattacharyya, M.K., Dunn, P. and Martin, C, (1992) Characterisation of cDNAs encoding two isoforms of granule-bound starch synthase which show differential expression in developing storage organs. *Plant J.* 2: 193–202.

Edwards, A., Marshall, J., Sidebottom, C., Visser, R.G.F., Smith, A.M. and Martin, C. (1995) Biochemical and molecular characterisation of a novel starch synthase from potato tubers. *Plant J.* 8: 283–294.

Edwards, A., Marshall, J., Denyer, K., Sidebottom, C., Visser, R.G.F., Martin, C. and Smith, A.M. (1996) Evidence that a 77-kilodalton protein from the starch of pea embryos is an isoform of starch synthase that is both soluble and granule-bound. *Plant Physiol.* 112: 89–97.

Edwards, A., Fulton, D., Hylton, C., Jobling, S., Gidley, M., Rössner, U., Martin, C. and Smith, A.M. (1999) A combined reduction in activity of starch synthases II and III of potato has novel effects on the starch of tubers. *Plant J.* (in press).

French, D. (1984) Organisation of starch granules. In: *Starch: Chemistry and Technology* (eds R.L. Whistler, J.N. BeMiller and J.F. Paschall). Academic Press, Orlando, FL, pp. 183–247.

Gallant, D.J., Bouchet, B. and Baldwin, P.M. (1997) Microscopy of starch: evidence of a new level of granule organization. *Carbohydr. Polymers* 32: 177–191.

Gao, M., Wanat, J., Stinard, P.S., James, M.G. and Myers, A.M. (1998) Characterisation of *dull*1, a maize gene coding for a novel starch synthase. *Plant Cell* 10: 399–412.

Guan, H.P. and Preiss, J. (1993) Differentiation of the properties of the branching enzymes from maize (*Zea mays*). *Plant Physiol.* 102: 1269–1273.

Guan, H.P., Kuriki, T., Sivak, M. and Preiss, J. (1995) Maize branching enzyme catalyses synthesis of glycogen-like polysaccharide in *Escherichia coli*. *Proc. Natl Acad. Sci. USA* **92**: 964–967.

Harn, C., Knight, M., Ramakrishnan, A., Guan, H.P., Keeling, P.L. and Wasserman, B.P. (1998) Isolation and characterization of the *zSSIIa* and *zSSIIb* cDNA clones from maize endosperm. *Plant Mol. Biol.* **37**: 639–649.

Hovenkamp-Hermelink, J.H.M., Jacobsen, E., Ponstein, A.S., Visser, R.G.F., Vos-Scheperkeuter, G.H., Bijmolt, E.W., de Vries, J.N., Witholt, B. and Feenstra, W.J. (1987) Isolation of an amylose-free starch mutant of the potato (*Solanum tuberosum* L.). *Theor. Appl. Genet.* **75**: 217–221.

Hylton, C.M., Denyer, K., Keeling, P.L., Chang, M.T. and Smith, A.M. (1996) The effect of *waxy* mutations on the granule-bound starch synthases of barley and maize endosperms. *Planta* **198**: 230–237.

James, M.G., Robertson, D.S. and Myers, A.M. (1995) Characterisation of the maize gene *sugary*1, a determinant of starch composition. *Plant Cell* **7**: 417–429.

Jane, J.L., Xu, A., Radosavljevic, M. and Seib, P.A. (1992) Location of amylose in normal starch granules. I. Susceptibility of amylose and amylopectin to cross-linking reagents. *Cereal Chem.* **69**: 405–409.

Jenkins, P.J. and Donald, A.M. (1995) The influence of amylose on starch granule structure. *Int. J. Biol. Macromol.* **17**: 315–321.

Jenkins, P.J., Cameron, R.E. and Donald, A.M. (1993) A universal feature in the structure of starch granules from different botanical sources. *Starch* **45**: 417–420.

Knight, M.E., Harn, C., Lilley, C.E., Guan, H.P., Singletary, G.W., Mu-Foster, C., Wasserman, B.P. and Keeling, P.L. (1998) Molecular cloning of starch synthase I from maize (W64) endosperm and expression in *Escherichia coli*. *Plant J.* **14**: 613–622.

van der Leij, F.R., Visser, R.G.F., Ponstein, A.S., Jacobsen, E. and Feenstra, W.J. (1991) Sequence of the structural gene for granule-bound starch synthase of potato (*Solanum tuberosum* L.) and evidence for a single point deletion in the *amf* allele. *Mol. Gen. Genet.* **228**: 240–248.

Lloyd, J.R., Bull, V.J., Hedley, C.L., Wang, T.L. and Ring, S.G. (1996) Determination of the effect of the *r* and *rb* mutations on the structure of amylose and amylopectin in pea *(Pisum sativum* L.) *Carbohydr. Polymers* **29**: 45–49.

Marshall, J., Sidebottom, C., Debet, M., Martin, C., Smith, A.M. and Edwards, A. (1996) Identification of the major starch synthase in the soluble fraction of potato tubers. *Plant Cell* **8**: 1121–1135.

Matsuo, T., Yano, M., Satoh, H. and Ohmura, T. (1987) Effects of *sugary* and *shrunken* mutant genes on carbohydrates in rice endosperm during the ripening period. *Jpn J. Breed.* **37**: 17–21.

Mouille, G., Maddelein, M.L., Libessart, N., Tagala, P., Decq, A., Delrue, B. and Ball, S. (1996) Preamylopectin processing: a mandatory step for starch biosynthesis in plants. *Plant Cell* **8**: 1353–1366.

Nakamura, Y., Umemoto, T., Takahata, Y., Komae, K., Amano, E. and Satoh, H. (1996) Changes in the structure of starch and enzyme activities affected by *sugary* mutations in developing rice endosperm. Possible role of starch debranching enzyme (R-enzyme) in amylopectin biosynthesis. *Physiol. Plant.* **97**: 491–498.

Oostergetel, G.T. and van Bruggen, E.F.J. (1993) The crystalline domains in potato starch granules are arranged in a helical fashion. *Carbohydr. Polymers* **21**: 7–12.

Pan, O. and Nelson, O.E. (1984) A debranching enzyme deficiency in endosperms of the *sugary-1* mutants of maize. *Plant Physiol.* **74**: 324–328.

Singletary, G.W., Banisadr, R. and Keeling, P. (1997) Influence of gene dosage on carbohydrate synthesis and enzyme activities in endosperm of starch-deficient mutants of maize. *Plant Physiol.* **113**: 293–304.

Smith, A.M. (1988) Major differences in isoforms of starch-branching enzyme between developing embryos of round- and wrinkled-seeded peas (*Pisum sativum* L.). *Planta* **175**: 270–279.

Smith, A.M. and Martin, C. (1993) Starch biosynthesis and the potential for its manipulation. In: *Biosynthesis and Manipulation of Plant Products* (ed. D. Grierson). Blackie, Glasgow, pp. 1–54.

Smith, A.M., Denyer, K. and Martin, C. (1997) The synthesis of the starch granule. *Annu. Rev. Plant Physiol. Plant Mol. Biol.* **48**: 67–87.

Sumner, J.B. and Somers, G.F. (1944) The water-soluble polysaccharides of sweet corn. *Arch. Biochem.* **4**: 7–9.

Takeda, Y., Guan, H.P. and Preiss, J. (1993) Branching of amylose by the branching isoenzymes of maize endosperm. *Carbohydr. Res.* **240**: 253–263.

Tomlinson, K., Lloyd, J.R. and Smith, A.M. (1997) Importance of isoforms of starch-branching enzyme in determining the structure of starch in pea leaves. *Plant J.* **11**: 31–43.

Tomlinson, K., Craig, J. and Smith, A.M. (1998) Major differences in isoform composition of starch synthases between leaves and embryos of pea (*Pisum sativum* L.). *Planta* **204**: 109–119.

Visser, R.G.F., Somhorst, I., Kuipers, G.J., Ruys, N.J., Feenstra, W.J. and Jacobsen, E. (1991) Inhibition of expression of the gene for granule-bound starch synthase in potato by antisense constructs. *Mol. Gen. Genet.* **225**: 289–296.

Waigh, T.A., Jenkins, P.J. and Donald, A.M. (1996) Quantification of water in carbohydrate lamellae using SANS. *Faraday Discuss.* **103**: 325–337.

Waigh, T.A., Hopkinson, I. and Donald, A.M. (1997) An analysis of the native structure of starch granules with x-ray microfocus diffraction. *Macromolecules* **30**: 3813–3820.

Waigh, T.A., Perry, P., Reikel, C., Gidley, M.J. and Donald, A.M. (1998) Chiral side-chain liquid-crystalline polymeric properties of starch. *Macromolecules* **31**: 7980–7984.

Zeeman, S.C., Umemoto, T., Lue, W.L., Au-Yeung, P., Martin, C., Smith, A.M. and Chen, J. (1998) A mutant of *Arabidopsis thaliana* lacking a chloroplastic isoamylase accumulates both starch and phytoglycogen. *Plant Cell* **10**: 1699–1711.

8

From enzyme activity to flux control: a quest to understand starch deposition in developing cereal grains

Peter L. Keeling

1. Introduction

In a single year our planet produces enough starch to meet over 80% of human caloric needs. All of this is derived from a molecule consisting of glucose units held together by simple α-1,4 linkages branched with simple α-1,6 bonds. It is a remarkable statistic that, averaged over a year, every second over 2×10^{29} α-1,4 linkages are synthesized in cereal crops alone. Since the earliest reports by Cardini and Leloir that ADPglucose and starch synthase are involved in the process of starch synthesis, great progress has been made. Yet, there remains much that we still do not understand because many simple questions remain only partially answered. In this paper I will focus on the enzymes of starch biosynthesis and how they interconnect to build a pathway that controls the flux of carbon into starch. In particular, I will focus on what is known of the pathway and some of the control mechanisms involved in starch synthesis in developing cereal grain.

The developing cereal grain is a remarkable factory dedicated to the survival of the species. It is composed of the embryo, pericarp and endosperm. The endosperm tissue contains the starch-based energy-store, able to withstand winters, droughts and fungal attack. For endosperm tissue in cereal grains, the 10–15 days after pollination sees a frantic series of nuclear-divisions culminating in a coenocyte which then rapidly forms cell walls (Olsen *et al.*, 1992, 1995). No further cell division occurs in this tissue. Next, starch synthesis is initiated as a wave of transcription and translation of the pathway enzymes. For the next 20–40 days, depending on the temperature and species, there is an unrelenting period of starch deposition (the so-called linear grain-filling period). Despite all kinds of environmental fluctuations the seed will fill out

Plant Carbohydrate Biochemistry, edited by J.A. Bryant, M.M. Burrell and N.J. Kruger.
© 1999 BIOS Scientific Publishers Ltd, Oxford.

every endosperm cell with starch granules. Starch synthesis ceases and the now fully developed grain becomes mature as it dries out ready to be stored. Cereal grains are remarkably consistent in size, weight, density, shape and germination potential. This consistency is achieved through a very precisely choreographed process. We know much about the pathway and the various enzymes involved. We know much about starch granule structure. Yet, for all that we know, there is much to be learned. This seemingly simple process hides a system of great interwoven complexity.

2. Pathway of starch synthesis

It all seems so simple. Sucrose made in the leaves is transported around the plant and delivered to the developing seed. There, a series of enzymes make ADPglucose available for starch synthesis and starch granules are built. It is important to remember that when we began this work, the biochemical pathway of starch synthesis was, and still is, being elucidated. For the purposes of this report I shall use the pathway of starch synthesis recently described by Denyer *et al.* (1996b). Starting with sucrose, the endosperm cells produce UDPglucose and fructose by the action of sucrose synthase. Next, the sugar phosphates (glucose-6-phosphate, glucose-1-phosphate and fructose-6-phosphate) are formed, which are readily interconverted by phosphoglucomutase and phosphoglucose isomerase. Glucose-1-phosphate is a substrate for ADPglucose pyrophosphorylase (AGPase) which makes the key intermediate ADPglucose. Next, a series of enzymes make amylopectin and amylose from ADPglucose. These include, but may not be limited to, starch synthases, branching enzymes and debranching enzymes. Amylopectin and amylose collectively assemble to become a starch granule, the essential storage component of cereal grain. Although we can describe much of the kinetics of individual enzymes in the pathway, there is much about this seemingly simple process which we do not understand.

In particular, we do not understand enough about the final steps in starch deposition. There is a complexity in a single starch granule which we do not adequately understand at the molecular level. Furthermore, we do not understand properly how the enzymes of starch deposition function together to form the complex molecular configuration of amylopectin.

3. Rate and duration of grain filling

The amount of starch accumulated is a function of the rate as well as the duration of grain filling. *Figure 1* presents data representing the average starch content of cereal grain taken from three different commercially available field-grown corn hybrids. After a short lag of a few days, when endosperm cell division is proceeding, the process of grain filling begins. This is known as the linear phase of grain filling. It usually lasts for 20–40 days and then seemingly finishes abruptly. During the next phase, no more starch is deposited and the seed dries down. Ultimately, in combination with the number of seeds per plant, plant yield is therefore determined by the average seed weight which is a function of the rate of grain filling and grain filling duration.

Considering the apparent simplicity of the linear rate of grain filling and the abrupt end to this process, one might expect there to be a similar simplicity in the expression of the enzymes which determine endosperm starch biosynthesis. Shown in *Figure 2* are activities of selected enzymes in extracts from developing maize endosperm. It is

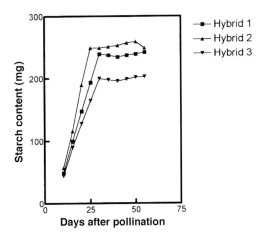

Figure 1. Starch content in developing endosperm of three different commercial maize hybrids.

clear that all of these enzymes appear to be expressed together in co-ordination with the onset of starch deposition. The activity of each enzyme increases rapidly until about 20 days after pollination when it generally reaches a maximum. This coincides with the time of mid-grain filling. Next the enzymes generally decline in activity over the extended period of grain filling when the seeds are drying down.

Contrasting *Figure 1* with *Figure 2*, it is not at all clear what gives rise to such a simple linear phase of grain filling, nor what so abruptly causes the process to simply stop and hence the seed to begin to dry down. It could be argued that plant senescence may have occurred such that no more photoassimilate is being transferred to the seed. However, we have measured sucrose, glucose and fructose content throughout grain development and there are clearly sugars still available after the seed has reached maturity and is no longer filling (data not shown). Thus, it seems likely that processes within the grain can be the only determinants of grain filling duration.

Despite our interest in understanding the duration of grain filling, for the remainder of this discussion I shall focus attention only on the linear phase of grain filling. We have studied this extensively over the years in a quest to decipher a biochemical explanation for its seemingly simple linearity. The primary premise being pursued is that this process

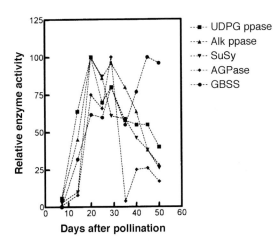

Figure 2. Enzyme activities extracted from maize endosperm tissue. UDPG ppase, UDPglucose pyrophosphorylase; Alk ppase, alkaline pyrophosphatase; SuSy, sucrose synthase; AGPase, ADPglucose pyrophosphorylase; GBSS, granule-bound starch synthase.

of linear grain filling is limited by metabolic regulation in the seed (so-called sink-limitation). Surely, linear grain filling must be linked to some simple rate-limiting or flux-controlling enzyme? When we began this quest in the early 1980s the prevailing doctrine was that pathways were rate-limited by specific enzymes. Over the last two decades we have come to learn that pathways are not limited by a single rate-limiting enzyme, but may instead have several flux-controlling steps which share in determining flux or flow through the pathway. Here then, I will describe our efforts to identify, initially, the rate-limiting enzymes in the pathway and how this has subsequently become a quest to quantify flux control coefficients for several enzymes in the pathway.

4. Control of starch synthesis during grain filling

4.1 *Identifying rate-liming steps*

The concept of 'rate-limiting steps' pervaded the biochemical literature until a more mathematical approach was invented (see Section 5, below). However, in the early 1980s, when we first started our efforts to understand starch deposition, the concept of flux control was in its infancy and so here I will summarize our preliminary efforts to understand the mechanisms of control of starch synthesis. To identify rate-limiting steps it was argued that such steps would be most likely to be associated with: (i) the irreversible enzymes in which the forward reaction rate far exceeded the backward rate through that enzymic step; (ii) the least catalytically active enzymes; (iii) the early steps in the pathway; and (iv) those enzymes having substrate-inhibition kinetics or being allosterically regulated, and those enzymes which become inactive. It turns out that much of this early doctrine was misleading (see Chapters 1 and 2). Nevertheless, at the time we started our research efforts in this field these tools were all that were available. Furthermore, they enabled us to ask some simple questions and study this process further.

4.2 *Equilibrium constants and mass-action ratios*

Historically, it was proposed that rate-limiting steps could be identified by comparing the apparent equilibrium constant (K_{eq}) and mass-action ratio (MAR) for each of the enzymes. K_{eq} is the ratio of the concentrations of products and substrates of an individual reaction at equilibrium *in vitro* (K_{eq} = [products]/[substrates]). MAR is very similar, but involves measuring the concentration of substrates and products under steady-state conditions *in vivo* (MAR = [products]/[substrates]). It was generally believed that a rate-controlling enzyme would be one that had a MAR value several-fold lower than the K_{eq}. In contrast, if a reaction was near equilibrium then the MAR would be similar to the K_{eq} value. Whilst these measurements seem very simple and straightforward there is a technical barrier, namely the compartmentation of individual steps in any one pathway within different intracellular organelles. This is clearly the case for starch synthesis, where one part of the pathway is located in the cytosol and the other is in the amyloplast. Nevertheless, although haunted by this technical problem, we did make some measurements of the MAR in developing wheat endosperm and compare these with published values of K_{eq}. We concentrated our efforts on those enzyme steps where we believed we could measure metabolite levels with minimal error caused by intracellular compartmentation. *Table 1* shows some

Table 1. *Measurements of 'rate-limiting' steps in wheat endosperm.*

Enzyme	Equilibrium constant	Mass-action ratio	'Irreversible' reaction	Maximum catalytic activity (relative to flux)
Sucrose synthase	0.2	0.6	No	19×
Hexokinase	2000	0.01	Yes	2×
UDPglucose pyrophosphorylase	0.15	0.1	No	900×
Phosphoglucomutase	0.06	4.3	No	?
Phosphoglucose isomerase	?	0.01	No	55×
ATPphosphofructokinase	1000	?	Yes	3×
PP$_i$-phosphofructokinase	1200	?	Yes?	?
Triose phosphate isomerase	0.2	?	No	50×
ADPglucose pyrophosphorylase	1	?	No	30×
Starch synthase	350	?	Yes	2×
Branching enzyme	?	?	Yes	?

interesting values for certain enzymes, notably hexokinase, but left us unable to measure the MAR values for the enzymes believed at the time to be located in the plastid compartment. These results were limited, but nevertheless left us with some potential candidate enzymes and others that were less likely to be major control points.

4.3 *Enzyme activities in relation to flux*

Next we measured enzyme activities of many enzymes in the pathway in developing wheat endosperm at mid grain filling when we knew that starch synthesis was in its linear phase. By comparing enzyme activity with the actual daily rate of flux of carbon into starch we were able to calculate an activity ratio for each enzyme. This provided an indication of which enzymes were the most likely to be regulatory. However, these measurements can be criticized on the grounds of being derived from *in vitro* assays in which the true intracellular conditions are not being represented. Furthermore, they still did not provide unequivocal quantitative evidence for the role of individual enzymes in regulation.

4.4 *Allosteric regulation* in vitro

Many enzyme activities are modified by some of the substrates and products of the biochemical pathway in which they serve. Such allosteric properties have been associated with enzymes which were thought to have key regulatory roles. These metabolites inhibit or activate an enzyme by interacting with allosteric sites on the enzyme (usually remote from the catalytic site). Within the pathway of starch biosynthesis other researchers have focussed particularly on the allosteric properties of AGPase because this enzyme has been proposed by many to be a key rate-liming enzyme (Preiss and Sivak, 1996; see also Chapter 9). However, this is not the only enzyme that displays allosteric properties. For example, other candidate enzymes might include the hexokinases and phosphofructokinases. Unfortunately, these allosteric properties create a special problem for us in trying to identify candidate rate-controlling enzymes. Namely, it becomes virtually impossible to relate *in vitro* measurements of enzyme activities to *in vivo* flux through the pathway.

Thus, despite many efforts to measure enzyme activities and substrate/product concentrations we were only a little wiser. What was needed was a more holistic and mathematical approach (Stitt, 1995) in which the pathways are treated more as systems, and changes in flux are related to small changes in key enzymes in the pathway. This analysis relies on using mutants as well as using a transgenic approach to evaluate individual enzymes one at a time. Such an analysis became possible with the advent of metabolic control analysis.

5. Metabolic control analysis

5.1 *Flux control analysis*

Metabolic control analysis (Heinrich and Rapoport, 1974; Kacser and Burns, 1974) has been applied to many pathways since its introduction. However, it has only sparingly been applied to developing storage organs which accumulate starch such as those in the seed or the developing tuber. The approach is based on quantifying the fractional change in activity of an enzyme ($\delta E/E$) and comparing this with fractional change in flux through the pathway ($\delta J/J$). The flux control coefficient (C) is calculated from the two fractional changes expressed as a single ratio of $\delta J/J/\delta E/E$. In our studies we used the equations developed by Torres *et al.* (1986) to estimate the control coefficient. This approach requires a fractional change to be made in certain enzymes in a pathway in order to then measure the influence on flux through that pathway. In our studies we used two strategies. One was based on an observation made in wheat endosperm that the enzyme soluble starch synthase may be selectively inactivated using moderately high temperatures. The other exploited mutants in enzymes of starch biosynthesis in combination with gene dosage to effect small changes in the activities of branching enzyme, AGPase and sucrose synthase in corn endosperm. Whilst these studies did not include all the enzymes involved in starch biosynthesis, they represented what we believed were some key steps in the pathway.

5.2 *Soluble starch synthase from wheat endosperm*

Several years ago we observed that the soluble starch synthases of plants have unusual kinetic properties at moderately elevated temperatures (Keeling *et al.*, 1993, 1994). In particular we noticed that raising the temperature to 40°C and above caused an irreversible loss in starch synthase activity. When this temperature was applied *in vivo* it was possible to induce a seemingly specific loss in starch synthase activity without affecting other enzymes in the pathway. Using a combination of time and temperature we were able to inactivate this activity to varying degrees and use this to measure the control strength of starch synthase. Although this technique can be criticized on the grounds that there may have been some other part of the pathway that was adversely affected by high temperature, no clear evidence has been presented supporting this view. Nevertheless, it is prudent to argue that this possibility should not be overlooked. As a result of this work we found that the control coefficient for wheat endosperm starch synthase is quite high (0.82).

5.3 *Gene-dosage analysis in corn endosperm*

Another approach to measuring flux control comes from gene-dosage analysis. Cereal endosperm tissue is triploid, inheriting two doses of each allele from the maternal plant and one dose from the male parent. This unusual genetic basis provides a unique opportunity to study the effects of individual enzymes on the pathway of starch deposition.

By carefully crossing male and female plants it is possible to create a gene-dosage series of mutant and wild-type alleles. *Figure 3* shows how we can artificially create a series of geneotypes containing four different doses of the waxy gene (*WxWxWx*, *WxWxwx*, *wxwxWx* and *wxwxwx*). Interestingly, this kind of approach was used many years ago in pioneering work in Boyer's, Nelson's and Tsai's laboratories (Boyer *et al.*, 1976; Chourey and Nelson, 1976; Tsai, 1974). Tsai's group found that granule-bound starch synthase (GBSS) activity is related linearly to increasing dosage of wild-type allele. In contrast, when amylose content was quantified this was not related to gene-dosage in a linear fashion. This indicates that over-expressing GBSS would not be likely to significantly increase amylose content (*Figure 4*).

In our research we were interested in studying other mutants of starch synthesis, namely *brittle2*, *shrunken2*, *amylose extender* and *dull*. Furthermore, such mutants are important in helping us understand the links between genes, structure and functionality (Keeling, 1997). Such mutants have been reported by others to affect specific enzymes such as AGPase, sucrose synthase, branching enzyme (Hedman and Boyer, 1982), GBSS (Shure *et al.*, 1983) and soluble starch synthase (Gao *et al.*, 1998; Harn *et al.*, 1998). Just as in the example of using waxy mutants, we found that there was a linear relationship between gene dosage and enzyme activity. Interestingly, we found that other enzymes in the pathway of starch synthesis were several-fold over-expressed in the mutant backgrounds. However, as far as we could determine, for each mutant the only decrease in activity was the one associated with the enzyme encoded by the gene.

Our data show that the average flux control coefficients in corn endosperm are 0.22 for AGPase, 0.12 for branching enzyme and 0.09 for sucrose synthase (*Table 2*).

6. Studies of the individual enzymes in the starch synthesis pathway

In designing a molecular approach to improving starch yield and starch quality, it is essential to know on which enzymes to focus costly research efforts. Based on all of the evidence presented above, we have proposed that the most significant enzymes involved in starch biosynthesis are the starch synthases. Additionally, because of other evidence (not presented here) concerning the roles of branching and debranching

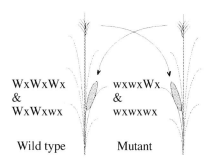

WxWxWx
&
WxWxwx

Wild type

wxwxWx
&
wxwxwx

Mutant

Figure 3. Artificial creation of a series of genotypes containing four different doses of the waxy gene.

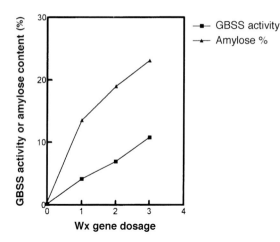

Figure 4. Granule-bound starch synthase (GBSS) activity and amylose content with different waxy (Wx) gene dosages.

enzymes in determining starch structure we cannot ignore these enzymes. Starch synthase (SS, EC 2.4.1.21) elongates α-1,4 glucan chains by transferring glucose from ADPglucose to the non-reducing end of an acceptor chain. Branching enzyme (BE, EC 2.4.1.18) hydrolyses an elongated α-1,4 glucan chain and simultaneously transfers it to an acceptor chain to form an α-1,6 linkage. Although it has been shown that the rate of glycogen synthesis in bacteria and leaf starch synthesis in plants is allosterically regulated by ADPglucose synthesis in response to the intracellular metabolite levels (Preiss and Romeo, 1994; Preiss and Sivak, 1996), this has never been unequivocally proven in storage organs. Thus in our research, we have not concentrated efforts on this enzyme. In other storage organs (e.g. potato tuber) it could be more important as evidenced by reports of increased starch formation when over-expressing this enzyme (Stark *et al.*, 1992).

6.1 *Studies of purified starch synthases and branching enzymes*

Several isoforms of SS and BE have been identified in maize, rice, potato and pea (Baba *et al.*, 1993; Macdonald and Preiss, 1985; Marshall *et al.*, 1996; Mu *et al.*, 1994). In comparison with BE, less is known about the specificities and functions of the various forms of SS. Since amylose production in a number of species such as maize, rice and potato depends on the GBSS encoded by the *waxy* gene, it is generally accepted that

Table 2. Summary of flux control coefficients for enzymes of starch synthesis in cereal endosperm

Enzyme	Flux control coefficient
Branching enzyme (*ae1*)	0.071–0.137
ADPglucose pyrophosphorylase (*bt2*)	0.045–0.195
Sucrose synthase (*sh1*)	0.005–0.014
Starch synthase (from wheat)	0.82[a]

All data from Singletary *et al.* (1997) except [a] from Keeling *et al.* (1993).

GBSS is primarily responsible for the synthesis of amylose (Martin and Smith, 1995; Preiss and Sivak, 1996). Researchers studying the *waxy* mutation in pea, corn and *Chlamydomonas reinhardtii* have suggested that GBSS is also involved in amylopectin synthesis (Delrue *et al.*, 1992; Denyer *et al.*, 1996a). Although it has been reported that SSII plays a major role in the synthesis of intermediate size glucans of amylopectin in *Chlamydomonas* (Fontaine *et al.*, 1993), SSII protein has not been purified and characterized from *Chlamydomonas*. Although some evidence for the functions of SS in higher plants has appeared, more research is needed. Antisense technology has been used to study the functions of SSII and SSIII in potato; however, the results have been inconclusive (Abel *et al.*, 1996; Marshall *et al.*, 1996). Although transgenic potato plants containing antisense SSIII produced smaller starch granules, no change in chain length distribution was observed (Abel *et al.*, 1996; Marshall *et al.*, 1996). Thus it is not certain how individual forms of SS control starch structure. In corn, mutations affecting the activities of BE and SS point to key roles of these enzymes in starch assembly and suggest that some of these enzymes may interact with one another to be fully active.

6.2 *Cloning of starch synthases and branching enzymes*

To date five SS genes have been isolated from maize endosperm (Gao *et al.*, 1998; Harn *et al.*, 1998; Knight *et al.*, 1998; Shure *et al.*, 1983). The identities of these, and other genes associated with starch synthesis in maize are summarized in *Table 3*. However, many of the corresponding enzymes have not been purified to homogeneity, a task which is apparently difficult due to their low abundance, instability and association with each other and amylolytic enzymes.

6.3 *Studies of the properties of enzymes of starch biosynthesis over-expressed in* Escherichia coli

Developing a proper understanding the structure–function relationships of SS has been hampered by the unavailability of the individual enzymes in a purified form. However, cloning of the genes encoding the various enzymes has opened the way for specific studies of these enzymes either individually or in specific combinations. Simple enzyme kinetic studies have now been completed for some of the maize SS isoforms (Imparl-Radosevich *et al.*, 1999).

Table 3. Enzymes and genes associated with starch metabolism in developing maize endosperm

Enzyme	Form 1	Form 2	Form 3	Form 4
ADPglucose pyrophosphorylase	*bt2*	*sh2*	*agp1*	*Agp2*
Soluble starch synthase	SSI	SSIIa, *su2*	SSIIb	SSIII, *dull*
Granule-bound starch synthase	GBSS, *wx*	—	—	—
Branching enzyme	BEI	BEIIa	BEIIb, *ae*	—
Amylogenin	Amylogenin	—	—	—
Debranching enzyme	Isoamylase, *su1*	Pullulanase	—	—
Disproportionating enzyme	—	—	—	—

Comparison of SS sequences from different sources shows a distinguishing feature: a unique N-terminal extension (Baba *et al.*, 1993; Imparl-Radosevich *et al.*, 1998; Martin and Smith, 1995). For example, comparison of the protein sequences deduced from the cDNAs of maize GBSS, maize SSI and *E. coli* glycogen synthase (Imparl-Radosevich *et al.*, 1998) reveals that maize SSI contains an N-terminal extension of 93 amino acids. Similar N-terminal extensions are found in the other maize SS (Gao *et al.*, 1998; Harn *et al.*, 1998). Computer modelling of pea SS (Martin and Smith, 1995) has led to the suggestion that this N-terminal arm is 'flexible'. However, there is little sequence similarity between the N-terminal extensions of SS isozymes from different species, making it difficult to understand their biochemical function(s). In order to investigate the functions of the maize SSI N-terminal extension, cDNA coding two different N-terminally truncated SSI forms (SSI-2, SSI-3) were individually expressed in *E. coli*, purified and kinetically characterized for comparison to the full-length SSI-1 (Imparl-Radosevich *et al.*, 1998). The full-length recombinant SSI-1 and both truncated forms of the enzyme exhibited similar properties to the native SSI isolated from maize endosperm. Similar experiments conducted the maize SSIIa and SSIIb and their truncated forms have given essentially the same result (Imparl-Radosevich *et al.*, 1999). Our results clearly demonstrate that the catalytic centre of the starch synthases is not located in its N-terminal extension. However, N-terminal truncation decreased the enzyme affinity for amylopectin, with the K_m for amylopectin of the truncated SSI-3 being about 60–90% higher than that of the full-length SSI-1. These results suggest that the N-terminal extension in SSI may not be directly involved in enzyme catalysis but may instead regulate the enzyme association with α-glucans. Additionally, the N-terminal extension may play a role in determining the localization of SSI to specific portions of the starch granule, or it may regulate its interactions with other enzymes involved in starch synthesis.

7. Conclusions

Initially it all seemed so simple. Sugars are delivered to the developing seed where they are converted into starch. The process requires a few biochemical steps involving enzymes which must be there in adequate amounts to support rapid starch accumulation in the seed. However, underneath this apparent simplicity lies a host of still poorly understood enzymic activity and system of genetic control mechanisms. The mechanisms of control of starch deposition are surely simple, but they still evade full discovery despite decades of intense research around the world. Thus there remain many unanswered and interconnected questions. What determines the rate and duration of starch synthesis? Could starch biosynthesis be source or sink limited? Which enzymes are the most obvious candidates for controlling flux? Do we really understand the biochemical pathway properly? How does a biochemical pathway interact with changing environmental conditions? Have we cloned all the enzymes of starch synthesis? Can the biochemical pathways of a plant be improved? Can starch structure be usefully manipulated? The questions appear to be quite simple and yet are surely endless. Throughout this chapter I have emphasized the apparent simplicity of this process, yet I hope I have revealed some of the complexity that lies within.

Acknowledgements

I am very grateful to many colleagues, over many years who have contributed much to this research; Ian Bridges, Tony Fentem, Jeff Foxon, Sheila Attenborough, Karen

Holt, Lois Catt, Mike Bayliss, John Wood, David Holt, Philippa Bacon, Mary Knight, Wolfgang Schuch, Mark Olive and Chris Jackson (Zeneca Seeds, Runcorn, UK), Tom ap Rees, Gale Entwistle and Huw Tyson (Cambridge University, Cambridge, UK), George Singletary, Roshie Banisadr, John Greaves, Norman Cloud, Ming Chang, Ed Wilhelm and Doris Rimathe (Garst Seeds, Slater, IA, USA), Alison Smith, Kay Denyer and Cathie Martin (John Innes Institute, Norwich, UK), Bruce Wasserman, Chee Harn and Chen Mu) (Rutgers University, Brunswick, USA), Hanping Guan, Jennifer Radosevich, Ping Li, Angela McKean, Zhong Gao, Francie Dunlap, Jingdong Sun, Deb Nichols, Padma Commuri, Ming Chang, Ed Wilhelm, Darcy Breyfogle, Dana Rewoldt and Amy Hohnstein (ExSeed Genetics, Ames, IA, USA).

Finally, I especially acknowledge Tom ap Rees, an important friend and mentor, who was a significant influence on my life and work.

References

Abel, G.J.W., Springer, F., Willmeitzer, L. and Kossman, J. (1996) Cloning and functional analysis of a cDNA encoding a novel 139 kDa starch synthase from potato (*Solanum tuberosum* L.). *Plant J.* **10**: 981–991.

Baba, T., Nishihara, M., Mizuno, K., Kawasaki, T., Shimada, H., Kobayashi, E., Ohnishi, S., Tanaka, K.-I. and Arai, Y. (1993) Identification, cDNA cloning, and gene expression of soluble starch synthase in rice (*Oryza sativa* L.) immature seeds. *Plant Physiol.* **103**: 565–573.

Boyer, C.D., Garwood, D.L. and Shannon, J.C. (1976) Interaction of the amylose extender and waxy mutants of maize. *J. Heredity* **67**: 209–214.

Chourey, P.S. and Nelson, O.E. (1976) The enzymatic deficiency conditioned by the *shrunken-1* mutations in maize. *Biochem. Genet.* **14**:1041–1055.

Delrue, B., Fontaine, T., Routier, F., Decq, A., Wieruszeski, J.M., Van der Koornhyse, N., Maddelein, M.L., Fournet, B. and Ball, S. (1992) Waxy *Chlamydomonas reinhardii*: monocellular algal mutants defective in amylose biosynthesis and granule-bound starch synthase activity accumulate a structurally modified amylopectin. *J. Bacteriol.* **174**: 3612–3620.

Denyer, K., Clarke, B., Hylton, C., Tatge, H. and Smith, A.M. (1996a) The elongation of amylose and amylopectin chains in isolated starch granules. *Plant J.* **10**: 1135–1143.

Denyer, K., Dunlap, F., Thorbjornsen, T., Keeling, P.L. and Smith, A.M. (1996b) The major form of ADP-glucose pyrophosphorylase activity in maize (*Zea mays* L.) endosperm is extraplastidial. *Plant Physiol.* **112**: 779–785.

Fontaine, T., D'Hulst, C., Maddelein, M.-L., Routier, F., Pepin, T.M., Decq, A., Wieruszeski, J.-M., Delrue, B., Van den Koornhuyse, Bossu, J.-P., Fournet, B. and Ball, S. (1993) Towards an understanding of the biogenesis of the starch granule. Evidence that chlamydomonas soluble starch synthase II controls the synthesis of intermediate size glucans of amylopectin. *J. Biol. Chem.* **268**: 16223–16230.

Gao, M., Wanat, J., Stinard, P.S., James, M.G. and Myers, A.M. (1998) Characterisation of *dull1*, a maize gene coding for a novel starch synthase. *Plant Cell* **10**: 339–412.

Harn, C., Knight M., Ramakrishnan, A., Guan, H.-P., Keeling, P.L. and Wasserman, B.P. (1998) Isolation and characterization of the Ss2 starch synthase cDNA clones from maize endosperm. *Plant Mol. Biol.* **37**: 639–649.

Hedman, K.D. and Boyer, C.D. (1982) Gene dosage at the amylose-extender locus of maize: effects on the levels of starch branching enzyme. *Biochem. Genet.* **20**: 483–492.

Heinrich, R. and Rapoport, T.A. (1974) A linear steady-state treatment of enzymatic chains. General properties, control and effector strength. *Eur. J. Biochem.* **42**: 89–95.

Imparl-Radosevich, J.M., Li, P., Zhang, L., McKean, A.L., Keeling, P.L. and Guan, H. (1998) Purification and characterization of maize starch synthase I and its truncated forms. *Arch. Biochem. Biophys.* **353**: 64–72.

Imparl-Radosevich, J.M., Li, P., Zhang, L., McKean, A.L., Gao, Z., Keeling, P.L. and Guan, H. (1999) Characterization of maize starch synthase IIa and IIb and comparison to other maize starch synthases. *Arch. Biochem. Biophys.* (in press).

Kacser, H. and Burns, J.A. (1974) The control of flux. *Symp. Soc. Exp. Biol.* **27:** 65–104.

Keeling, P.L. (1997) Starch biotechnology: technical barriers to starch improvement. In: *Starch: Structure and Functionality* (eds P.J. Frazier, P. Richmond and A.M. Donald). The Royal Society of Chemistry, London, pp. 180–187.

Keeling, P.L., Bacon, P.J. and Holt, D.C. (1993) Elevated temperature reduces starch deposition in wheat endosperm by reducing the activity of soluble starch synthase. *Planta* **191:** 342–348.

Keeling, P.L., Banisadr, R., Barone, L., Wasserman, B.P. and Singletary, G.W. (1994) Effects of temperature on enzymes in the pathway of starch biosynthesis in developing wheat and maize grain. *Aust. J. Plant Physiol.* **21:** 807–827.

Knight, M., Harn, C., Lilley, C.E.R., Guan, H.-P., Singletary, G. W., Mu-Forster, C., Wasserman, B.P. and Keeling, P.L. (1998) Molecular cloning of starch synthase I from *Zea mays* endosperm and expression in *E. coli*. *Plant J.* **14:** 613–622.

Macdonald, F.D. and Preiss, J. (1985) Partial purification and characterization of granule-bound starch synthases from normal and waxy maize. *Plant Physiol.* **78:** 849–852.

Marshall, J., Sidebottom, C., Debet, M., Martin, C., Smith, A.M. and Edwards, A. (1996) Identification of the major starch synthase in the soluble fraction of potato tubers. *Plant Cell* **8:** 1121–1135.

Martin, C. and Smith, A.M. (1995) Starch biosynthesis. *Plant Cell* **7:** 971–985.

Mu, C., Harn, C., Ko, Y., Singletary, G.W., Keeling, P.L. and Wasserman, B.P. (1994) Association of a 76 kDa polypeptide with soluble starch synthase I activity in maize (cv B73) endosperm. *Plant J.* **6:** 151–159.

Olsen, O.-A., Potter, R.H. and Kalla, R. (1992) Histo-differentiation and molecular biology of developing cereal endosperm. *Seed Sci. Res.* **2:** 117–131.

Olsen, O.-A., Brown, R.C. and Lemmon, B.E. (1995) Pattern and process of cell wall formation in developing cereal endosperm. *BioEssays* **17:** 803–812.

Preiss, J. and Romeo, T. (1994) Molecular biology and regulatory aspects of glycogen biosynthesis in bacteria. *Prog. Nucl. Acid Res. Mol. Biol.* **47:** 299–329.

Preiss, J. and Sivak, M.N. (1996) Starch synthesis in sinks and sources. In: *Photoassimilate Distribution in Plants and Crops: Source–Sink Relationships* (eds E. Zamski and A.A. Schaffer). Marcel Dekker, New York, pp. 63–95.

Shure, M., Wessler, S. and Fedoroff, N. (1983) Molecular identification and isolation of the *waxy* locus in maize. *Cell* **35:** 225–233.

Singletary, G.W., Banisadr, R. and Keeling, P.L. (1997) Influence of gene dosage on carbohydrate synthesis and enzymatic activities in endosperm of starch-deficient mutants of maize. *Plant Physiol.* **113:** 293–304.

Stark, D.M., Timmerman, K.P., Barry, G., Preiss, J. and Kishore, G.M. (1992) Regulation of the amount of starch in plant tissues by ADP-glucose pyrophosphorylase. *Science* **258:** 287–292.

Stitt, M. (1995) The use of transgenic plants to study the regulation of plant carbohydrate metabolism. *Aust. J. Plant Physiol.* **22:** 635–646.

Torres, N.U., Mateo, F., Melendez-Hevia, E. and Kacser, H. (1986) Kinetics of metabolic pathways. A system in vivo to study the control of flux. *Biochem J.* **234;** 169–174.

Tsai, C.-Y. (1974) The function of the *waxy* locus in starch synthesis in maize endosperm. *Biochem. Genet.* **11:** 83–96.

9

Allosteric regulation and reductive activation of ADPglucose pyrophosphorylase

Jack Preiss, Miguel Ballicora and Yingbin Fu

1. Introduction

ADPglucose pyrophosphorylase (AGPase; EC 2.7.7.27) catalyses an important regulatory step in the biosynthesis of starch in plants and of glycogen in bacteria (Preiss, 1988, 1991, 1997; Preiss and Sivak, 1996). This enzyme catalyses the synthesis of ADPglucose and PP_i from glucose-1-phosphate (G1P) and ATP. The product ADPglucose, serves as the activated glucosyl donor in α-1,4-glucan synthesis. AGPase from higher plants is heterotetrameric, encoded by two different genes (Preiss and Sivak, 1996; Smith-White and Preiss, 1992), while the enzyme from enterobacteria and cyanobacteria is homotetrameric in structure. The small subunit of higher plant AGPases is highly conserved, whereas the similarity among different large subunits is lower (Smith-White and Preiss, 1992). It has been speculated that the two plant subunits were originally derived from the same gene. This gene was duplicated during evolution, and then the two polypeptides diverged in sequence. Both subunits are required for optimal activity at least for the *Arabidopsis thaliana* (Li and Preiss, 1992) and potato tuber (Iglesias *et al.*, 1993; Ballicora *et al.*, 1995) AGPases.

Studies based on a wide range of sources have shown that AGPase is regulated by effectors derived from the dominant carbon assimilation pathway in the organism. The enzyme from higher plants (Preiss, 1988, 1991), green algae (Ball *et al.*, 1991; Sanwal and Preiss, 1967), cyanobacteria (Iglesias *et al.*, 1991) is mainly activated by 3-phosphoglycerate (PGA) and inhibited by P_i. AGPases from enteric bacteria are activated by frutose-1,6-bisphosphate while inhibited by AMP (Preiss and Romeo, 1989). These allosteric effects have been shown to functional physiologically in *Escherichia coli* and in *Salmonella typhimurium*, in which mutants having their AGPases affected in their allosteric properties have altered glycogen levels compared to the wild-type organism (Preiss and Romeo, 1989). Likewise, a *Chlamydomonas rheinhardtii* mutant containing an AGPase that cannot be activated by PGA is starch deficient whether the organism is grown autotrophically during the day or at night on acetate (Ball *et al.*,

1991). Similarly, Giroux *et al.*, 1996 showed that a maize endosperm mutant having an AGPase insensitive to P_i inhibition compared to normal maize endosperm AGPase accumulates 15% more starch than normal maize.

The potato (*Solanum tuberosum* L.) tuber AGPase consists of two different sub-units of 51 and 50 kDa (Okita *et al.*, 1990) as do those from other higher plant tissues. The cDNAs encoding the large and small subunits of the potato tuber AGPase have been expressed in *E. coli*, which yielded a recombinant heterotetrameric enzyme with similar properties to the native enzyme purified from potato tuber (Ballicora *et al.*, 1995). It was found that a homotetrameric enzyme comprised of only small subunits exhibited catalytic activity, with different allosteric regulatory properties from the heterotetrameric enzyme. In contrast, the large subunit by itself has negligible cat-alytic activity (Ballicora *et al.*, 1995). It was thus proposed that the major function of the large subunit is to modulate the sensitivity of the small subunit to allosteric regu-lation by P_i and PGA, and the major function of the small subunit is catalysis. However, the precise role of the small and large subunits is not well understood with respect to the substrate, allosteric effector binding sites, the precise role of the differ-ent subunits with respect to the binding sites and the activation by dithiothreitol (DTT) of the potato tuber AGPase.

Chemical modification and site-directed mutagenesis studies have identified that Lys-195 in AGPase from *E. coli* is involved in binding of the substrate G1P (Parsons and Preiss, 1978; Hill *et al.*, 1991). This residue is highly conserved in the bacterial enzymes as well as in both the small and large subunits of plant AGPases (Preiss and Sivak, 1996). However, no studies have been done to investigate the function of this highly conserved Lys residue present in the plant AGPases in the two different sub-units. The expression system for potato tuber AGPase provides a useful tool to char-acterize the role of the corresponding Lys residues in the small subunit, Lys-198, as well as in the large subunit, Lys-213, and to see if they retain the same function as in the homotetrameric *E. coli* enzyme.

The first structural study on the regulatory site of the plant enzymes was performed on the AGPase purified from spinach (*Spinacia oleracea*) leaves. After chemical modifi-cation with pyridoxal 5′-phosphate, protection with effectors and sequencing of the peptides covalently modified, lysines involved in the allosteric regulation were identi-fied. Those residues were involved either in the binding of the activator (PGA) or the inhibitor (P_i) (Ball and Preiss, 1994; Morell *et al.*, 1988). Three of those lysines are highly conserved in the enzymes from higher plants and cyanobacteria (*Table 1*; s-II, l-I, l-II). Two of these are equivalent (s-II, l-II), one corresponding to the large subunit and the other to the small subunit (Lys-440 in the spinach leaf small subunit and Lys-441 in the potato tuber small subunit). The third lysine labelled by pyridoxal 5′-phosphate was pre-sent in the large subunit (l-I; *Table 1*). The enzyme from *Anabaena*, which is comprised of only one type of subunit, conserves these lysines in the primary structure (sites I and II; *Table 1*). Both Lys-382 and Lys-419 were shown to be involved in the activation by PGA with chemical modification and site-directed mutagenesis techniques (Charng *et al.*, 1994; Sheng *et al.*, 1996). However, no studies were performed to replace these residues on an enzyme consisting of two different subunits, which would be important in order to understand the role that each subunit plays in the higher plant enzymes. In the present study, site-directed mutagenesis was used to determine whether the lysine residues of both subunits are involved in the activation by PGA or in the inhibition by P_i of the potato tuber AGPase. The effects of substitution of the lysine residues in the

Table 1. Sequence alignment of the plant and cyanobacterial AGPase regulatory sites

Accession number	Source	Tissue	Site	Site
Cyanobacterial			I	II
			382	419
Z11539	*Anabaena*		DTIIRRAIIDKNARIG	IVVVLKNAVITDGTII
Plants: small subunit			s-I	s-II
			404	441
L33648	*Solanum tuberosum*	Tuber	NCHIKRAIIDKNARIG	IVTVIKDALIPSGIII
x83500	*Spinacia oleracea*	Leaf	NSHIKRAIIDKNARIG	IVTVIKDALIPSGTVI
x76941	*Vicia faba* (VfAGPP)	Cotyledons	NSHIRRAIIDKNARIG	IVTVIKDALIPSGTVI
x96764	*Pisum sativum*	Cotyledons	NSHIKRAIIDKNARIG	IVTVIKDALIPSGTVI
z79635	*Ipomoea batatas* (psTL1)	Tuberous root/leaf	NSHIKRAIIDKNARIG	IVTIIKDALIPSGTII
L41126	*Lycopersicon esculentum*	Fruit	NCLYKRAIIDKNARIG	IVTVIKDALIPSGIVI
J04960	*Oryza sativa*	Endosperm	NCHIRRAIIDKNARIG	IVTVIKDALLLAEQLYE
M31616	*Oryza sativa*	Leaf	NCHIRRAIIDKNARIG	IVTVIKDALLLAEQLYE
X66080	*Triticum aestivum*	Leaf	NSHIKRAIIDKNARIG	IVTVIKDALLPSGTVI
Z48562	*Hordeum vulgare*	Starchy endosperm	NSHIKRAIIDKNARIG	IVTVIKDALLPSGTVI
Z48563	*Hordeum vulgare*	Leaf	NSHIKRAIIDKNARIG	IVTVIKDALLPSGTVI
	Zea mays (Brittle 2)	Endosperm	NSCIRRAIIDKNARIG	IVTVIKDALLPSGTVI
S72425	*Zea mays*	Leaf	NSHIRKAIIDKNARIG	IVTVIKDALLPSGTVI
Algae: large subunits			l-I	l-II
x91736	*Chlamydomonas reinhardtii*		NSVITNAIIDKNARVG	ILVIDKDALVPTGTTI
Plants: large subunits			l-I	l-II
x6118	*Solanum tuberosum*	Tuber	417 NTKIRKCIIDKNAKIG	455 IIIILEKATIRDGTVI
	Spinacia oleracea[a]	Leaf	...IKDAIIDKNAR...	ITVIFKNATIKDGVV
x74982	*Solanum tuberosum* Desiree	Leaf	NTKIQNCIIDKNAKIG	ITVIMKNATIKDGTV
u81033	*Lycopersicon esculentum*		NTKIRKCIIDKNAKIG	IIIISEKATIRDGTVI

continued overleaf

Table 1. Continued

Accession number	Source	Tissue	Site	Site
x96766	*Pisum sativum*	Cotyledons	NTKIKNCIIDKNAKIG	ITIIMEKATIEDGTVI
x78900	*Beta vulgaris*	Tap root	NTKIKNCIIDKNAKIG	TIILKNATIQDGLVI
x14348	*Triticum vulgaris aestivum,* cv. Mardler	Leaf	NTSIQNCIIDKNARIG	ITVVLKNSVIADGLVI
x14349	*Triticum aestivum*	Endosperm	NTKISNCIIDMNARIG	IVVIQKNATIKDGTVV
z38111	*Zea mays*	Embryo	NTKISNCIIDMNCQGW	IVVVLKNATIKDGTVI
s48563	*Zea mays* (*Sh*-2)	Endosperm	NTKIRNCIIDMNARIG	IVVILKNATINECLVI
u66041	*Oryza sativa*	Endosperm	NTKIRNCIIDMNARIG	IVVILKNATNATIKHG
u66876	*Hordeum vulgare*	Leaf	NTSIQNCIIDKNARIG	ITVVLKNSVIADGLVI
x67151	*Hordeum vulgare*	Starchy endosperm	NTKISNCIIDMNARIG	IVVIQKNATIKDGTVV

[a]Sequence determined by amino acid sequencing (Ball and Preiss, 1994).

large and small subunit are not equivalent. Mutating Lys-404 and Lys-441 on the small subunit decreased the affinity for the activator PGA and the ability of the enzyme to be inhibited by P_i. However, the effects were less pronounced when the homologous mutations were performed on the large subunit.

Previous results have shown that DTT activated the potato tuber AGPase three-fold and increased the affinity of the enzyme for ATP and ADPglucose (Sowokinos, 1981; Sowokinos and Preiss, 1982). It was suggested that key sulphydryl groups were present at the catalytic and/or allosteric sites. The mechanism of DTT stimulation, however, is not known. This activation by DTT seems to be similar to the regulation observed for several chloroplast enzyme activities which are controlled by reversible thiol/disulphide interchange (Buchanan, 1980; Wolosiuk *et al.*, 1993). During photosynthetic electron transport in the light, covalent redox modification is mediated by a redox chain, the ferredoxin–thioredoxin system, leading to reductive activation of several stromal target enzymes, for example, fructose-1,6-bisphosphatase (FBPase), NADP-malate dehydrogenase and phosphoribulokinase (Scheibe, 1991). Evidence will be presented to show that the activation is due to synergism involving the enzyme interacting with both DTT and its substrates. A reduction of the intermolecular disulphide bridge between Cys-12 of the two small subunits of potato tuber AGPase is involved in the activation process.

The expression level of all the mutant enzymes used in this study was similar to that of the wild type based on the results of immunoblotting (Ballicora *et al.*, 1998; Fu *et al.*, 1998a, 1998b). Wild-type and mutant enzymes of potato tuber AGPase were identified by immunoblotting with antibody prepared against the spinach leaf AGPase (Fu *et al.*, 1998a). It had been shown that the small subunit of the potato tuber enzyme cross-reacts significantly with the spinach leaf AGPase antibody (Okita *et al.*, 1990). The apparent sizes of these mutant polypeptides were the same as that of the wild type, having a molecular mass of about 50 kDa. Mutant enzyme $S_{K198R}L_{wt}$ was purified to more than 85% homogeneity, as estimated from about 4 μg protein electrophoresed on SDS–PAGE. All the other enzymes were purified to greater than 95% homogeneity except as noted.

2. Characterization of the glucose-1-phosphate binding site

2.1 *Kinetic characterization of $S_{K198}L_{wt}$ mutant enzymes*

The apparent affinity for G1P decreased dramatically when Lys-198 in the small subunit was mutated to either Arg, Ala or Glu. The $S_{0.5}$ values for G1P of the $S_{K198R}L_{wt}$, $S_{K198A}L_{wt}$ and $S_{K198E}L_{wt}$ enzymes were about 135-, 400- and 550-fold higher than that of the wild-type enzyme, respectively (*Table 2*). Substitution of Lys-198 in the small subunit by Ala and Glu resulted in such large $S_{0.5}$ changes that they could not be accurately determined (Fu *et al.*, 1998a). The highest G1P amount used in the assay mixture was 40 mM due to a solubility problem, and this value is only about two times higher than the $S_{0.5}$ of the Ala mutant and slightly higher than the $S_{0.5}$ of the Glu mutant. Thus, the $S_{0.5}$ value is only approximate for the Ala- and Glu-198 mutants. Since the V_{max} of the mutants of residue Lys-198 in the small subunit is not affected to a great extent, it does not have a role in the rate-determining step of catalysis. This is consistent with the observation on the corresponding Lys-195 of the *E. coli* AGPase (Hill *et al.*, 1991). However, replacement with Glu not only caused the largest increase of $S_{0.5}$ value for G1P, but also substantially decreased the catalytic efficiency relative to the wild-type enzyme.

The $S_{0.5}$ value of ADPglucose was increased 2.5–10-fold. The kinetic constant for the other substrate, PP$_i$, was increased 3–6-fold for the mutant enzymes (Fu *et al.*,

Table 2. *The apparent affinity of G1P for the potato tuber wild-type and mutant AGPases. Reactions were performed in the synthesis direction. Data represent the average of two identical experiments ± average difference of the duplicates*

AGPase	$S_{0.5}$ for G1P (mM)	V_{max} (U mg^{-1})[a]
Wild type	0.057 ± 0.003	48 ± 1
$S_{K198R}L_{wt}$	7.7 ± 0.1	24 ± 1
$S_{K198A}L_{wt}$	22.0 ± 2.5	46 ± 3
$S_{K198E}L_{wt}$	31.1 ± 2.7	1.7 ± 0.1
$S_{wt}L_{K213R}$	0.044 ± 0.002 (1.1)	27 ± 1
$S_{wt}L_{K213A}$	0.037 ± 0.001 (1.0)	25 ± 1
$S_{wt}L_{K213E}$	0.036 ± 0.001 (0.9)	31 ± 1
$S_{K198R}L_{K213R}$	5.6 ± 0.1 (1.5)	24 ± 1

[a]One unit of enzyme activity is expressed as the amount of enzyme required to form 1 mole of ADPglucose per min at 37°C assayed in either synthesis or pyrophosphorolysis direction.
G1P, glucose-1-phosphate; AGPase, ADPglucose pyrophosphorylase.

1998a). However, these changes are relatively small in comparison with the change of the $S_{0.5}$ value for G1P. The various mutations at residue 198 in the small subunit caused little or no alteration in the apparent affinities for the other substrates (ATP and Mg^{2+}) and activator, PGA (Fu *et al.*, 1998a). The data suggest that the conformations of those ligand-binding sites are relatively unchanged. The 2–7-fold increase of the $I_{0.5}$ value for the inhibitor, P_i, is also relatively small (Fu *et al.*, 1998a).

2.2 *Kinetic characterization of $S_{wt}L_{K213}$ mutant enzymes*

The apparent affinity for G1P was not affected when Lys-213 in the large subunit was replaced with either Arg, Ala or Glu (*Table 2*). These effects are in sharp contrast to the effect caused by mutations on Lys-198 in the small subunit (*Table 2*; Fu *et al.*, 1998a). In general, mutations on Lys-213 of the large subunit caused small changes (less than four-fold) to the kinetic constants for substrates (ATP, Mg^{2+}), activator, PGA and inhibitor, P_i (Fu *et al.*, 1998a).

2.3 *Kinetic characterization of $S_{K198R}L_{K213R}$ mutant enzyme*

When both Lys-198 in the small subunit and Lys-213 in the large subunit were replaced with Arg, the $S_{0.5}$ value for G1P was about 100-fold higher than that of the wild-type enzyme (*Table 2*). Considering the 135-fold increase of the $S_{0.5}$ value in mutant enzyme $S_{K198R}L_{wt}$, the double mutation did not cause a further decrease in the apparent affinity of G1P over the single mutation. In either direction of assay, the V_{max} of the double mutant enzyme is essentially the same as the single mutant enzyme, $S_{K198R}L_{wt}$.

The double mutation did not cause much alteration in the apparent affinities for Mg^{2+} and P_i (Fu *et al.*, 1998a). However, the $A_{0.5}$ value for PGA increased 11-fold relative to wild type. This effect seemed to be additive since the $A_{0.5}$ value for both $S_{K198R}L_{wt}$ and $S_{wt}L_{K213R}$ increased three-fold. The six-fold increase of $S0.5$ value for PP_i was similar to the effect seen in the single mutant enzyme, $S_{K198R}L_{wt}$. The other minor effects were four-fold increases of the $S_{0.5}$ values for both ATP and ADPglucose (Fu *et al.*, 1998a).

The Lys-195 region (FVEKP) of *E. coli* AGPase is not only conserved in potato tuber AGPase (FAEKP), but also identical to the mannose-1-phosphate binding site of phosphomannose isomerase/GDPmannose pyrophosphorylase (PMI/GMP) (Marolda and Volvano, 1993) as well as other sugar nucleotide pyrophosphorylases. This motif or a closely related sequence (GVEKP, IVEKY, KVIKP, FKEKP) is found in many enzymes with the common characteristic of catalysing the synthesis of an NDP-sugar from sugar phosphate and NTP (*Table 3*). Therefore, FVEKP is part of a sugar phosphate-binding motif for this class of sugar nucleotide pyrophosphorylases. Of course, other sequences must be involved in the sugar specificity, for example for mannose-1-phosphate or G1P.

3. Studies of the allosteric sites

3.1 *Effect of the mutations on the pyrophosphorolysis direction*

Using partially purified enzymes, it has been reported that in the presence of the large subunit from potato tuber, the small subunit has a higher affinity for the activator

Table 3. Conservation of the sugar phosphate-binding motif in sugar nucleotide pyrophosphorylases

Enzymes	Source	Sugar-1-phosphate-binding motif	References
AGPase	*Escherichia coli*	I I E F V E K P A N	Preiss and Sivak (1996); Hill *et al.* (1991)
GMP	*Escherichia coli*	RT*****NL	Marolda and Valvano (1993)
GMP	*Salmonella typhimurium*	VAE*****DI	Jiang *et al.* (1991)
UGP	*Escherichia coli*	PMVG****KA	Hossain *et al.* (1994)
UGP	*Bacillus subtilis*	VKN*****PK	Varnó *et al.* (1993)
UGP	*Solanum tuberosum* L.	TLKI***Y*	Katsube *et al.* (1990)
UGP	*Dictyostelium discoideum*	ETNK*I**YK	Ragheb and Dottin (1987)
CGP	*Yersinia pseudotuberculosis*	VRS*K***KG	Thorson *et al.* (1994)
PMI/GMP	*Pseudomonas aeruginosa*	VQS*****DE	May *et al.* (1994)
PMI/GMP	*Xanthomonas campestri*	VER*****LA	Köplin *et al.* (1992)

Lys-195 in *Escherichia coli* AGPase and Lys-175 in *P. aeruginosa* PMI/GMP, which were shown to bind to glucose-1-phosphate and mannose-1-phosphate, respectively, are underlined.
*signifies the same amino acid as in the *E. coli* AGPase sequence.
AGPase, ADPglucose pyrophosphorylase; GMP, GDPmannose pyrophosphorylase; UGP, UDPglucose pyrophosphorylase; CGP, CDPglucose pyrophosphorylase; PMI, phosphomannose isomerase.

PGA both in synthesis and pyrophosphorolysis directions (Ballicora *et al.*, 1995). In the present study, using proteins purified to homogeneity, we observed that the homotetramer comprised only of the small subunit (S_4) and the heterotetramer (L_2S_2) had similar specific activities at saturated concentrations of PGA in the pyrophosphorolysis reaction. They were 55 and 48 U mg^{-1} and the $A_{0.5}$ for PGA was 900 and 2.2 μM, respectively (*Table 4*). When the activity was analysed in the absence of PGA, the activity of the small subunit alone is negligible. However, when expressed in the presence of the large subunit, the activity in the absence of activator is 13.3 U mg^{-1}, only 3.6-fold lower than the maximal activity at saturated concentrations of PGA (*Table 4*). The presence of the large subunit in the quaternary structure is very important regarding the allosteric activation. Thus, it was of interest to determine the effect of mutations on the putative allosteric sites of the large subunit.

Lys-455 on the large subunit was mutated and it was found that the apparent affinity for PGA decreased. $A_{0.5}$ values were three- and five-fold higher when the residue was replaced by Ala and Glu, respectively. Greater changes were observed when large subunit Lys-417 was mutated. In this case, the $A_{0.5}$ values are nine- and 12-fold higher when Lys-417 was replaced by Ala or Glu, respectively (*Table 4*). The catalytic ability of the enzyme was not dramatically affected by these four mutations. A bigger reduction was observed, however, when the activity was assayed in the absence of PGA. The activation-fold increased from 3.6 (wild type) to 6, 22, 18 and 50 for the mutants K455A, K417A, K417E and K455E, respectively (*Table 4*). Even though these mutated large subunits produced heterotetrameric enzymes with lower affinity for the activator than the heterotetrameric wild type, the mutated large subunits still increased the apparent affinity of PGA for the small subunit when both subunits were expressed together.

Table 4. Activation of the potato tuber AGPase by PGA in the pyrophosphorolysis direction. Escherichia coli AC70R1-504 cells were co-transformed with plasmids encoding either the large or small subunits, mutated or wild type. The enzymes were expressed and purified as described by Ballicora et al. (1998). The specific activities of the enzymes are determined at saturating concentrations of activator and substrates

Large subunit	Small subunit	$A_{0.5}$ (PGA) (μM)	Ratio mutant/ wild type	Specific activity (U mg^{-1})	Activation[a] (-fold)
None	wt	900	–	55	>500
wt	wt	2.2	1	48	3.6
K455A	wt	6	3	47	6
K455E	wt	10	5	51	50
K417A	wt	20	9	13	22
K417E	wt	27	12	14	18
K417M	wt	7	3.5	45	14
wt	K441A	120	54	32	2.9
wt	K441E	420	191	39	3.1
wt	K404A	6800	3090	12	130
K417A	K441A	390	177	7	80
K417E	K441E	6500	2955	5	96

[a]The fold-activation is the ratio of the activated activity at saturating activator concentration to the activity in the absence of activator. The K417E, K441E double mutant enzyme is only about 40% pure so the specific activity is estimated to be higher.
AGPase, ADPglucose pyrophosphorylase; PGA, 3-phosphoglycerate.

When the wild-type small subunit was expressed along with the mutated large subunits K455A, K455E, K417A and K417E, the $A_{0.5}$ for PGA was reduced from 900 mM to 6, 10, 20 and 27 μM, respectively (*Table 4*).

Mutations of the homologous residues on the small subunit were done to determine their effects on the apparent affinity for the activator. Lys-441 when replaced by alanine or glutamic acid, increased the $A_{0.5}$ for PGA 54- and 191-fold, respectively. The specific activities of the purified mutant enzymes at saturated concentrations of activator were 32 U mg^{-1} for the Ala and 39 U mg^{-1} for the Glu mutant, not much different than the wild-type enzyme. Interestingly, the activity of these mutants in the absence of PGA did not change much. Thus, the maximal activation is 2.9 for the Ala and 3.1 for the Glu mutant (*Table 4*). The biggest decrease in apparent affinity was obtained when Lys-404 on the small subunit was replaced by Ala. The $A_{0.5}$ for PGA increased to 6.8 mM, representing a 3090-fold increase over the wild type (*Table 4*). Although the mutant, K404A, was severely affected in its allosteric properties, its maximal specific activity was 12 U mg^{-1}, only four times lower than the wild-type activity. However, the pyrophosphorolysis activity was very dependent on the allosteric activator, as PGA activated the K404A mutant enzyme 130-fold (*Table 4*).

Two double mutants were prepared by co-transforming with plasmids that encoded subunits with a single mutation. The $A_{0.5}$ for PGA of the $L_{K417A}S_{K441A}$ and $L_{K417E}S_{K441E}$ enzymes were much higher than the wild type, and both were also higher than the $A_{0.5}$ of each single mutant. The effects of these mutations thus seemed to be additive. The

$A_{0.5}$ for PGA of $L_{K417A}S_{K441A}$ was 177-fold higher than the wild type. That is close to the value expected (162-fold) if the L_{K417A} mutation increased three-fold and the S_{K441A} mutation further contributed with a 54-fold increase as they did when each single mutant was analysed (*Table 4*). At the same time, the $A_{0.5}$ for PGA of $L_{K417E}S_{K441E}$ was 2955 times higher that the wild type. If each mutation contributed to the same extent that they did in the single mutants (five- and 191-fold increase of the $A_{0.5}$ for PGA) an increase of 955-fold would be expected. In this case, the effect observed with the double mutant is higher but in the same range. The activation by PGA was very high for both $L_{K417A}S_{K441A}$ and $L_{K417E}S_{K441E}$ mutants, 80- and 96-fold, respectively (*Table 4*).

A role on the catalytic efficiency can be ruled out for these residues (L_{K417} and S_{K441}). The specific activities of the double mutants were lower than the wild type; however, this decrease is not significant when compared to the effect on the apparent affinity for the activator (*Table 4*).

3.2 *Effect of the mutations on the inhibition by* P_i

It was previously shown for spinach leaf AGPase that two lysine residues, in similar sequences as potato tuber AGPase small subunit Lys-441 and large subunit Lys-417, were protected by both PGA and P_i against chemical modification by pyridoxal phosphate (Ball and Preiss, 1994). Since the mutations of those residues of the potato tuber AGPase altered the activation properties of the enzyme, it was of interest to determine whether the P_i inhibition constant was also altered. Thus, the mutants were analysed in the pyrophosphorolysis direction and, inhibition studies were performed in the presence and absence of the activator.

The activity was analysed in the absence of PGA at different concentrations of P_i for determination of the $I_{0.5}$. None of the mutations performed in the large subunit decreased the affinity for the inhibitor, P_i (Ballicora *et al.*, 1998). On the contrary, those mutants were equally or more sensitive to P_i inhibition than the wild type. The $I_{0.5}$ for the wild type was 74 μM whereas the $I_{0.5}$ of the mutants K455A, K455E, K417A and K417E were 19, 20, 76 and 22 μM, respectively (*Table 5*). Conversely, when Lys-441 and Lys-404 on the small subunit were mutated, the mutants became virtually insensitive to P_i inhibition. The $I_{0.5}$ of all mutants K441A, K441E and K404A were more than 200-fold higher than the wild type (*Table 5*). These results suggest that Lys-441 and Lys-404 residues of the small subunit are important for the inhibition by P_i whereas Lys-417 and Lys-455 on the large subunit play little or no role.

The effect of P_i in presence of the activator, PGA, was also studied. P_i can reverse the activation caused by PGA and higher concentrations of PGA can overcome the P_i inhibition. P_i in the reaction mixture shifts the PGA activation curve toward higher concentrations, thus increasing the $A_{0.5}$. To study this effect, an activation curve was performed for each mutant in the presence of 0.5 mM P_i. The resultant $A_{0.5}$ was then compared to the one determined in the absence of inhibitor (*Table 5*). When the wild-type enzyme was analysed, 0.5 mM P_i increased the $A_{0.5}$ for PGA, 25-fold. All the mutations performed in the large subunit produced enzymes that showed similar results. P_i increased the $A_{0.5}$ for PGA, 27-, 28-, 20- and 22-fold for the mutants K455A, K455E, K417A and K417E, respectively. None of the mutations performed on the small subunit, K441A, K441E and K404A, exhibited the same effect. Activation curves in the presence or absence of P_i were very similar. The $A_{0.5}$ values were little affected, increasing only 2-, 1.9- and 1.2-fold for the mutants K441A, K441E and K404A, respectively (*Table 5*).

Table 5. *Inhibition of the potato tuber ADPglucose pyrophosphorylase by P_i in the pyrophosphorolysis direction.* Escherichia coli *AC70R1-504 cells were co-transformed with plasmids encoding either the large or small subunits, mutated or wild type. The enzymes were expressed and purified as described by Ballicora* et al. *(1998). The specific activities of the enzymes are determined at saturating concentrations of activator and substrates*

Large subunit	Small subunit	$I_{0.5}$		$A_{0.5}$	
		P_i (μM)	Ratio mutant/wild type	PGA (μM) (+0.5 mM P_i)	Ratio +P_i/–P_i
wt	wt	74	1	56	25
K455A	wt	19	0.25	160	27
K455E	wt	20	0.25	280	28
K417A	wt	76	1.0	400	20
K417E	wt	22	0.3	600	22
K417M	wt	20	0.25	198	28
wt	K441A	32 000	432	240	2.0
wt	K441E	16 000	216	800	1.9
wt	K404A	48 000	650	8000	1.2

PGA, 3-phosphoglycerate.

3.3 *Effect of the mutations on the kinetic constants and stability*

Although the allosteric properties were altered when Lys-417, Lys-455 (large subunit), Lys-404 and Lys-441 (small subunit) were replaced by either Ala or Glu acid, the affinities for Mg^{2+} and PP_i did not change dramatically in any of the mutants tested (Ballicora et al., 1998). The $S_{0.5}$ for Mg^{2+} remained between 2.6 and 4.2 mM and the K_m for PP_i did not increase more than two-fold for any of the mutants relative to the wild type. The apparent affinity for ADPglucose did not change for most of the mutants except for two. The double mutant $L_{K417E}S_{K441E}$ and single mutant $L_{wt}S_{K404A}$ had, respectively, a K_m of 0.60 and 0.90 mM for ADPglucose and were 2.7- and 4.1-fold higher, respectively, than the wild type. However, those changes are small compared to the striking changes in allosteric properties observed for these two mutants (*Tables 4* and *5*). Thus, the mutations had little, if any effect, on the catalytic site and specifically altered the regulatory properties.

3.4 *Effect of the mutations on the synthesis direction*

The activation in the reaction of synthesis was also studied since this is the direction of the reaction *in vivo* in the pathway of the starch biosynthesis.

When Lys-455 on the large subunit was mutated to Ala or Glu, the $A_{0.5}$ for PGA increased two- and eight-fold, respectively (Ballicora et al., 1998). Greater increases in $A_{0.5}$ were obtained mutating Lys-417 on the large subunit. The $A_{0.5}$ of mutants K417A and K417E were three and 13 times higher than the wild-type $A_{0.5}$ of 0.10 mM (Ballicora et al., 1998). Therefore, replacing of the Lys residues with Ala or Glu in the synthesis direction also affected the activation of the enzyme. These mutations on the large subunit had an $A_{0.5}$ for PGA lower than the $A_{0.5}$ (3.5 mM) of the small subunit expressed alone (Ballicora et al., 1998). As in the pyrophosphorolysis direction, mutations on the small subunit caused a bigger decrease in the apparent affinity for the activator. Mutant K441A and K441E had an $A_{0.5}$ for PGA, 32 and 83 times higher than the wild type. At the same time, mutant K404 could not be activated by PGA even by raising the concentration up to 50 mM (Ballicora *et al*., 1998).

Double mutants also showed that effects of those mutations were somewhat additive. The $A_{0.5}$ for PGA of $L_{K417A}S_{K441A}$ was 6.0 mM, which is 60 times higher than the wild type. This is in the range of what is expected if one mutation contributes with an increase of three-fold and another of 32-fold as they do when the single mutants are analysed. The $A_{0.5}$ of $L_{K417E}S_{K441E}$ could not be measured because the enzyme was not activated by PGA even up to 50 mM. In this case, at the highest concentrations assayed, the PGA partially inhibits $L_{K417E}S_{K441E}$. In this double mutant, if both Glu mutations contribute similarly as they do in the single mutants (with an increase of 13- and 83-fold) an $A_{0.5}$ of ~100 mM would have been expected. At such high concentrations, it is possible that PGA is also interacting non-specifically with other sites of the enzyme causing inhibition.

3.5 *Characterization of the* $L_{K417}MS_{wt}$ *mutant*

The residue Lys-417 in the potato tuber AGPase is highly conserved among the large subunits from plant enzymes. However, there are some exceptions such as in the large subunits of the maize and barley endosperm AGPases (*Table 1*). Since this residue could be involved in the regulation of the enzyme, a question arose about how relevant a change would be in the allosteric properties? As shown in *Tables 4* and *5*, the mutant $L_{K417M}S_{wt}$ showed a significant but small effect on the activation by PGA and no effect on the inhibition by P_i. Thus, some substitutions at this position such as Met and Ala could still give an enzyme with the ability to be regulated allosterically *in vivo*. The $L_{K417M}S_{wt}$ mutant was activated by PGA in the pyrophosphorolysis and synthesis directions. In synthesis, the $A_{0.5}$ was 450 μM and in pyrophosphorolysis, 7 μM, only 4.5 and 3.2 times higher than the wild type. At the same time, specific activities did not change dramatically. In the pyrophosphorolysis direction, it was 45 U mg^{-1} and in synthesis, 12 U mg^{-1}. The inhibition by P_i was not altered significantly. The sensitivity toward P_i in the absence of activator increased from 74 to 20 μM but the ability to shift the PGA activation curve remained the same. In the presence of 0.5 mM P_i, the $A_{0.5}$ for PGA in the pyrophosphorolysis direction increased 28-fold. Under the same conditions, the wild-type enzyme showed an increase of 25-fold. The most altered characteristic between $L_{K417M}S_{wt}$ and the wild type was that the activation-fold in pyrophosphorolysis increased from 3.6- to 14-fold because the enzyme was lower in activity in the absence of the activator. This alteration may have very little physiological effect.

4. Reductive activation of potato tuber ADPglucose pyrophosphorylase by ADPglucose and dithiothreitol or by ATP, glucose-1-phosphate and Ca²⁺

When measured in the absence of activator, PGA, the catalytic activity of potato tuber AGPase was found to increase with time showing non-linear kinetics (Fu *et al*., 1998b). Various combinations of effectors were tested for their ability to activate the enzyme during a preincubation at 37°C. As seen in *Table 6*, both ADPglucose and DTT were required to give about 10-fold activation of the potato tuber enzyme. In the

Table 6. *Activation of potato tuber AGPase by ADPglucose and DTT or ATP, G1P and* Ca^{2+}

Additions	ADPglucose formed (nmol per 2 min)
(A)	
Control	1.5
None	1.5
2 mM ADPglucose	2.5
3 mM DTT	0.3
2 mM ADPglucose + 3 mM DTT	12
(B)	
Control	2.5
None	0.4
1.5 mM ATP	0.7
0.5 mM G1P	0.2
1.5 mM ATP + 0.5 mM G1P	1.4
1.5 mM ATP + 2 mm Ca^{2+}	1.1
1.5 mM ATP + 2 mm Ca^{2+} + 0.5 mM G1P	10.3

AGPase, ADPglucose pyrophosphorylase; DTT, dithiothreitol; G1P, glucose-1-phosphate.
(A) Enzyme (1.7 µg) was incubated with 100 mM Hepes, pH 8.0, 0.2 mg ml^{-1} bovine serum albumin, and with different pyrophosphorolysis substrates in a final volume of 8 µl for 30 min at 37°C. The synthesis reaction was started by adding 192 µl assay mixture into the incubated solution and continued at 37°C for 2 min. The control experiment was carried out without adding effectors and omitting the preincubation step. (B) Enzyme (2 µg) was preincubated with a mixture that contained 80 mM glycylglycine, pH 8.0, 0.2 mg ml^{-1} bovine serum albumin, 3 mM DTT and with different combinations of synthesis substrates in a final volume of 80 µl for 30 min. The synthesis reaction was started by adding 60 µl incubated solution to 140 µl of assay mixture and continued at 37°C for 2 min. A control experiment was carried out without adding effectors and the activation step was omitted. Since the enzymatic reaction would slowly take place when the substrate components, ATP, G1P and Ca^{2+}, were present in the preincubation, the value (10.3 nmol per 2 min) was obtained after the subtraction of 9.6 nmol ADPglucose produced in the 30 min preincubation. Whenever Ca^{2+} was used, 1.6 mM EGTA was also included in the assay mixture.

absence of DTT, ADPglucose could slightly activate the enzyme (close to two-fold). However, when DTT was included in the preincubation mixture in the absence of ADPglucose, about 70% activity was lost after a 30 min preincubation. To further examine the effect of DTT, the rate of ADPglucose synthesis was measured in the presence or absence of DTT. When DTT was eliminated from the assay mixture, the enzyme was kept in a low activity form. The conversion to a high activity form only took place when DTT was present. This may suggest that reduction of a disulphide bridge(s) is involved in the activation process.

Different combinations of ATP, Glc-1-P, Ca^{2+} and DTT were also tested for their effect on the activation of potato tuber AGPase. DTT was included in all the combinations since it was required for the activation. Since catalysis would take place when the three effectors (ATP, G1P and Mg^{2+}) were present together, Ca^{2+} was used as a substitute for Mg^{2+} to separate the activation process from catalysis. Experiments showed that Ca^{2+} could replace Mg^{2+} as a co-factor for the potato tuber AGPase at about one-tenth of the rate seen with Mg^{2+}, and the apparent affinity of Ca^{2+} for the enzyme ($S_{0.5}$ = 1.8 mM) is similar to that of Mg^{2+} ($S_{0.5}$ = 2.0 mM). After preincubation, EGTA and Mg^{2+} were added to start the assay. Since EGTA has a very high affinity for Ca^{2+} and a very poor affinity for Mg^{2+}, Ca^{2+} in the assay mixture was efficiently chelated and Mg^{2+} would be the synthesis reaction cation. This metal exchange method was successfully used in the study of activation of chloroplast FBPase (Hertig and Wolosiuk, 1993; Wolosiuk *et al.*, 1993). As shown in *Table 6*, the enzyme was only activated when all three effectors were present at the same time. With DTT present in cases where one of the other two effectors were not present enzyme activity actually decreased from the control value in the 30-min preincubation.

Since 9.6 nmol ADPglucose was produced when all three effectors were present in the preincubation (see *Table 6*), another experiment was conducted to differentiate the activation from ADPglucose and that from ATP, G1P and Ca^{2+} in the preincubation. Even before ADPglucose was produced in the preincubation (2–6 min), the enzyme was already activated (Fu *et al.*, 1998b). Thus ATP, G1P, Ca^{2+} could activate potato tuber AGPase without prior formation of ADPglucose.

4.1 *Reduction of an intermolecular disulphide bridge during activation*

To determine if a reduction occurred in the activation process, 5,5′-dithiobio(2-nitro-benzoic acid) (DTNB) was used to quantitate the available sulphydryl groups of the potato tuber enzyme under activated or non-activated conditions. The activation of potato tuber AGPase was accompanied by an increase of about 2.5 new sulphydryl groups per tetrameric enzyme over the non-activated form. This difference correlates with the reduction of a disulphide bridge in the activation.

Electrophoretic evidence data (Fu *et al.*, 1998b) have also been obtained that show the disulphide linkage is between the Cys residues of the small subunit. Since proteins with disulphide bridges often exhibit altered migration on SDS–PAGE under non-reducing conditions, both the activated and non-activated potato tuber AGPases were subjected to SDS–PAGE under non-reducing conditions. The activated protein migrated as a single band with molecular mass of about 50 kDa. This was in agreement with previous studies on native and cloned AGPase (Ballicora *et al.*, 1995; Okita *et al.*, 1990) that showed the molecular masses of the small and large subunit were 50 and 51 kDa, respectively. The non-activated enzyme migrated as two bands corresponding to molecular masses of 50 and 100 kDa. Both bands were transferred to a ProBlott membrane and their N-terminal sequences were determined. For the 100 kDa band, it was: AVSDSQN; for the 50 kDa band, AVSVITT. The former was the same as the N-terminal sequence of the small subunit (Ballicora *et al.*, 1995), and the latter was the same as that of the large subunit deduced from cDNA sequence (Nakata *et al.*, 1991) except that the first methionine was processed in both cases. Thus, the 100 kDa band was the dimer of the small subunit. This result indicated the existence of an intermolecular disulphide bridge between the small subunits of the potato tuber enzyme, which was reduced during activation. Under reducing conditions, both the activated and non-activated enzyme migrated as a single band of 50–51 kDa.

4.2 *Identification of the intermolecular disulphide bridge*

There are 28 cysteine residues in potato tuber AGPase (six in the small subunit and eight in the large subunit). Prior to locating the intermolecular disulphide bridge between the small subunits, the total number of disulphide bridges of potato tuber AGPase was determined by [^{14}C]iodoacetic acid labelling (Fu *et al.*, 1998b). The non-activated (oxidized) enzyme was first denatured with urea to expose all free sulphydryl groups, which were blocked by subsequent addition of iodoacetamide. Then the protein was reduced with DTT before being labelled with [^{14}C]iodoacetic acid. In this way, only the oxidized (disulphide) groups would be labelled. It was found that 2.7 sulphydryl groups were labelled per tetrameric protein (Fu *et al.*, 1998b). When this procedure was applied to the activated enzyme, 0.7 sulphydryl group was labelled per tetrameric protein, apparently from non-specific labelling. This indicates that there is only one disulphide bridge in the potato tuber AGPase.

To determine the location of the intermolecular disulphide bridge, both the activated and non-activated enzyme were labelled with [^{14}C]iodoacetic acid and then digested with trypsin (Fu *et al.*, 1998b). The digests were separated by reversed-phase HPLC. One major radioactive fraction was obtained for the activated enzyme. After further purification by HPLC, sequence analysis showed that its N-terminal sequence corresponded to Ala2-Ser18 (A in *Table 7*) in the small subunit of potato tuber AGPase. One labelled carboxymethylcysteine was identified at cycle 11. For the non-activated enzyme, the overall labelling was low, suggesting most sulphydryl groups were buried in the protein. The sequences for three other minor radioactive peaks were not determined due to their low level of labelling.

To eliminate the possibility that the labelling was due to the unmasking of buried sulphydryl groups in the activated enzyme, a reverse labelling experiment which would specifically label the oxidized (disulphide) groups was performed for the non-activated enzyme. Only one major radioactive peak was obtained. Its N-terminal sequence was determined after further HPLC separation. The major sequence corresponded to Ala2-Pro15 in the small subunit (B in *Table 7*). One carboxymethylcysteine was identified at cycle 11. When this procedure was performed on the activated enzyme, no significant labelling could be observed.

Table 7. *Sequence analysis of [^{14}C]iodoacetic acid-labelled tryptic peptides and 4-vinylpyridine-labelled AGPase*

Peptides or proteins	Amino acid sequences
A	*AVSDSQNSQTCm-CLDPDAS*
B	*AVSDSQNSQTCm-CLDP*
C	*AVSDSQNSQTPE-CLD*
	AYSVITTENDTQT
D	*AVSDSQNSQTX*[a] *L*
	AYSVITTENDTQ

A and B refer to purified labelled fractions obtained by reverse-phase HPLC. C and D refer, respectively, to the activated and non-activated potato tuber AGPases labelled with 4-vinylpyridine. Both C and D contain two N-terminal sequences corresponding to the small and large subunits. The sequence corresponding to each subunit was deduced from the known sequences of the two subunits (Ballicora *et al.*, 1995; Nakata *et al.*, 1991). Italic sequences were assigned to the small subunit. The labelled cysteine residues are underlined: Cm-C, carboxymethylcysteine; PE-C, s-b-(4-pyridylethyl)-cysteine. Cm-C and PE-C were identified by comparison with the elution times of the two standard cysteine derivatives as determined separately. Besides the major sequence shown in the table, there was also a minor sequence corresponding to Ala196-Lys208 in the small subunit.
[a]X indicates that no amino acid could be detected in the cycle.

In order to avoid any ambiguity, sequence analysis was performed on 4-vinylpyridine-labelled whole protein since the N-terminal cysteines of the small subunits were implicated in forming the disulphide bridge. The result shows two sequences corresponding to both the small and large subunit. Based on the known sequences of the two subunits, the sequence of each subunit could be deduced. As shown in *Table 7*, Cys-12 in the small subunit of the activated enzyme was labelled by 4-vinylpyridine. On the contrary, sequence analysis did not detect any residue at the same position in the small subunit of the non-activated enzyme, suggesting the presence of non-derivatizable sulphydryls (unmodified cysteine is not stable during Edman degradation). The data confirm the results obtained by direct labelling (A in *Table 7*) and reverse labelling (B in *Table 7*) by [^{14}C]iodoacetic acid. These results demonstrate that the cysteine residues located at position 12 of the two small subunits are linked together by a disulphide bridge in the non-activated (oxidized) potato tuber AGPase. The disulphide linkage was reduced during the activation.

4.3 *Production and purification of Cys-12 mutant enzymes*

The expression of mutant AGPase cDNAs was confirmed by resolving the crude extract proteins on SDS–PAGE. Potato tuber AGPases were identified by immunoblotting with antibody against spinach leaf AGPase that has been shown to be reactive with the potato tuber enzyme (Okita *et al.*, 1990). The two mutant enzymes, $S_{C12S}L_{wt}$ and $S_{C12A}L_{wt}$, were produced at a level similar to the wild-type enzyme based on the intensity of the immunoblotting. Their apparent sizes were the same as that of the wild type.

The mutant and wild-type enzymes were subjected to SDS–PAGE under reducing and non-reducing conditions and transferred to nitrocellulose membranes. Immunoblotting results showed that the mutant proteins migrated as a single band under non-reducing condition, while the wild type migrated as two bands corresponding to molecular masses of 50 and 100 kDa (Fu *et al.*, 1998b). Under reducing conditions, the mutant and wild-type enzymes migrated as a single band. Thus, mutagenesis indeed eliminated the intermolecular disulphide in potato tuber enzyme. This observation confirms the results obtained from chemical modification approaches.

4.4 *Activation characteristics of Cys-12 mutant enzymes*

Substitution of Cys-12 in the small subunit by either Ser or Ala eliminated the requirement of DTT for the activation of the potato tuber AGPase (Fu *et al.*, 1998b). Another striking difference between the mutant and wild-type enzymes was the time course of activation. The wild type needed about 17.5 min to reach maximal activity, the two mutant enzymes were fully activated within 10 s. Thus, the mutant enzymes show the same activation characteristics as the reduced wild type (Fu *et al.*, 1998b).

5. Concluding remarks and discussion

This chapter deals with three properties of the potato tuber AGPase, the nature of the G1P binding site, the characteristics of the activator binding sites of the catalytic (small) and regulatory (large) subunits and the nature of the reductive activation of the enzyme by DTT.

We can conclude that Lys-198 of the small subunit of potato tuber AGPase is primarily involved in G1P binding. The 135–550-fold increases of the $S_{0.5}$ value for G1P when this residue was replaced by other amino acids explains the high conservation of this Lys in plant and bacterial AGPases. The Lys residue is probably required for the proper substrate binding to AGPase under physiological concentrations of G1P. From the the modest effect on V_{max} values and the kinetic constants for ATP, Mg^{2+}, PGA and P$_i$ (Fu et al., 1998a), Lys-198 is probably neither involved in the rate-limiting step of the catalytic mechanism nor responsible for maintaining the native conformation of the enzyme. As the substitution of Lys-198 varied from basic to neutral to acidic amino acid, the apparent affinities for G1P decreased. There seems to be a highly specific requirement for a Lys residue in terms of its charge, size and shape to be present in the active site to allow optimal binding of substrate. Even the most conservative substitution of an Arg resulted in a mutant enzyme with 135-fold lower apparent affinity for G1P, suggesting that charge alone is insufficient to account for proper interaction with the substrate. Arg, being a slightly larger amino acid than Lys, may sterically interfere with substrate binding.

In contrast to the effects observed for the mutations of the small subunit Lys-198, mutations of Lys-213 of the large subunit had no effect on the $S_{0.5}$ of G1P. When both residues were replaced by Arg, the effect on the apparent affinity for G1P was similar to that obtained with the single Arg substitution of the small subunit, ruling out a direct role of Lys-213 in binding of the substrate. A sequence search on the large subunit of tuber AGPase revealed no consensus sequence other than the region surrounding Lys-213. Therefore, it is unlikely that G1P binds to an alternative site on the large subunit. This seems to be consistent with the proposed function of this subunit, that is, modulating the allosteric regulation of the small subunit by PGA and P$_i$, with no direct role in catalysis. Still, there is a possibility that G1P may bind to the large subunit, but with no catalysis after the binding event. In any case, the data provide further evidence that the main function of the small subunit is catalysis as suggested by a previous study (Ballicora et al., 1995).

The Lys-195 region (FVEKP) of E. coli AGPase is not only conserved in potato tuber AGPase (FAEKP), but also identical to the mannose-1-phosphate binding site of PMI/GMP (May et al., 1994). Furthermore, this motif or a closely related sequence (GVEKP, IVEKY, KVIKP, FKEKP) is found in many enzymes with the common characteristic of catalysing the synthesis of an NDP-sugar from sugar phosphate and NTP (Table 3). Therefore, FVEKP may be part of a sugar phosphate-binding motif for this class of sugar nucleotide pyrophosphorylases. Of course, other sequences must dictate the specificity for different sugar phosphates.

The study of the allosteric regulation of the AGPase has always been a very important topic because this enzyme catalyses a key step in the pathway of the synthesis of glycogen in bacteria and starch in plants (Preiss, 1984, 1988, 1991, 1997, 1999; Preiss and Romeo, 1989).

The activator site Lys residues (large Lys-417, large Lys-455, small Lys-441) were mutated conservatively to Arg. However, very little change was observed on the affinity for the activator. The $A_{0.5}$ for PGA in the synthesis direction did not increase more than two-fold (data not shown). The mutant enzymes of neutral (Ala) and negative (Glu) amino acid replacements were expressed, purified and characterized kinetically. Much greater effects on the affinity for activator were found when the Lys residues were replaced by Ala and Glu. The $A_{0.5}$ for PGA of these mutants increased up to 8.3

mM (83-fold higher than the wild type) when Lys-441 on the small subunit was mutated to Glu (*Tables 4* and *5*). This indicated that the most important characteristic about those Lys residues was the positive charge that probably interacts with the negative charge of the phosphate and/or carboxyl groups of the activator, PGA. At the same time, replacing the Lys-404 on the small subunit (homologous to large Lys-417) to Ala was enough to make this mutant insensitive to the activation in the synthesis direction (Ballicora *et al.*, 1998). When the pyrophosphorolysis reaction was analysed, this mutant had an $A_{0.5}$ that was 3090-fold higher than the wild type. In all mutants at the activator sites, the kinetic constants for the substrates ADPglucose, PP_i and the co-factor Mg^{2+} did not change significantly (Ballicora *et al.*, 1998). The most significant difference between these mutants was that mutations on the small subunit showed a much lower affinity for the PGA than the mutations performed on the large subunit (*Tables 4* and *5*). Thus, both of these sites in the small subunit appear to be more important for regulation than the homologous lysines in the large subunit.

It has been thought that the large subunit plays the most important role in the regulation since its presence in the heterotetramer increases the affinity for the activator PGA (Ballicora *et al.*, 1995). One of the possibilities that has been considered was that the large subunit provides a regulatory site with higher affinity. However, performing mutations in the large subunit of the putative amino acids involved in the binding of the activator did not support that idea. Those large subunit mutants still increased the affinity for PGA when they were combined with the wild-type small subunit. This effect was observed whether the reaction was in the synthesis or the pyrophosphorolysis direction (*Table 4*; Ballicora *et al.*, 1998). Thus, the large subunit of the potato tuber AGPase increases the affinity for the activator but the integrity of its own regulatory site residues is not required to cause the effect. Mutations on the homologous sites on the small subunit showed larger effects on the regulation of the enzyme. All this suggests that the role of the large subunit is to *modulate* the regulatory properties of the small subunit rather than providing a more efficient allosteric site for activation. The small subunit not only has a functional catalytic site but also has a functional yet relatively inefficient regulatory site (with a low affinity for the activator). In presence of the large subunit, this site in the small subunit is more efficient as an activator.

The inhibition by phosphate showed a very clear difference between the residues on the small and large subunit. When Lys-441 and Lys-404 on the small subunit were mutated to either Ala or Glu, the inhibition by P_i in absence of the activator almost disappeared. The $I_{0.5}$ increased more than two orders of magnitude (*Table 5*). These mutants also lost the ability to shift the activation curve by the presence of inhibitor. Thus, P_i cannot effectively compete with the activator. Nevertheless, when the mutations were performed on the large subunit the effect was very different. None of the mutants on the Lys-417 or the Lys-455 could increase the $I_{0.5}$ for P_i in absence of PGA. In these mutants, P_i shifted the activation curve to the same extent as the wild type (*Table 5*). Therefore, Lys-404 (site s-I) and Lys-441 (site s-II) are very important for inhibition by P_i. The homologous sites on the large subunit do not seem to have the same role. It is interesting that in the spinach leaf enzyme the P_i can prevent the binding of pyridoxal 5'-phosphate to the large subunit site (l-I) (Ball and Preiss, 1994). This can be explained by P_i binding and partially blocking the access of the activator or chemical analogue (pyridoxal 5'-phosphate) to that residue. Recently, another P_i binding site was found on the enzyme from *Anabaena* (Sheng and Preiss, 1997). Since this enzyme is a homotetramer, it would be interesting to know if this residue plays a different role in the large or small subunit of the heterotetrameric higher plant enzyme.

The results obtained in the present study strongly suggest that the different subunits have very different regulatory functions in the heterotetrameric AGPase. Previously, it has been suggested that they have different roles in catalysis. The small subunit seemed to have all the catalytic activity of the heterotetramer, as the large subunit expressed alone did not show AGPase activity (Ballicora *et al.*, 1995; Iglesias *et al.*, 1993). Site-directed mutagenesis experiments on the G1P site confirmed the idea that the role of the large subunit on catalysis is minimal (Fu *et al.*, 1998a). We propose that the main role of the large subunit is to interact with the small subunit and modulate the properties of the activator site. The structure of the putative activator sites in the large subunit would not be as important as the overall interaction of the large subunit that allows the small subunit to increase its affinity for the activator. This effect could be the result of an induced conformational change or the direct interaction between regulatory sites. This hypothesis is consistent with the Lys residues identified as part of the activator site showing a higher conservation in the small subunit than the large subunit of the plant enzymes sequenced so far.

It has been suggested that Lys-417 (site l-I, *Table 1*) in the large subunit is not conserved and may provide different grades of activation for PGA to different AGPases from different sources (Martin and Smith, 1995). However, a change in this residue can explain only slight decreases in affinity for the PGA. Mutating Lys-417 to Met, a residue that is present in this position in some enzymes, showed little changes in regulatory properties (*Tables 4* and *5*). Thus, enzymes bearing a Met instead of a Lys at this position can still have all the properties needed to regulate the synthesis of starch *in vivo*. If an AGPase from a higher plant does not show the classical regulation by PGA and P_i, it is not because there is a Met in this site. Other structural reasons have to be found.

Thus, the Lys residues on the small subunit are more important than the respective homologous residues in the large subunit. Still, some variations could exist in the relative importance of these two sites in the small subunit. In the potato tuber AGPase, site s-I (small Lys-404) seems to be more important than site s-II (small Lys-441). In other enzymes the opposite effect cannot be discarded. For instance, in the spinach leaf enzyme, site s-I is not labelled by pyridoxal 5'-phosphate (Ball and Preiss, 1994). This could be due to s-II being more important for the regulation than s-I or that pyridoxal 5'-phosphate does not bind for steric reasons. For instance, when in the enzyme from *Anabaena*, Lys-419 (homologous to s-II and l-II) was mutated to Arg, Lys-382 (homologous to s-I and l-I) became the preferred site for labelling (Charng *et al.*, 1994).

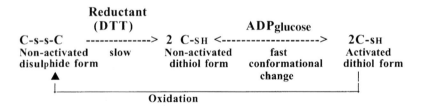

Figure 1. A proposed model of the reductive activation mechanism of potato tuber ADPglucose pyrophosphorylase. Only the small (catalytic) subunits, C, are shown. DTT, dithiothreitol.

The results presented show that DTT activation of the potato tuber AGPase proceeds via a reduction of the intermolecular disulphide bridge between the small subunits and a subsequent conformational change induced by the substrates. The observation that all three ligands ATP, G1P and Ca^{2+} (Mg^{2+}) are required to be present to have an equivalent activation effect as ADPglucose on the enzyme is consistent with an ordered binding mechanism as previously shown for AGPases from *E. coli* (Haugen and Preiss, 1979), *Rhodospirillum rubrum* (Paule and Preiss, 1971) and barley leaf (Kleczkowski *et al.*, 1993). ATP: Mg^{2+} binds first, and then G1P binds. Mg^{2+} was required for the binding of ATP, but not for the binding of ADPglucose (Haugen and Preiss, 1979). Therefore, all substrates, ATP, G1P and Ca^{2+} (Mg^{2+}) are needed to bind the catalytic sites in contrast to ADPglucose. Both the ATP site and G1P site are required to be occupied in order to induce the conformational change following the reduction step.

Reduction of an intermolecular disulphide bridge resulted in a shift of the dimer band of the small subunit to its monomer position in SDS gels. The enzyme needs to be reduced first in order for the conformational change to take place (Fu *et al.*, 1998b). For some enzymes such as chloroplast FBPase, the rate of the reduction process is strongly accelerated by specific conformational changes induced by modulators (Ballicora and Wolosiuk, 1994). However, inclusion of ADPglucose did not show acceleration of the reduction rate of potato tuber AGPase. This further indicates that the activation of potato tuber enzyme proceeds in a sequential order with the reduction occurring first.

It is important to distinguish the activation induced by the substrates plus DTT from that induced by the physiological activator, PGA. All the studies on the reductive activation of potato tuber AGPase were performed in the absence of PGA. When PGA was included in the reaction mixture, the enzyme showed linear kinetics and the specific activity was about 13-fold higher in the ADPglucose synthesis reaction and 2.5-fold higher in the pyrophosphorolysis reaction than that from the reductive activation.

Intermolecular disulphide bridges are often involved in maintaining the quaternary structure of proteins. Besides its involvement in the activation of potato tuber AGPase, the intermolecular disulphide bridge between the small subunits is apparently also involved in maintaining the enzyme in a correct folding state. The potato tuber AGPase became unstable when DTT was present in the preincubation mixture (Fu *et al.*, 1998b). Once the disulphide bridge was reduced, either ADPglucose or ATP, G1P plus Ca^{2+} are needed to protect the enzyme from inactivation. Consistent with this observation is that reduced wild-type and mutant enzymes are heat labile at 60°C for 5 min, while the wild-type enzyme is stable at this condition (Ballicora *et al.*, 1995).

Mutation of Cys-12 in the small subunit into either Ala or Ser yielded mutants with instantaneous activation rates as the wild type in the reduced state. This suggested that the role of Cys-12 was neither related to its hydrophobicity nor its hydrogen-bonding capacity but specifically to its ability to form a disulphide bridge. Sequence alignment of all the plant AGPases available indicates that Cys-12 and its surrounding sequence, -S-Q-T-C-L-D-P-, is conserved in the small subunit of enzymes from all dicot plants (e.g. spinach, *Vicia faba*, *Beta vulgaris*, *Pisum sativum*, *Arabidopsis thaliana*). It is also conserved in the small subunit of one monocot plant enzyme, that from barley leaf. However, AGPase from spinach leaf could not be activated by ADPglucose and DTT. By analysing the spinach leaf AGPase on SDS–PAGE under non-reducing condition, it was also found that an intermolecular disulphide bridge existed between its small subunits. Reduction of this disulphide bridge made the spinach leaf enzyme heat-labile as in the case of the potato tuber enzyme. Information regarding the reductive activation of the other AGPases with the conserved Cys is not available.

Several chloroplast enzymes are regulated by reversible thiol–disulphide interchange mediated by a light-controlled ferredoxin–thioredoxin system (Wolosiuk *et al.*, 1993). Interestingly, the same potato AGPase small subunit gene is expressed both in tubers (non-photosynthetic tissue) and leaves (photosynthetic tissue; Nakata *et al.*, 1994). However, the expression level in leaves is significantly lower than that in tubers. It is not clear if the same potato tuber AGPase is also expressed in potato leaves. When reduced thioredoxin from *Spirulina* was substituted for DTT to activate the potato tuber AGPase, no significant effect could be observed. There is still a possibility that a different isozyme of thioredoxin may be active. However, the physiological importance of the reductive activation phenomenon in this enzyme is still unclear, as *in vivo* the enzyme may be continuously exposed to the activator PGA. Nevertheless, the possibility cannot be discarded that an endogenous reductant plays a role in the fine regulation of the potato tuber enzyme.

An activation mechanism of potato tuber AGPase is proposed (*Figure 1*). The intermolecular disulphide bridge between the small subunits locks the protein in a non-activated conformation. Reduction frees the enzyme, and subsequent binding of ADPglucose induces a rapid conformational change of the enzyme to the activated state. Removal of ADPglucose converts the enzyme back to its non-activated dithiol form, while reoxidation of the intermolecular disulphide bridge converted the enzyme back to its non-activated disulphide conformation. For clarity, only the small subunits are shown, but it must be kept in mind that the reduction of the intermolecular disulphide bridge probably leads to a rearrangement of the small and large subunits during the activation.

Acknowledgement

This research was supported in part by Department of Energy Grant DE-FG02-93ER20121.

References

Ball, K. and Preiss, J. (1994) Allosteric sites of the large subunit of the spinach leaf adenosine diphosphate glucose pyrophosphorylase. *J. Biol. Chem.* **269:** 24706–24711.

Ball, S., Marianne, T., Dirick, L., Fresnoy, M., Delrue, B. and Decq, A. (1991) A *Chlamydomonas reinhardtii* mutant is defective for 3-phosphoglycerate activation and orthophosphate inhibition of ADPglucose pyrophosphorylase. *Planta* **185:** 17–26.

Ballicora, M.A. and Wolosiuk, R.A. (1994) Enhancement of the reductive activation of chloroplast fructose-1,6-bisphosphatase by modulators and protein pertubants. *Eur. J. Biochem.* **222:** 467–474.

Ballicora, M.A., Laughlin, M.J., Fu, Y., Okita, T.W., Barry, G.F. and Preiss, J. (1995) Adenosine 5'-diphosphate-glucose pyrophosphorylase from potato tuber. Significance of the N-terminus of the small subunit for catalytic properties and heat stability. *Plant Physiol.* **109:** 245–251.

Ballicora, M.A., Fu, Y., Nesbitt, N.M. and Preiss, J. (1998) ADP-glucose pyrophosphorylase from potato tuber. Site-directed mutagenesis studies of the regulatory sites. *Plant Physiol.* **118:** 265–274.

Buchanan, B.B. (1980) Role of light in the regulation of chloroplasts. *Annu. Rev. Plant Physiol.* **31:** 341–374.

Charng, Y., Iglesias, A.A. and Preiss, J. (1994) Structure–function relationships of cyanobacterial ADP-glucose pyrophosphorylase: site-directed mutagenesis and chemical modification of the activator-binding sites of ADP-glucose pyrophosphorylase from *Anabaena* PCC 7120. *J. Biol. Chem.* **269:** 24107–24113.

Fu, Y., Ballicora, M.A., Leykam, J.F. and Preiss, J. (1998a) Mechanism of reductive activation of potato tuber ADP-glucose pyrophosphorylase. *J.Biol. Chem.* **273:** 25045–25062.

Fu, Y., Ballicora, M.A. and Preiss, J. (1998b) Mutagenesis of the glucose-1-phosphate binding site of potato tuber ADP-glucose pyrophosphorylase. *Plant Physiol.* **117:** 989–996.

Giroux, M.J., Shaw, J., Barry, G., Cobb, B.G., Greene, T., Okita, T. and Hannah, L.C. (1996) A single mutation that increases maize seed weight. *Proc. Natl Acad. Sci. USA* **93:** 5824–5829.

Haugen, T.H. and Preiss, J. (1979) Biosynthesis of bacterial glycogen. The nature of the binding of substrates and effectors to ADP-glucose synthases. *J. Biol. Chem.* **254:** 127–136.

Hertig, C.M. and Wolosiuk, R.A. (1983) Studies on the hysteretic properties of the chloroplast fructose-1,6-bisphosphatase. *J. Biol. Chem.* **258:** 984–989.

Hill, M.A., Kaufmann, K., Otero, J. and Preiss, J. (1991) Biosynthesis of bacterial glycogen: mutagenesis of a catalytic site residue of ADPglucose pyrophosphorylase from *Escherichia coli. J. Biol. Chem.* **266:** 12455–12460.

Hossain, S.A., Tanizawa, K., Kazuta, Y. and Fukui, T. (1994) Overproduction and characterization of recombinant UDP-glucose pyrophosphorylase from *Escherichia coli* K-12. *J. Biochem. (Japan)* **115:** 965–972.

Iglesias, A.A., Kakefuda, G. and Preiss, J. (1991) Regulatory and structural properties of the cyanobacterial ADP-glucose pyrophosphorylase. *Plant Physiol* **97:** 1187–1195.

Iglesias, A.A., Barry, G.F., Meyer, C., Bloksberg, L., Nakata, P.A., Greene, T., Laughlin, M.J., Okita, T.W., Kishore, G.M. and Preiss, J. (1993) Expression of the potato tuber ADP-glucose pyrophosphorylase in *Escherichia coli. J. Biol. Chem.* **268:** 1081–1086.

Jiang, X.M., Neal, B., Santiago, F., Lee, S.J., Romana, L.K. and Reeves, P.R. (1991) Structure and sequence of the rfb (O antigen) gene cluster of *Salmonella* serovar typhimurium (strain LT2). *Mol. Microbiol.* **5:** 695–713.

Katsube, T., Kazuta, Y., Mori, H., Nakano, K., Tanizawa, K. and Fukui, T. (1990) UDP-glucose pyrophosphorylase from potato tuber: cDNA cloning and sequencing. *J. Biochem. (Tokyo)* **108:** 321–326.

Kleczkowski, L.A., Villand, P., Preiss, J. and Olsen, O. (1993) Kinetic mechanism and regulation of ADP-glucose pyrophosphorylase from barley (*Hordeum vulgare*) leaves. *J. Biol. Chem.* **268:** 6228–6233.

Köplin, R., Arnold, W., Hotte, B., Simon, R., Wang, G. and Pühler, A. (1992) Genetics of xanthan production in *Xanthomonas campestris*: the *xan* A and *xan* B genes are involved in UDPglucose and GDPmannose biosynthesis. *J. Bacteriol.* **174:** 191–199.

Li, L. and Preiss, J. (1992) Characterization of ADPglucose pyrophosphorylase from a starch deficient mutant of *Arabidopsis thaliana. Carbohydr. Res.* **227:** 227–239.

Marolda, C.L. and Valvano, M.A. (1993) Identification, expression, and DNA sequence of the GDP-mannose biosynthesis genes encoded by the O7 rfb gene cluster of strain VW187 (*Escherichia coli* O7: K1). *J. Bacteriol.* **175:** 148–158.

Martin, C. and Smith, A.M. (1995) Starch biosynthesis. *Plant Cell* **7:** 971–985.

May, T.B., Shinabarger, D., Boyd, A. and Chakrabarty, A.M. (1994) Identification of amino acid residues involved in the activity of phosphomannose isomerase-guanosine 5′-diphospho-D-mannose pyrophosphorylase. A bifunctional enzyme in the alginate biosynthetic pathway of *Pseudomonas aeruginosa. J. Biol. Chem.* **269:** 4872–4877.

Morell, M., Bloom, M. and Preiss, J. (1988) Affinity labeling of the allosteric activator site(s) of spinach leaf ADPglucose pyrophosphorylase. *J. Biol. Chem.* **263:** 633–637.

Nakata, P.A., Greene, T.W., Anderson, J.M., Smith-White, B.J., Okita, T.W. and Preiss, J. (1991) Comparison of the primary sequences of two potato tuber ADP-glucose pyrophosphorylase subunits. *Plant. Mol. Biol.* **17:** 1089–1093.

Nakata, P.A., Anderson, J.M. and Okita, T.W. (1994) Structure and expression of the potato ADP-glucose pyrophosphorylase small subunit. *J. Biol. Chem.* **269:** 30798–30807.

Okita, T.W., Nakata, P.A., Anderson, J.M., Sowokinos, J., Morell, M. and Preiss, J. (1990) The subunit structure of potato tuber ADP-glucose pyrophosphorylase. *Plant Physiol.* **93:** 785–790.

Parsons, T.F. and Preiss, J. (1978) Biosynthesis of bacterial glycogen. Isolation and characterization of the pyridoxal-P allosteric activator site and the ADP-glucose protected pyridoxal-P binding site of *Escherichia coli* B ADP-glucose synthase. *J. Biol. Chem.* **253:** 7638–7645.

Paule, M.R. and Preiss, J. (1971) Biosynthesis of bacterial glycogen X. The kinetic mechanism of adenosine diphosphoglucose pyrophosphorylase from *Rhodospirillum rubrum*. *J. Biol. Chem.* **246:** 4602–4609.

Preiss, J. (1984) Bacterial glycogen synthesis and its regulation. *Annu. Rev. Microbiol.* **38:** 419–458.

Preiss, J. (1988) Biosynthesis of starch and its regulation. In: *The Biochemistry of Plants*, Vol. 14 (ed. J. Preiss). Academic Press, San Diego, CA, pp. 181–254.

Preiss, J. (1991) Biology and molecular biology of starch synthesis and its regulation. *Surv. Plant Mol. Cell Biol.* **7:** 59–114.

Preiss, J. (1997) Modulation of starch synthesis. In: *A Molecular Approach to Primary Metabolism in Higher Plants* (eds C. Foyer and P. Quick). Taylor & Francis, London and Washington, DC, pp. 81–104.

Preiss, J. (1999) The chemistry and molecular biology of plant starch synthesis. In: *Starch: Chemistry and Technology*, 3rd Edn (eds R.L. Whistler and J.N. BeMiller). Academic Press, New York, in press.

Preiss, J. and Romeo, T. (1989) Physiology, biochemistry and genetics of bacterial glycogen synthesis. *Adv. Microb. Physiol.* **30:** 183–238.

Preiss, J. and Sivak, M. (1996) Starch synthesis in sinks and sources. In: *Photoassimilate Distribution in Plants and Crops* (eds E. Zamski and A. Schaffer). Marcel Dekker, New York, pp. 63–96.

Ragheb, J.A. and Dottin, R.P. (1987) Structure and sequence of a UDP glucose pyrophosphorylase gene of *Dictyostelium discoideum*. *Nucleic Acids Res.* **15:** 3891–3906.

Sanwal, G.G. and Preiss, J. (1967) Biosynthesis of starch in *Chlorella pyrenoidosa*. II. Regulation of ATP: alpha-D-glucose 1-phosphate adenyl transferase (ADP-glucose pyrophosphorylase) by inorganic phosphate and 3-phosphoglycerate. *Arch. Biochem. Biophys.* **119:** 454–459.

Scheibe, R. (1991) Redox modulation of chloroplast enzymes. A common principle for individual control. *Plant Physiol.* **96:** 1–3.

Sheng, J. and Preiss, J. (1997) Arginine294 is essential for the inhibition of *Anabaena* PCC 7120 ADP-glucose pyrophosphorylase by phosphate. *Biochemistry* **36:** 13077–13084.

Sheng, J., Charng, Y. and Preiss, J. (1996) Site-directed mutagenesis of Lysine382, the activator binding site of ADP-glucose pyrophosphorylase from *Anabaena* PCC 7120. *Biochemistry* **35:** 3115–3121.

Smith-White, B. and Preiss, J. (1992) Comparison of proteins of ADP-glucose pyrophosphorylase from diverse sources. *J. Mol. Evol.* **34:** 449–464.

Sowokinos, J.R. (1981) Pyrophosphorylases in *Solanum tuberosum* II. Catalytic properties and regulation of ADP-glucose and UDPglucose pyrophosphorylases activities in potatoes. *Plant Physiol.* **68:** 924–929.

Sowokinos, J.R. and Preiss, J. (1982) Pyrophosphorylases in *Solanum tuberosum*. III. Purification, structural and catalytic properties of potato tuber ADP-glucose pyrophosphorylase. *Plant Physiol.* **6:** 1459–1466.

Thorson, J.S., Kelly, T.M. and Liu, H.W. (1994) Cloning, sequencing, and overexpression in *Escherichia coli* of the α-D-Glc-1-P cytidylyltransferase gene isolated from *Yersinia pseudotuberculosis J. Bacteriol.* **176:** 1840–1849.

Varnó, D., Boylan, S.A., Okamoto, K. and Price, C.W. (1993) *Bacillus subtilis gtaB* encodes UDP-glucose pyrophosphorylase and is controlled by stationary-phase transcription factor sigma B. *J. Bacteriol.* **175:** 3964–3971.

Wolosiuk, R.A., Ballicora, M.A. and Hagelin, K. (1993) The reductive pentose phosphate cycle for photosynthetic CO_2 assimilation. Enzyme modulation. *FASEB J.* **7:** 622–637.

.

10

Regulation of Rubisco

Martin A.J. Parry, Jane E. Loveland and P. John Andralojc

1. Introduction

Ribulose-1,5-bisphosphate carboxylase/oxygenase (Rubisco; EC 4.1.1.39), the world's most abundant protein, catalyses the initial step of photosynthetic carbon metabolism which is the carboxylation of ribulose-1,5-bisphosphate (RuBP) by CO_2. In higher plants, the enzyme is a hexadecamer composed of eight large, chloroplast-encoded, catalytic subunits and eight small, nuclear-encoded subunits. Rubisco also catalyses a competing and wasteful reaction with oxygen which is examined in detail elsewhere within this book (see Chapter 11). Rubisco has a low affinity for CO_2 and a slow catalytic rate (e.g. 3 s^{-1} for each wheat Rubisco catalytic site) and so plants require extraordinarily large amounts of this enzyme (up to 60% soluble leaf protein) to photosynthesize rapidly. A recent comprehensive review of catalytic mechanism and structure is given in Cleland *et al.* (1998). The present review focuses on the regulation of Rubisco in higher plants.

Plants are subjected to marked fluctuations in their physical environment. Rubisco activity is regulated in response to these environmental changes and this helps to regulate flux through the photosynthetic carbon reduction cycle. Such regulation of Rubisco activity is important in matching the leaf's capacity for RuBP regeneration (a function of the light-dependent reduction of NADP$^+$ and photophosphorylation) with the prevailing rate of RuBP utilization (by carboxylation, photorespiration and – more remotely – starch and sucrose synthesis). Thus, regulation of Rubisco prevents any imbalance between regeneration and utilization of RuBP, helping to ensure that CO_2 fixation is efficient over a wide range of ambient growth conditions (Geiger and Servaites, 1994). Under some conditions Rubisco can exert a high level of control over photosynthetic rate (e.g. the flux control coefficient is large at high irradiance or in plants grown with low N supply (Hudson *et al.*, 1992; Stitt *et al.*, 1991)).

In the long term, carboxylation rates *in vivo* may be altered by changes in the amount of Rubisco protein through changes in the rate of biosynthesis and/or degradation of the enzyme. Whilst the amounts of *rbc*S and *rbc*L transcripts may change within minutes (Krapp *et al.*, 1993; Sheen, 1990), the amount of Rubisco present responds less rapidly, over days (e.g. where Rubisco is selectively degraded during stress (Desimone *et al.*, 1996; Eckardt and Pell, 1995; Mehta *et al.*, 1992)).

Plant Carbohydrate Biochemistry, edited by J.A. Bryant, M.M. Burrell and N.J. Kruger.

In the short term, carboxylation rates *in vivo* may be altered by modulating the activity of the enzyme far beyond that determined by substrate availability alone (Badger *et al.*, 1984; Mott *et al.*, 1984). It is this regulation that is the subject of this review. The mechanisms involved are summarized in *Figure 1*.

2. Rubisco carbamylation

Rubisco is only active when the amino side-chain of lysine 201 within the large subunit is carbamylated with CO_2 and when Mg^{2+} becomes co-ordinated to this carbamate to form a ternary complex at the catalytic site (Lorimer and Miziorko, 1980; Lorimer *et al.*, 1976). This process is ordered and reversible and the equilibria between the two forms of Rubisco (non-carbamylated, E, and carbamylated ternary complex, ECM) depends on the concentrations of CO_2 and Mg^{2+} (*Figure 1*).

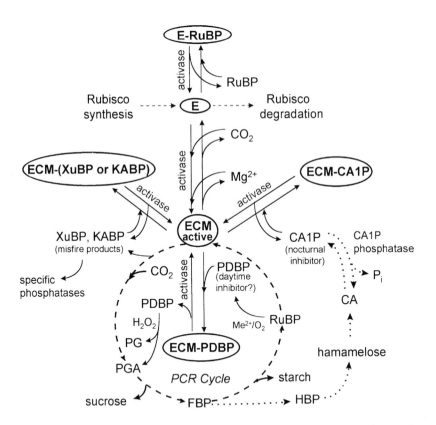

Figure 1. *Diagram showing integration of Rubisco regulation with the photosynthetic carbon reduction (PCR) cycle. The only active form of Rubisco is ECM (carbamylated ternary complex) shown at the centre of the diagram, all other forms are catalytically inactive. In all cases Rubisco activation is mediated by Rubisco activase. E, non-carbamylated Rubisco; RuBP, ribulose-1,5-bisphosphate; XuBP, xylulose-1,5-bisphosphate; KABP, 3-keto-arabinitol-1,5-bisphosphate; CA1P, 2-carboxy arabinitol-1-phosphate; PDBP, D-glycero-2,3-pentodiulose-1,5-bisphosphate; FBP, fructose-1,6-bisphosphate; PGA, 3-phosphoglycerate; PG, 2-phosphoglycolate; HBP, hamamelose 2',5-bisphosphate; CA, 2-carboxyarabanitol. PCR cycle is illustrated with dashed arrows; CA1P biosynthesis with dotted arrows.*

Illumination increases both the Mg^{2+} concentration and the pH of the stroma, promoting activation. However, the concentration of CO_2 in the stroma of most C_3 plants is usually lower in the light than in the dark, limiting activation. Whilst photosynthetic rate and the extent of Rubisco carbamylation respond to a wide range of irradiances (Machler and Nosberger, 1980; Perchorowicz et al., 1981), the increase in stromal pH and Mg^{2+} concentrations saturate at much lower levels of irradiance than does Rubisco activity, so these cannot be the only factors determining Rubisco activity. Furthermore, the concentration of CO_2 required to produce maximal activition in vitro is at least 83 μM (equivalent to 10 mM HCO_3^- at pH 8.2) which is almost an order of magnitude higher than the CO_2 concentration in vivo. Thus, in many cases spontaneous carbamylation alone cannot account for the in vivo activity of Rubisco.

The extent of carbamylation of Rubisco in vivo may be determined by comparing the catalytic activity, measured immediately following extraction in the cold, with the activity measured after incubation with saturating concentrations of CO_2 and Mg^{2+}, to carbamylate lysine 201 in vacant catalytic sites. These measurements are referred to as 'initial' and 'total' activities, respectively.

3. Rubisco activase

3.1 Background

Full activation in vivo requires an additional protein called Rubisco activase (Figure 1). This was first identified by analysis of the rca mutant of Arabidopsis thaliana which required high CO_2 for survival, due to the absence of Rubisco activase (Salvucci et al., 1985; Sommerville et al., 1982). Rubisco activase is also found in some cyanobacteria (Li et al., 1993) and in all the higher plants and green algae examined (Salvucci et al., 1987). Rubisco activase has been estimated to reduce the K_m (CO_2) for Rubisco carbamylation by a factor of 4 (Portis et al., 1986). In vitro, spinach Rubisco activase has a K_m (CO_2) of 8 μM (Werneke et al., 1988a)

Two isoforms of Rubisco activase have been identified but, as explained below, the subunit composition of the active enzyme is unclear. Rubisco activase-catalysed activation of Rubisco requires ATP and a Rubisco–ligand complex (e.g. E.RuBP) (Robinson and Portis, 1988a, 1988b; Streusand and Portis, 1987). During the activation process, Rubisco activase hydrolyses ATP, but not in a manner that is tightly coupled with Rubisco activation; indeed, Rubisco activase catalyses the hydrolysis of ATP even in the absence of Rubisco. However, both the hydrolysis of ATP and the activation of Rubisco are inhibited by ADP (Wang and Portis, 1992).

RuBP binds much more tightly to the non-carbamylated form of Rubisco (Jordan and Ogren, 1983) than to the carbamylated, active form (Laing and Christeller, 1976). The tight binding of RuBP to non-carbamylated Rubisco develops in a time-dependent manner but only occurs at six of the eight available sites per Rubisco holoenzyme (Jordan and Chollet, 1983). RuBP is probably the most common ligand to associate with non-carbamylated Rubisco. In the absence of Rubisco activase, the carbamylation of Rubisco in the presence of RuBP is nearly impossible (Bassham et al., 1978; Gutteridge et al., 1982; Jordan and Chollet, 1983). Non-carbamylated Rubisco can, at least in vitro, exist in several different forms (e.g. depending on temperature) (Gutteridge et al., 1982; Schmidt et al., 1984) for which the rate of dissociation of RuBP differs (Jordan and Chollet, 1983); the physiological relevance of this has yet to be established.

An additional function for Rubisco activase was proposed by Sanchez de Jimenez *et al.* (1995) who suggested that it could function as chaperonin and restore activity to heat denatured Rubisco. However, Eckardt and Portis (1997) were unable to confirm this finding. Furthermore, they demonstrated that Rubisco activase was more sensitive to heat degradation than Rubisco. This conclusion was supported by others (Feller *et al.*, 1998) who found that even moderately high temperatures (e.g. 30–45°C) inhibited Rubisco activase-mediated activation of Rubisco but had no effect on total Rubisco activity in wheat and cotton plants. Therefore any role for Rubisco activase in ameliorating heat-induced damage to Rubisco is unlikely to be of physiological significance (Eckardt and Portis, 1997).

3.2 *Rubisco activase reaction mechanism*

A model for Rubisco activase action in which activase promotes the release of RuBP from inactive enzyme was first proposed by Robinson (Robinson *et al.*, 1988). Studies of transgenic tobacco plants with an antisense gene directed against the mRNA of Rubisco activase have provided further information about its function (Mate *et al.*, 1996). Ligand binding to Rubisco is a biphasic process, the initial binding at the active site being followed by a tightening of the association as a result of the closure of active site polypeptide loops over the ligand, sequestering it from the aqueous environment. Current models (Mate *et al.*, 1996; Salvucci and Ogren, 1996) suggest that activase binds to ligand-bound Rubisco and in so doing causes the flexible loops closed over the ligand to open, leading to the release of both the ligand and activase. In this way, Rubisco activase can facilitate the release of ligand from both non-carbamylated Rubisco (E) or activated Rubisco (ECM). However, release of RuBP from activated Rubisco is a natural consequence of catalytic turnover, being released as product. As a result, Rubisco activase predominantly acts on E–RuBP complexes rather than on ECM–RuBP complexes. The non-carbamylated form of Rubisco (E) – but not E–RuBP – is susceptible to carbamylation and so Rubisco activase activity leads to an increase in the amount of carbamylated enzyme (Mate *et al.*, 1996). It is envisaged that this mechanism will also work for other ligands that bind at the catalytic site. This is essential since many phosphorylated metabolites in the stroma can also modulate Rubisco activity (e.g. positive effectors: 6-phosphogluconate and fructose-1,6-bisphosphate (FBP) or negative effector: ribose-5-phosphate) not only by stabilization of the non-carbamylated or carbamylated enzyme (Badger and Lorimer, 1981; Jordan *et al.*, 1983) but also by competitive inhibition of carboxylation/oxygenation. During Rubisco activase-mediated activation, there is general agreement that ATP hydrolysis promotes vital conformational changes. Mate *et al.* (1996) suggest that free Rubisco activase is 'primed' by prior conversion to an active conformation by a process involving ATP hydrolysis, whereas Salvucci and Ogren (1996) propose that ATP hydrolysis causes productive changes in conformation of Rubisco activase whilst associated with Rubisco. Further study is needed to clarify this point.

The details of the mechanism by which Rubisco activase promotes the release of the bound ligand remain unknown. To date, disappointingly little direct evidence for an interaction between Rubisco and Rubisco activase has been produced. Disuccinimidyl suberate has been used to cross-link the subunits of Rubisco activase with the Rubisco large subunit, indicating a close association between the two (Yokota and Tsujimoto, 1992). In addition, although there was a preliminary report of a complex visualized by electron microscopy (Büchen-Osmond *et al.*, 1992) further structural evidence remains elusive. Despite the limited direct evidence for interaction between Rubisco and Rubisco activase we can remain confident that such interactions must occur. Indirect evidence has come from examining the specificity of Rubisco activase from diverse sources. Tobacco Rubisco activase does not markedly facilitate activation of Rubisco from non-solanaceous species (Wang *et al.*, 1992). A comparison of the Rubisco large subunits revealed residues on the outside of Rubisco that were unique to Rubisco of the *Solanaceae*. Site-directed mutagenesis was used to change the cDNA encoding one such residue in the large subunit of *Chlamydomonas* Rubisco to that in tobacco (P89R). *Chlamydomonas* Rubisco is not normally activated by tobacco Rubisco activase, but once the change had been made the resultant enzyme could be activated by tobacco Rubisco activase (Larson *et al.*, 1997). Further new approaches are required to study the interactions between Rubisco and Rubisco activase.

3.3 *Rubisco activase genes and structure*

One or two nuclear *rca* genes encode Rubisco activase. The genes encode a mature protein of about 420 residues. In many plant species two isoforms of about 41 and 46 kDa are found, due to alternative mRNA splicing of the same or similar primary transcripts (Werneke *et al.*, 1988b). The two forms differ only at the C-terminus (Shen *et al.*, 1991). When expressed separately in *Escherichia coli* the specific activity of the smaller isoform for Rubisco activation was 1.5–3-fold greater than that of the larger isoform. Some species such as tobacco only have the smaller isoform. Aggregation of Rubisco activase subunits into high molecular mass complexes has made it difficult to determine accurately a molecular mass or subunit stoichiometry for catalytic function. As yet there is no evidence to support a specific stoichiometry. Nevertheless, association of the subunits enhances both ATP hydrolysis and the Rubisco activation activity, for which reason a multimer is thought to be the functional form *in vivo* (Yang *et al.*, 1993).

Recent *in vitro* experiments have shown that the larger isoform has greater thermal stability than the smaller one and that incorporation of the larger isoform in heteromultimers increased the thermal stability relative to the homomultimeric form of Rubisco activase (Crafts-Brandner, 1997). The physiological significance of this proposal was examined by comparing the temperature response of wild-type *Arabidopsis* with a transgenic line expressing only the small isoform, but no evidence for increased thermal tolerance was obtained (Kelly *et al.*, 1998).

Conserved features (*Figure 2*) include two consensus nucleotide binding regions (Werneke *et al.*, 1988b). A series of single amino acid substitutions in one of these consensus sequences (the purine nucleotide-binding, p-loop: GGKGQGKS) altered both the ATP hydrolytic and Rubisco activation activities of Rubisco activase (Shen and Ogren, 1992). In addition, a conserved lysine residue has been shown to be required for efficient Rubisco activation (Salvucci and Klein, 1994), and two conserved aspartate residues to be involved both in the precise positioning of the γ-phosphate group of the ATP/Mg^{2+} complex and in the interaction between Rubisco activase subunits (van de Loo and Salvucci, 1998). These conserved residues are indicated in *Figure 2*, which shows the aligned sequences of 10 higher plant Rubisco

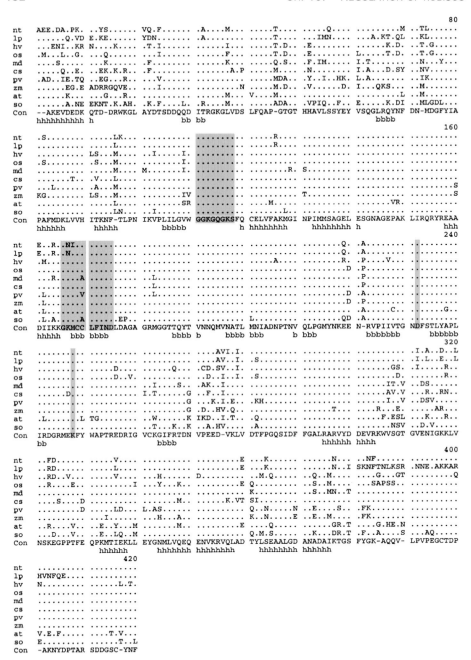

Figure 2. Amino acid sequence alignment for Rubisco activase from 10 higher plant species. The
transit peptide sequence is not shown. Cucumber (cs, Preisig-Mueller and Kindl, 1992);
Arabidopsis (at, Werneke and Ogren, 1989); apple (md, Watillon et al., 1993); rice (os, To et al.,
1996); Phaseolus vulgaris (pv, Ryan et al., 1998); spinach (so, Werneke et al., 1988b); maize
(zm, Ayala-ochoa et al., 1998); barley (hv, Rundle and Zielinski, 1991); tomato (lp, Zhu et al.,
1997); and tobacco (nt, Rodermel, 1992). The consensus sequence (Con) is also illustrated. Only
residues which differ from the consensus sequence are shown explicitly. Conserved regions/residues
discussed in the text are shaded. Secondary structure (α-helix, h and β-sheet, b) is indicated only
where agreement was obtained between the algorithms of Chou and Fasman (1974) and Garnier
et al. (1978).

activases, the corresponding consensus sequence and a secondary structural prediction. Whilst homology between the higher plant sequences remains quite high (around 80%, *Figure 2*) their homology with *Anabaena* (Li *et al.*, 1993) and *Chlamydomonas* Rubisco activase is much lower, the latter being truncated at the N-terminus and extended at the C-terminus (Roesler and Ogren, 1990).

3.4 *Regulation of Rubisco activase activity*

To effectively regulate Rubisco *in vivo* Rubisco activase activity must itself respond to changes in irradiance and other environmental conditions but it is still not clear how Rubisco activase is regulated.

High catalytic activity requires the association of Rubisco activase subunits. Since Mg^{2+} is necessary for association to occur (van de Loo and Salvucci, 1998) changes in stromal Mg^{2+} may modulate activity *in vivo*. However, the increase in stromal Mg^{2+} concentrations is maximal at relatively low irradiance (Machler and Nosberger, 1980; Perchorowicz *et al.*, 1981) and so cannot provide the necessary control of Rubisco activase activity over a wide range of irradiances.

Some additional control could be provided by the availability of ATP (substrate) and ADP (inhibitor) making activase particularly responsive to the ATP:ADP ratio. However, whilst a reasonable correlation between ATP:ADP ratio and Rubisco activation is found under certain conditions, for example P_i depletion (Sharkey *et al.*, 1986), ATP:ADP regulation of Rubisco activase cannot satisfactorily explain why enzymic activity increases with irradiance, since the ratio of ATP:ADP frequently decreases with increasing irradiance (Brooks *et al.*, 1988; Dietz and Heber, 1986). An added level of control exists, since the affinity of Rubisco activase for both ATP and ADP is itself dependent upon the ambient Mg^{2+} and pH (Wang and Portis, 1991).

Like Rubisco itself, Rubisco activase activity is also stimulated by high concentrations of some sugar phosphates (e.g. FBP and RuBP) (Parry *et al.*, 1988; Yokota and Tsujimoto, 1992) and thus the concentrations of a variety of intermediates of the photosynthetic carbon reduction cycle may contribute indirectly to the regulation of Rubisco activity (Geiger and Servaites, 1994).

There are conflicting reports as to the additional requirement for ΔpH across the thylakoid membrane for activation of Rubisco activase, even when ATP is supplied externally (Campbell and Ogren, 1990, 1992; Parry *et al.*, 1988). Similarly, measurements of Rubisco activase activity have generated conflicting results. Using very low concentrations of protein Lan *et al.* (1992) concluded that activation of Rubisco activase is light dependent. However, results from Campbell and Ogren (1992) using a 25-fold greater concentration of protein (which should have enhanced Rubisco activase activity) did not support this hypothesis.

The extent to which the activity of Rubisco activase is regulated in response to changes in irradiance by factors other than ATP:ADP ratios and bisphosphate availability has yet to be fully evaluated and should be a focus for future research.

4. Rubisco inhibitors

A number of tight binding inhibitors of Rubisco have been identified, such as the non-physiological inhibitor 2-carboxy-D-arabinitol-1,5-bisphosphate (CABP) and the naturally occurring inhibitor, 2-carboxy-D-arabinitol-1-phosphate (CA1P; *Figure 1*).

They are all phosphorylated compounds which resemble intermediates of catalysis. From crystallographic studies at high resolution, structural information for Rubisco with and without various ligands bound at the catalytic site has been obtained (for a list of structures, see Table 1 in the review by Cleland *et al.* (1998). By analogy with substrate binding to Rubisco, inhibitor binding is also accompanied by conformational changes involving flexible polypeptide elements which restrict access to the catalytic site subsequent to inhibitor binding (Gutteridge and Gatenby, 1995; Taylor and Andersson, 1996).

4.1 *Misfire products*

Two additional tight binding inhibitors, 3-ketoarabinitol-1,5-bisphosphate (KABP) and xylulose-1,5-bisphosphate (XuBP), are produced by higher plant Rubiscos in non-productive side reactions during catalysis (*Figure 1*). The reactions occur at low frequency, for example XuBP at a frequency of 1 in 400, and result from misprotonation of the enediol intermediate at C2 (KABP) or at C3 (XuBP) (Edmonson *et al.*, 1990a, 1990b). These inhibitors are often referred to as misfire products.

Just as it does for E–RuBP complexes, Rubisco activase is able to greatly decrease the inhibition by many such tight binding inhibitors (Robinson and Portis, 1988a). After release by the action of Rubisco activase, the degradation of these compounds to a non-inhibitory form requires specific phosphatases (Larson and Portis, 1994) (*Figure 1*).

4.2 *2-Carboxy-D-arabinitol-1-phosphate (CA1P)*

At low irradiances the naturally occurring, branched chain, sugar phosphate CA1P may directly control activity by acting as a potent inhibitor of Rubisco (*Figure 1*). CA1P closely resembles an intermediate in the carboxylation reaction, 2-carboxy-3-ketoarabinitol-1,5-bisphosphate (Berry *et al.*, 1987; Gutteridge *et al.*, 1986) and inhibits Rubisco by binding tightly to activated enzyme (ECM), thereby preventing substrate binding.

CA1P quantitation has been achieved directly by measurement of CA1P purified from leaf extracts (Andralojc *et al.*, 1994; Moore *et al.*, 1991) or indirectly, being inferred from measurements of total Rubisco activity (Sage *et al.*, 1990, 1993; Servaites *et al.*, 1986) or both (Sage *et al.*, 1993). It is our opinion that a direct assay incorporating an internal standard should be the method of choice. Although direct measurements are the most reliable, they are more labour-intensive, and so not widely used. Estimating CA1P from measurements of total Rubisco activity may be misleading since many factors (and inhibitors) other than CA1P may contribute to the measured activity.

Since CA1P is produced at low irradiances or in the dark, its production in different species has been inferred by comparison of the activity of Rubisco extracted from leaves sampled either pre-dawn (pd) or at midday (md) (Servaites *et al.*, 1986). The activities are commonly expressed as the ratio, pd/md. The amount of CA1P present in leaves is dependent upon the plant species and the length of the dark period. The lowest recorded pd/md ratios were measured in *Phaseolus vulgaris* (pd/md = 0.1). *P. vulgaris* has subsequently been shown to contain more CA1P at night than any other species (Moore *et al.*, 1991). In species such as potato, soybean and tobacco there is enough CA1P present in the dark to block only about 50% of the active sites of Rubisco (Salvucci, 1989). However, CA1P has been detected in the leaves of the majority of plants tested under low light or dark conditions (Moore and Seemann, 1994; Moore *et al.*, 1991; Servaites *et al.*, 1986), although not all species accumulated photosynthetically significant amounts (Moore *et al.*, 1991).

During the day, even in CA1P-producing species like *P. vulgaris*, decreases in CO_2 concentration reduced the amount of carbamylated Rubisco. As this loss of activity was readily reversed by pre-incubation with CO_2 and Mg^{2+}, Rubisco activity was being controlled by the extent of carbamylation and not inhibitor production (Farineau *et al.*, 1988).

The physiological relevance of CA1P to Rubisco regulation has been questioned. The photosynthetic carbon reduction cycle is strongly regulated at a number of other steps and even Rubisco itself is regulated by Rubisco activase-mediated changes in carbamylation. At night, the substrate RuBP would also become a limiting factor, due to a reduction in the activity of phosphoribulokinase (Wirtz *et al.*, 1982). Additional regulation involving CA1P would therefore appear unnecessary. Strong evidence is emerging that CA1P protects Rubisco from proteolysis by a variety of proteases *in vitro* (Khan *et al.*, 1998) which suggests that, *in vivo*, CA1P may decrease the rate of Rubisco degradation. It is envisaged that conformational changes associated with CA1P binding protect Rubisco from proteolysis, by decreasing the accessibility of proteases to cleavage sites on the enzyme. During the day such protection would be achieved through catalysis, but at night the concentrations of RuBP would be too low for this (Servaites *et al.*, 1991).

CA1P phosphatase. CA1P is completely degraded within a few minutes of bright irradiance by a specific phosphatase, 2-carboxyarabinitol-1-phosphate phosphatase (CA1P-phosphatase) which hydrolyses CA1P to 2-carboxy-D-arabinitol (CA) and P_i (Charlet *et al.*, 1997; Gutteridge and Julien, 1989; Holbrook *et al.*, 1989; Kingston-Smith *et al.*, 1992; Salvucci *et al.*, 1988) (*Figure 1*). CA1P-phosphatase only hydrolyses free CA1P after it has been removed from Rubisco by the action of Rubisco activase (Robinson and Portis, 1988). No other physiologically relevant compounds have been shown to be effective substrates for CA1P-phosphatase (Charlet *et al.*, 1997; Kingston-Smith *et al.*, 1992). It is interesting to speculate whether naturally occurring, branched chain sugar phosphates (e.g. hamamelose phosphates) that are similar to CA1P are substrates for CA1P-phosphatase.

CA1P-phosphatase activity has been found in all the species tested irrespective of their ability to accumulate photosynthetically significant amounts of CA1P (Charlet *et al.*, 1997; Gutteridge and Julien, 1989; Holbrook *et al.*, 1989; Kingston-Smith *et al.*, 1992; Salvucci *et al.*, 1988) and so CA1P-phosphatase, like CA1P, is likely to be present in most if not all species. There are substantial conflicts in the reported characteristics of CA1P-phosphatase isolated by different groups (e.g. effects of ionic strength and dithiothreitol on activity) even where the enzyme has been extracted from the same species (Charlet *et al.*, 1997; Gutteridge and Julien, 1989; Holbrook *et al.*, 1989; Kingston-Smith *et al.*, 1992; Salvucci *et al.*, 1988). However, there is general agreement that several other metabolites increase CA1P-phosphatase activity (e.g. FBP, RuBP, 3-phosphoglycerate (PGA) and NADPH). In a comprehensive study, Charlet *et al.* (1997) showed that positive effectors of catalytic activity contained a phosphate group in close proximity to a second phosphate or carboxyl group. Although structurally similar to CA1P these positive effectors did not serve as substrates for the enzyme. Changes in the concentrations of these effectors in response to environmental conditions may serve to regulate activity *in vivo*, for example NADPH concentrations are closely correlated with irradiance.

CA1P biosynthesis. CA1P is only synthesized at very low irradiances or in the dark (*Figure 1*). In tobacco, CA1P seems to be synthesized relatively slowly (over hours). However, *P. vulgaris* accumulates CA1P within minutes of the transition from high to low irradiances (Kobza and Seemann, 1989). Details of the pathway for the *de novo* synthesis of CA or CA1P are still incomplete. It has been demonstrated that during low irradiance (less than 100 μmol m^{-2} s^{-1}) up to 8% of recently assimilated carbon is incorporated collectively into CA1P and CA in *P. vulgaris* (Andralojc *et al.*, 1994). This high level of incorporation indicates that photosynthetic carbon reduction cycle intermediates are likely to be close precursors of CA and CA1P. A pulse-chase experiment using $^{14}CO_2$ has shown that a pre-existing pool of CA1P precursors acts as the immediate source of carbon skeletons for CA1P synthesis and that recently assimilated carbon only finds its way into CA1P after a lag of 1–2 h in the dark (Andralojc *et al.*, 1994).

Of the photosynthetic carbon reduction cycle intermediates, the most likely to be a precursor of CA1P is FBP. *In vivo*, FBP has been shown to undergo an intramolecular rearrangement to form hamamelose-2',5-bisphosphate (HBP) (Beck *et al.*, 1968). HBP is structurally closely related to CA1P (Beck, 1982). Consistent with the role of FBP as a precursor of CA1P is the increased incorporation of newly assimilated radiolabel into CA1P at low irradiances (Andralojc *et al.*, 1994), conditions under which FBP concentrations would be relatively high (Sassenrath-Cole and Pearcy, 1994). HBP has been shown to be dephosphorylated to hamamelose *in vivo* (Beck *et al.*, 1971). These interconversions may be universal in higher plants, since hamamelose is likely to be ubiquitous amongst higher plants (Sellmair *et al.*, 1977).

Hamamelose is readily converted by leaf discs into CA (and thence into CA1P) and so is a plausible precursor of CA1P (Andralojc *et al.*, 1996; Martindale *et al.*, 1997). Even in species that do not accumulate photosynthetically significant amounts of CA1P at night, for example *Triticum aestivum*, the capacity exists for the conversion of hamamelose into CA (Andralojc *et al.*, 1996).

Radioisotope dilution assays have shown that, in both light- and dark-adapted leaves, there is a large pre-existing pool of CA (Moore *et al.*, 1992) which may serve as the immediate precursor of CA1P (Moore and Seemann, 1992). This could explain why the appearance of recently assimilated carbon as CA1P lags behind the first appearance of (CA-derived) CA1P (Andralojc *et al.*, 1994).

Recently, enzyme activities which may support alternative routes of CA1P biosynthesis have been identified (Andralojc *et al.*, 1998). A putative pathway for CA1P biosynthesis is incorporated in *Figure 1* (dotted arrows).

Intracellular location. In a controversial report, one group has questioned the interaction *in vivo* between CA1P and Rubisco (Anwaruzzaman *et al.*, 1996). However, there is a substantial body of evidence from studies with (i) protoplasts (Salvucci and Anderson, 1987); (ii) non-aqueous fractionation experiments (Moore *et al.*, 1995); and (iii) the use of low concentrations (25–50 mM) of SO_4^{2-} ions to prevent both the dissociation of CA1P from Rubisco and the binding of free CAIP to Rubisco, during extraction and assay (Moore and Seemann, 1994). All suggest a chloroplastic location for both CA1P and CA1P-phosphatase and indicate that Rubisco does bind CA1P *in vivo*. Given the high affinity of CA1P for carbamylated Rubisco ($K_d = 3.2 \times 10^{-8}$ M; Berry *et al.*, 1987) it is difficult to believe that there is no interaction between these two chloroplast components.

4.3 *Daytime inhibitors*

There is strong evidence for an additional tight binding inhibitor of Rubisco that is produced during the day. The presence of this inhibitor in a wide range of higher plants, both those that do and those that do not have CA1P, can be inferred by comparing measurements of total and maximal Rubisco activity. Measurements of maximal activity exploit the ability of SO_4^{2-} ions (200 mM or above) to displace inhibitors bound to Rubisco and, following removal of the SO_4^{2-} ions, permit the measurement of maximum catalytic potential (Parry *et al.*, 1997). Maximal activity measures the activity of all the active sites whether or not they were carbamylated or occupied by tight binding inhibitors immediately prior to extraction and thus provides the best available measurement of maximum catalytic potential. In contrast, total activity only measures the activity of sites that were either carbamylated or readily carbamylated by incubation with activating co-factors and cannot exceed measurements of maximal activity (*Table 1*).

The amount of this inhibitor in leaves is dependent upon the plant species and the environmental conditions, but does not appear to be related to the ability to produce significant amounts of CA1P (*Table 1*). Indeed, under growth room conditions in the light, species such as potato and French bean (*P. vulgaris*) contained little of this inhibitor but *Petunia* and barley contained sufficient inhibitor to block over 40 and

Table 1. *A comparison of total and maximal Rubisco activities in extracts from plants grown at 300 μmol m^{-2} s^{-1}*

Species	Rubisco activity (nmol min^{-1} mg protein^{-1})	
	Total	Maximal
Species with photosynthetically significant amounts of CA1P		
French bean	685 ± 6[a]	738 ± 17[b]
Potato	656 ± 7[a]	673 ± 24[a]
Sugar beet	460 ± 4[a]	578 ± 39[b]
Tobacco	444 ± 20[a]	634 ± 26[b]
Petunia	505 ± 40[a]	926 ± 44[b]
Species with insignificant amounts of CA1P		
Pea	526 ± 18[a]	611 ± 23[b]
Wheat	845 ± 18[a]	1059 ± 36[b]
Barley	678 ± 21[a]	913 ± 38[b]

Values are the means of five independent estimations ± SE, different superscript characters indicate statistically significant differences between columns at $p \leq 0.05$. CA1P, 2-carboxy-D-arabinitol-1-phosphate.

34% of the Rubisco active sites, respectively (Parry *et al.*, 1997). Thus, reports that the Rubisco k_{cat} (i) varies with irradiance; (ii) may be decreased late in the day in response to water stress (Medrano *et al.*, 1997; Parry *et al.*, 1993) or by CO_2 enrichment (Lawlor *et al.*, 1995); or (iii) increased by warmer growth temperature (Bowes, 1995) are consistent with such inhibitors regulating Rubisco activity in response to environmental changes.

Only small amounts of the daytime inhibitor have been isolated from leaves (Keys *et al.*, 1995) and it has yet to be identified. It has been shown to be distinct from the misfire product, XuBP (Keys *et al.*, 1995; Parry *et al.*, 1997) although many of its properties are consistent with those of D-*glycero*-2,3-pentodiulose-1,5-bisphosphate (PDBP), a compound produced *in vitro* by the oxidation of RuBP (Kane *et al.*, 1998) (*Figure 1*). If present *in vivo*, it is likely that PDBP would be metabolized to 3-phosphoglycerate and 2-phosphoglycolate. Any contribution of this process to photorespiration (as opposed to Rubisco oxygenase activity) is open to question.

4.4 *Oxyanions*

Rubisco activity is also affected by the oxyanions HPO_4^{2-} (P_i) and SO_4^{2-}. In addition to being competitive inhibitors, these oxyanions were able to increase the activity of Rubisco without a concomitant increase in carbamylation (Parry *et al.*, 1985). The oxyanion-dependent increase in activity was much greater (five-fold) when only one or two active sites per holoenzyme were carbamylated than when all sites were carbamylated (two-fold). This effect is thought to be due to interactions between the active sites. Thus, in the absence of complete Rubisco carbamylation, disproportionately high Rubisco activities would be observed *in vivo* in the presence of P_i (10 mM) (Parry *et al.*, 1985). Therefore, fluctuations in stromal P_i concentrations should have a significant effect on Rubisco activity. Despite the availability of a number of high-resolution structures for Rubisco, no structural basis for this behaviour is evident. Nevertheless, similar interactions have been independently reported by a number of groups (Johal *et al.*, 1985; Parry *et al.*, 1985; Yokota and Tsujimoto, 1992).

5. Mechanisms and physiological relevance

Under most conditions more of the control of photosynthetic carbon metabolism resides with Rubisco than any other enzyme of photosynthetic metabolism (Stitt, 1996). Regulation of Rubisco activity *in vivo* is complex, and the balance between the different mechanisms contributing to regulation may vary not only from species to species but also in relation to the environment (Farineau *et al.*, 1988; Salvucci and Anderson, 1987; Servaites *et al.*, 1986).

On the transition from darkness to light, Rubisco activase exerts a high degree of control over the rate of activation and hence catalytic activity. This is achieved via the release either of RuBP from non-carbamylated Rubisco or of CA1P (and misfire products) from carbamylated enzyme, under which conditions Rubisco activity has been shown to be inversely correlated with either RuBP or CA1P bound to Rubisco. The flux control coefficient of Rubisco activase for the activation of Rubisco is unusually high (almost 1) (Hammond *et al.*, 1998).

In contrast, the transition from high to low irradiance is accompanied by a rapid decline in photosynthetic rate, due to diminishing supply of RuBP. Rubisco activase is thought not to play a direct role in the inactivation of Rubisco, although such inactivation is presumably inversely correlated to Rubisco activase activity. Both the decrease in Rubisco carbamylation and the synthesis of CA1P occur slowly (Hammond *et al.*, 1998).

There is increasing evidence that the carboxylation capacity of water-stressed plants can be reduced by a decrease in Rubisco k_{cat} rather than by a change in carbamylation state (Medrano *et al.*, 1997; Parry *et al.*, 1993). Such a reduction in k_{cat} occurs when inhibitors bind tightly to Rubisco. To what extent this is a regulatory response or merely reflects a failure of Rubisco activase to relieve such inhibition remains to be answered. Under water stress conditions, Rubisco activase may be reversibly deactivated by unfavourable changes in metabolite concentrations (e.g. a decrease in the ATP:ADP ratio) or the Rubisco activase:Rubisco ratio may be reduced because Rubisco activase is less stable than Rubisco (particularly the Rubisco–inhibitor complex). Moderately high temperatures have also been shown to decrease Rubisco activase activity in both cotton and wheat (Feller *et al.*, 1998).

Rubisco with bound inhibitor is more resistant to protein degradation (Khan *et al.*, 1998), which may be important in preventing premature Rubisco degradation. Interestingly, the breakdown of Rubisco was delayed in transgenic plants with reduced amounts of Rubisco activase. Under these conditions, inhibitors might not be efficiently removed from the active site and so could protect Rubisco from breakdown (He *et al.*, 1997).

Further experimentation is needed to clarify the interplay between Rubisco activase and Rubisco inhibitors in Rubisco regulation under the fluctuating environmental conditions plants experience in the field.

Acknowledgements

IACR receives grant-aided support from the Biotechnology and Biological Sciences Research Council of the United Kingdom.

References

Andralojc, P.J., Dawson, G.W., Parry, M.A.J. and Keys, A.J. (1994) Incorporation of carbon from photosynthetic products into 2-carboxyarabinitol-1-phosphate and 2-carboxyarabinitol. *Biochem. J.* **304**: 781–786.

Andralojc, P.J., Keys, A.J., Martindale, W., Dawson, G.W. and Parry, M.A.J. (1996) Conversion of D-hamamelose into 2-carboxy-D-arabinitol and 2-carboxyarabinitol 1-phosphate in leaves of *Phaseolus vulgaris* L. *J. Biol. Chem.* **271**: 26803–26810.

Andralojc, P.J., Keys, A.J. and Parry, M.A.J. (1998) Biosynthesis of the Rubisco inhibitor, 2-carboxy-D-arabinitol 1-phosphate (CA1P). In *XIth International Congress on Photosynthesis Abstracts*, SY9-P2.

Anwaruzzaman, Nakano, Y. and Yokota, A. (1996) Different location in dark-adapted leaves of *Phaseolus vulgaris* of ribulose 1,5-bisphosphate carboxylase/oxygenase and 2-carboxyarabinitol 1-phosphate. *FEBS Lett.* **388**: 233–227.

Ayala-ochoa, A., Loza-tavera, H. and Sanchez de Jimenez, E. (1998) *Zea mays* ribulose-1,5-bisphosphate carboxylase/oxygenase activase precursor (*rca*) mRNA, complete cds. *GenBank* AF084478.

Badger, M. and Lorimer, G. (1981) Interaction of sugar phosphates with the catalytic site of ribulose-1,5-bisphosphate carboxylase. *Biochemistry* **20**: 2219–2225.

Badger, M., Sharkey, T. and von Caemmerer, S. (1984) The relationship between steady state gas exchange of bean leaves and the level of carbon reduction cycle intermediates. *Planta* **160**: 305–313.

Bassham, J.A., Krohne, S. and Lendzian, K. (1978) *In vivo* control of carboxylation reaction. In: *Photosynthetic Carbon Assimilation* (eds H.W. Spiegelman and G. Hinds). Plenum, New York, pp. 77–93.

Beck, E. (1982) Branched-chain sugars. In: *Encyclopedia of Plant Physiology, New Series* (eds F.A. Loewus and W. Tanner). Springer-Verlag, Heidelberg, pp. 124–157.

Beck, E., Sellmair, J. and Kandler, O. (1968) Biosynthese der hamamelose. *Z. Pflanzenphysiol.* **58**: 434–451.

Beck, E., Stransky, H. and Furbringer, M. (1971) Synthesis of hamamelose-diphosphate by isolated spinach chloroplasts. *FEBS Lett.* **13**: 229–234.

Berry, J.A., Lorimer, G.H., Pierce, J., Seemann, J., Meek, J. and Freas, S. (1987) Isolation, identification and synthesis of 2-carboxyarabinitol 1-phosphate, a diurnal regulator of ribulose-bisphosphate carboxylase activity. *Proc. Natl Acad. Sci. USA* **84**: 734–738.

Bowes, G. (1995) Elevated atmospheric carbon dioxide:a return to the past? In: *Photosynthesis from Light to Biosphere*, Vol. 5 (ed. P. Mathis). Kluwer Academic Publishers, The Netherlands, pp. 761–766.

Brooks, A., Portis, A. and Sharkey, T. (1988) Effects of irradiance and methyl viologen treatment on ATP, ADP and activation of ribulose bisphosphate carboxylase in spinach leaves. *Plant Physiol.* **88**: 850–853.

Büchen-Osmond, C., Portis, A.R. and Andrews, T.J. (1992) Rubisco activase modifies the appearance of Rubisco in the electron microscope. In: *Research in Photosynthesis*, Vol. 3. (ed. N. Murata). Kluwer Academic Publishers, The Netherlands, pp. 653–656.

Campbell, W.J. and Ogren, W.L. (1990) Electron transport through photosystem 1 stimulates light activation of Ribulose bisphosphate carboxylase/oxygenase (Rubisco) by Rubisco activase. *Plant Physiol.* **94**: 479–484.

Campbell, W.J. and Ogren, W. (1992) Rubisco activase activity in spinach leaf extracts. *Plant Cell Physiol.* **36**: 215–220.

Charlet, T., Moore, B.D. and Seemann, J.R. (1997) Carboxyarabinitol 1-phosphate phosphatase from leaves of *Phaseolus vulgaris* and other species. *Plant Cell Physiol.* **38**: 511–517.

Chou, P.Y. and Fasman, G.D. (1974) Prediction of protein conformation. *Biochemistry* **13**: 222–245.

Cleland, W., Andrews, T.J., Gutteridge, S., Hartman, F. and Lorimer, G. (1998) Mechanism of Rubisco: the carbamate as general base. *Chem. Rev.* **98**: 549–561.

Crafts-Brandner, S.J., van de Loo, F.J. and Salvucci, M.E. (1997) The two forms of ribulose-1,5-bisphosphate carboxylase/oxygenase activase differ in sensitivity to elevated temperature. *Plant Physiol.* **114**: 439–444.

Desimone, M., Henke, A. and Wagner, E. (1996) Oxidative stress induces partial degradation of the large subunit of ribulose-1,5-bisphosphate carboxylase/oxygenase in isolated chloroplasts of barley. *Plant Physiol.* **111**: 789–796.

Dietz, K.J. and Heber, U. (1986) Light and CO_2 limitation of photosynthesis and states of reaction regenerating ribulose-1,5-bisphosphate or reducing glycerate 3-phosphate. *Biochim. Biophys. Acta* **767**: 392–401.

Eckardt, N. and Pell, E. (1995) Oxidative modification of Rubisco from potato foliage in response to ozone. *Plant Physiol. Biochem.* **33**: 273–282.

Eckardt, N. and Portis, A. (1997) Heat denaturation profiles for ribulose-1,5-bisphosphate carboxylase/oxygenase (Rubisco and Rubisco activase and the inability of Rubisco activase to restore activity to heat denatured Rubisco. *Plant Physiol.* **113**: 243–248.

Edmonson, D., Badger, M. and Andrews, T. (1990a) Slow inactivation of ribulose bisphosphate carboxylase during catalysis is caused by the accumulation of a slow tight binding inhibitor at the active site. *Plant Physiol.* **93**: 1390–1397.

Edmonson, D.L., Kane, H.J. and Andrews, T.J. (1990b) Substrate isomerization inhibits ribulose bisphosphate carboxylase-oxygenase during catalysis. *FEBS Lett.* **260**: 62–66.

Farineau, J., Susuki, A. and Morot-Gaudry, J.F. (1988) Changes in the activation state of Rubisco in bean leaves in relation to recovery of photosynthetic activity after various treatments. *Plant Sci.* **55**: 191–198.

Feller, U., Crafts-Bradner, S. and Salvucci, M. (1998) Moderately high temperatures inhibit ribulose-1,5-bisphosphate carboxylase/oxygenase (Rubisco) activase-mediated activation of Rubisco. *Plant Physiol.* **116**: 539–546.

Garnier, J., Osguthorpe, D.J. and Robson, B. (1978) Analysis of the accuracy and implications of simple methods for predicting the secondary structure of globular proteins. *J. Mol. Biol.* **120**: 97–120.

Geiger, D. and Servaites, J.C. (1994) Dynamics of self-regulation of photosynthetic carbon metabolism. *Plant Physiol. Biochem.* **32**: 173–183.

Gutteridge, S. and Gatenby, A.A. (1995) Rubisco synthesis, assembly, mechanism, and regulation. *Plant Cell* **7**: 809–819.

Gutteridge, S. and Julien, B. (1989) A phosphatase from chloroplast stroma of *Nicotiana tabacum* hydrolyses 2'-carboxyarabinitol 1-phosphate, the natural inhibitor of Rubisco to 2'-carboxyarabinitol. *FEBS Lett.* **254**: 225–230.

Gutteridge, S., Parry, M.A.J. and Schmidt, C.N.G. (1982) The reactions between active and inactive forms of wheat Rubisco and effectors. *Eur. J. Biochem.* **126**: 597–602.

Gutteridge, S., Parry, M.A.J., Burton, S., Keys, A.J., Mudd, A., Feeney, J., Servaites, J.C. and Pierce, J. (1986) A nocturnal inhibitor of carboxylation in leaves. *Nature* **324**: 274–276.

Hammond, E., Andrews, T., Mott, K. and Woodrow, I. (1998) Regulation of Rubisco activation in antisense plants of tobacco containing reduced levels of Rubisco activase. *Plant J.* **14**: 101–110.

He, Z., von Caemmerer, S., Hudson, G., Price, G., Badger, M. and Andrews, T.J. (1997) Ribulose bisphosphate carboxylase/oxygenase activase deficiency delays senescence of ribulose bisphosphate carboxylase/oxygenase but progressively impairs its catalysis during tobacco leaf development. *Plant Physiol.* **115**: 1569–1580.

Holbrook, G.P., Bowes, G. and Salvucci, M. (1989) Degradation of 2-carboxyarabinitol 1-phosphate by a specific chloroplast phosphatase. *Plant Physiol.* **90**: 673–678.

Hudson, G.S., Evans, J.R., von Caemmerer, S., Arvidsson, Y.B.C. and Andrews, T.J. (1992) Reduction of ribulose-1,5-bisphosphate carboxylase/oxygenase content by antisense RNA reduces photosynthesis in transgenic tobacco plants. *Plant Physiol.* **98**: 294–302.

Johal, S., Partridge, B. and Chollet, R. (1985) Structural characterization and the determination of negative cooperativity in the tight binding of 2'-carboxyaranitol bisphosphate to higher plant ribulose bisphosphate carboxylase. *J. Biol. Chem.* **260**: 9894–9904.

Jordan, D. and Chollet, R. (1983) Inhibition of ribulose bisphosphate carboxylase by substrate ribulose bisphosphate. *J.Biol. Chem.* **258**: 13752–13758.

Jordan, D.B. and Ogren, W.L. (1983) Species variation in kinetic properties of ribulose 1,5-bisphosphate carboxylase/oxygenase. *Arch. Biochem. Biophys.* **227**: 425–433.

Jordan, D., Chollet, R. and Ogren, W. (1983) Binding of phosphorylated effectors by active and inactive forms of ribulose-1,5-bisphosphate carboxylase. *Biochemistry* **22**: 3410–3418.

Kane, H., Wilkin, J., Portis, A. and Andrews, T.J. (1998) Potent inhibition of ribulose bisphosphate carboxylase by an oxidized impurity in ribulose 1,5-bisphosphate. *Plant Physiol.* **117**: 1059–1069.

Kelly, W., Whitmarsh, J. and Portis, A. (1998) Effect of two forms of Rubisco activase on the high temperature inhibition of photosynthesis in *Arabidopsis* plants. *Plant Physiol. (suppl.)* **117**: 688.

Keys, A.J., Major, I. and Parry, M.A.J. (1995) Is there another player in the game of Rubisco regulation? *J. Exp. Bot.* **46**: 1245–1251.

Khan, S., Andralojc, P. and Parry, M.A.J. (1998) Regulation of Rubisco:the protective nature of 2-carboxy-D-arabinitol 1-phosphate. *Plant Physiol. (suppl.)* **117**: 526.

Kingston-Smith, A.H., Major, I., Parry, M.A.J. and Keys, A.J. (1992) Purification and properties of a phosphatase in French bean (*Phaseolus vulgaris* L.) leaves that hydrolyses 2′-carboxy-D-arabinitol-1-phosphate. *Biochem. J.* **287**: 821–825.

Kobza, J. and Seemann, J. (1989) Regulation of ribulose-1,5-bisphosphate carboxylase activity in response to changes in irradiance. *Plant Physiol.* **89**: 918–924.

Krapp, A., Hoffman, B., Schafer, C. and Stitt, M. (1993) Regulation of the expression of *rbc*S and other photosynthetic genes by carbohydrates: a mechanism for sink regulation of photosynthesis. *Plant J.* **3**: 817–828.

Laing, W. and Christeller, J. (1976) A model for the kinetics of activation and catalysis of ribulose-1,5-bisphosphate carboxylase. *Biochem. J.* **159**: 563–570.

Lan, Y., Woodrow, I.E. and Mott, K.A. (1992) Light dependent changes in ribulose bisphosphate carboxylase activase activity in leaves. *Plant Physiol.* **99**: 304–309.

Larson, E.M. and Portis, A.R. (1994) Identification of xylulose bisphosphatase in spinach. *Plant Physiol.* **105**: 85S.

Larson, E.M., O'Brien, C.M., Zhu, G., Spreitzer, R.J. and Portis, A.R. (1997) Specificity for activase is changed by a Pro-89 to Arg substitution in the large subunit of ribulose-1,5-bisphosphate carboxylase/oxygenase. *J. Biol. Chem.* **272**: 17033–17037.

Lawlor, D.W., Delgado, E., Habash, D.Z., Driscoll, S.P., Mitchell, V.J., Mitchell, R.A.C. and Parry, M.A.J. (1995) Photosynthetic acclimation of winter wheat to elevated CO_2 and temperature. In: *Photosynthesis from Light to Biosphere*, Vol. 5 (ed. P. Mathis). Kluwer Academic Publishers, The Netherlands, pp. 989–992.

Li, L.-A., Gibson, J.L. and Tabita, F.R. (1993) The Rubisco activase (*rca*) gene is located downstream from the *rbc*S in *Anabaena* sp. strain CA and is detected in other *Anabaena / Nostoc* strains. *Plant Mol. Biol.* **21**: 753–764.

Lorimer, G. and Miziorko, H. (1980) Carbamate formation on the e-amino group of a lysyl residue as the basis for the activation of ribulose bisphosphate carboxylase by CO_2 and Mg^{2+}. *Biochemistry* **19**: 5321–5328.

Lorimer, G., Badger, M. and Andrews, T.J. (1976) The activation of ribulose-1,5-bisphosphate carboxylase by carbon dioxide and magnesium ions. *Biochemistry* **32**: 9018–9024.

Machler, F. and Nosberger, J. (1980) Regulation of ribulose bisphosphate carboxylase activity in intact wheat leaves by light, CO_2, and temperature. *J. Exp. Bot.* **31**: 1485–1491.

Martindale, W., Parry, M.A.J., Andralojc, P.J. and Keys, A.J. (1997) Synthesis of 2′-carboxy-D-arabinitol-1-phosphate in French bean (*Phaseolus vulgaris*, L.): a search for precursors. *J. Exp. Bot.* **48**: 9–14.

Mate, C.J., von Caemmerer, S., Evans, J.R. Hudson, G.S. and Andrews, T.J. (1996) The relationship between CO_2 assimilation rate, Rubisco carbamylation and Rubisco activase content in activase-deficient transgenic tobacco suggests a simple model of activase action. *Planta* **198**: 604–613.

Medrano, H., Parry, M.A.J., Socias, X. and Lawlor, D.W. (1997) Long term water stress inactivates Rubisco in subterranean clover. *Ann. Appl. Biol.* **131**: 491–501.

Mehta, R.A., Fawcett, T., Porath, D. and Mattoo, A. (1992) Oxidative stress causes rapid membrane translocation and *in vivo* degradation of ribulose-1,5-bisphosphate carboxylase. *J. Biol. Chem.* **267**: 2810–2816.

Moore, B.D. and Seemann, J.R. (1992) Metabolism of 2′-carboxyarabinitol in leaves. *Plant Physiol.* **99**: 1551–1555.

Moore, B.D. and Seemann, J.R. (1994) Evidence that 2-carboxyarabinitol 1-phosphate binds to ribulose-1,5-bisphosphate carboxylase *in vivo*. *Plant Physiol.* **105**: 731–737.

Moore, B.D., Kobza, J. and Seemann, J.R. (1991) Measurement of 2-carboxyarabinitol 1-phosphate in plant leaves by isotope dilution. *Plant Physiol.* **96**: 208–213.

Moore, B.D., Sharkey, T.D., Kobza, J. and Seemann, J.R. (1992) Identification and levels of 2′-carboxyarabinitol in leaves. *Plant Physiol.* **99**: 1546–1550.

Moore, B.D., Sharkey, T.D. and Seemann, J.R. (1995) Intracellular localization of CA1P and CA1P phosphatase activity in leaves of *Phaseolus vulgaris* L. *Photosynthesis Res.* **45**: 219–224.

Mott, J., Ogren, W., O'Leary, J. and Berry, J. (1984) Photosynthesis and ribulose 1,5-bisphosphate concentrations in intact leaves of *Xanthium stumarium* L. *Plant Physiol.* **76**: 968–971.

Parry, M.A.J., Schmidt, C.N.G., Cornelius, M.J., Keys, A.J., Millard, B.N. and Gutteridge, S. (1985) Stimulation of ribulose bisphosphate carboxylase activity by inorganic orthophosphate without an increase in bound activating CO_2: co-operativity between subunits of the enzyme. *J. Exp. Bot.* **170**: 1396–1404.

Parry, M.A.J., Keys, A.J., Foyer, C.H., Furbank, R.T. and Walker, D. (1988) Regulation of ribulose-1,5-bisphosphate carboxylase activity by the activase system in lysed spinach chloroplasts. *Plant Physiol.* **87**: 558–561.

Parry, M.A.J., Delgado, E., Vadell, J., Keys, A.J., Lawlor, D.W. and Medrano, H. (1993) Water stress and the diurnal activity of ribulose-1,5-bisphosphate carboxylase in field grown *Nicotiana tabacum* genotypes selected for survival at low CO_2 concentrations. *Plant Physiol. Biochem.* **31**: 113–120.

Parry, M.A.J., Andralojc, P.J., Parmar, S., Keys, A.J., Habash, D., Paul, M.J., Alred, R., Quick, W.P. and Servaites, J.C. (1997) Regulation of Rubisco by inhibitors in the light. *Plant Cell Environ.* **20**: 528–534.

Perchorowicz, J.T., Raynes, D.A. and Jensen, R.G. (1981) Light limitation of photosynthesis and activation of ribulose bisphosphate carboxylase in wheat seedlings. *Proc. Natl. Acad. Sci USA.* **78**: 2985–2989.

Portis, A.R., Salvucci, M. and Ogren, W. (1986) Activation of ribulose bisphosphate carboxylase/oxygenase at physiological concentrations by Rubisco activase. *Plant Physiol.* **82**: 967–971.

Preisig-Mueller, R. and Kindl, H. (1992) Sequence analysis of cucumber cotyledon ribulose bisphosphate carboxylase/oxygenase activase cDNA. *Biochim. Biophys. Acta* **1171**: 205–206.

Robinson, S.P. and Portis, A.R. (1988a) Release of nocturnal inhibitor, carboxyarabinitol-1-phosphate from ribulose-1,5-bisphosphate carboxylase/oxygenase. *FEBS Lett.* **233**: 413–416.

Robinson, S.P. and Portis, A.R. (1988b) Involvement of stromal ATP in the light activation of ribulose-1,5-bisphosphate carboxylase/oxygenase in intact isolated chloroplasts. *Plant Physiol.* **86**: 293–298.

Robinson, S.P., Streusand, V.J., Chatfield, J.M. and Portis, A.R. (1988) Purification and assay of Rubisco activase from leaves. *Plant Physiol.* **88**: 1008–1014.

Rodermel, S. (1992) *N. tabacum rca* gene for ribulose bisphosphate carboxylase activase. *GenBank Z14980.*

Roesler, K. and Ogren, W. (1990) Primary structure of *Chlamydomonas reinhardtii* ribulose 1,5-bisphosphate carboxylase/oxygenase activase and evidence for a single polypeptide. *Plant Physiol.* **94**: 1837–1841.

Rundle, S.J. and Zielinski, R. (1991) Organization and expression of two tandemly orientated genes encoding ribulose bisphosphate carboxylase/oxygenase activase in barley. *J. Biol. Chem.* **266**: 4677–4685.

Ryan, J., Andralojc, P.J., Willis, A., Gutteridge, S. and Parry, M.A.J. (1998) Cloning and sequencing of a *Phaseolus vulgaris* cDNA (accession number AF041068) encoding Rubisco activase (PGR98-044). *Plant Physiol.* **116**: 1192.

Sage, R. (1993) Light dependent modulation of ribulose-1,5-bisphosphate carboxylase/oxygenase activity in the genus *Phaseolus*. *Photosynthesis Res.* **35**: 219–226.

Sage, R., Sharkey, T.D. and Seemann, J.R. (1990) Regulation of ribulose-1,5-bisphosphate carboxylase activity in response to light intensity and CO_2 in the C_3 annuals *Chenopodium album* L. and *Phaseolus vulgaris* L. *Plant Physiol.* **94**: 1735–1742.

Sage, R., Reid, C.D., Moore, B.D. and Seemann, J.R. (1993) Long term kinetics of the light-dependent regulation of ribulose-1,5-bisphosphate carboxylase/oxygenase activity in plants with and without 2-carboxyarabinitol 1-phosphate. *Planta* **191**: 222–230.

Salvucci, M.E. (1989) Regulation of Rubisco activity *in vivo*. *Physiol. Plant.* **77**: 164–171.

Salvucci, M. and Anderson, J. (1987) Factors affecting the activation state of the level of total activity of ribulose bisphosphate carboxylase in tobacco protoplasts. *Plant Physiol.* **85**: 66–71.

Salvucci, M. and Klein, R.R. (1994) Site-directed mutagenesis of a reactive lysyl residue (Lys-247) of Rubisco activase. *Arch. Biochem. Biophys.* **314**: 178–185.

Salvucci, M. and Ogren, W. (1996) The mechanism of Rubisco activase: insights from studies of the properties and structure of the enzyme. *Photosynthesis Res.* **47**: 1–11.

Salvucci, M., Portis, A.R. and Ogren, W.L. (1985) A soluble chloroplast protein catalyzes ribulosebisphosphate carboxylase/oxygenase activation *in vivo*. *Photosynthesis Res.* **7**: 193–201.

Salvucci, M., Werneke, J., Ogren, W.L. and Portis, A.R. (1987) Purification and species distribution of Rubisco activase. *Plant Physiol.* **84**: 930–936.

Salvucci, M., Holbrook, G., Anderson, J. and Bowes, G. (1988) NADPH-dependent metabolism of ribulose bisphosphate carboxylase/oxygenase inhibitor 2-carboxyarabinitol 1-phosphate by a chloroplast protein. *FEBS Lett.* **231**: 197–201.

Sanchez de Jimenez, E., Medrano, L. and Martinez-Barajas, E. (1995) Rubisco activase: a possible new member of the molecular chaperone family. *Biochemistry* **34**: 2826–2831.

Sassenrath-Cole, G. and Pearcy, R. (1994) Regulation of photosynthetic induction state by the magnitude and duration of low light exposure. *Plant Physiol.* **105**: 1115–1123.

Schmidt, C., Gutteridge, S., Parry, M.A.J. and Keys, A.J. (1984) Inactive forms of wheat ribulose bisphosphate carboxylase. *Biochem. J.* **220**: 781–785.

Sellmair, J., Beck, E., Kandler, O. and Kress, A. (1977) Hamamelose and its derivatives as chemotaxonomic markers in the genus Primula. *Phytochemistry* **16**: 1201–1204.

Servaites, J.C., Parry, M.A.J., Gutteridge, S. and Keys, A.J. (1986) Species variation in the predawn inhibition of ribulose-1,5-bisphosphate carboxylase/oxygenase. *Plant Physiol.* **82**: 1161–1163.

Servaites, J.C., Shieh, W.-J. and Geiger, D.R. (1991) Regulation of photosynthetic carbon reduction cycle by ribulose bisphosphate and phosphoglyceric acid. *Plant Physiol.* **97**: 1115–1121.

Sharkey, T.J. Stitt, M., Heineke, D., Gehardht, R., Raschke, K. and Heldt, H. (1986) Limitation of photosynthesis by carbon metabolism. II. O_2 insensitive CO_2 uptake results from limitation to triose phosphate utilization. *Plant Physiol.* **81**: 1123–1129.

Sheen, J. (1990) Metabolic repression to transcription in higher plants. *Plant Cell* **2**: 1027–1038.

Shen, J.B. and Ogren, W.L. (1992) Alteration of spinach ribulose-1,5-bisphosphate carboxylase/oxygenase activities by site-directed mutagenesis. *Plant Physiol.* **99**: 1201–1207.

Shen, J.B., Orozco, E.M. and Ogren, W.L. (1991) Expression of the two isoforms of spinach ribulose 1,5-bisphosphate carboxylase activase and essentiality of the conserved lysine in the consensus nucleotide-binding domain. *J. Biol. Chem.* **266**: 8963–8968.

Sommerville, C.R., Portis, A.R. and Ogren, W.L. (1982) A mutant of *Arabidopsis thaliana* which lacks activation of RuBP carboxylase *in vivo*. *Plant Physiol.* **70**: 381–387.

Stitt, M. (1996) Metabolic regulation of photosynthesis. In: *Photosynthesis and the Environment* (ed. E.R. Baker). Kluwer Academic Publishers, The Netherlands, pp. 151–190.

Stitt, M., Quick, W.P., Schurr, U., Schulze, E.D., Rodermel, S. R. and Bogorad, L. (1991) Decreased ribulose-1,5-bisphosphate carboxylase-oxygenase in transgenic tobacco transformed with 'antisense' *rbc*S II. Flux control coefficients for photosynthesis in varying light, CO_2 and air humidity. *Planta* **183**: 555–566.

Streusand, V.J. and Portis, A.R. (1987) Rubisco activase mediates ATP-dependent activation of ribulose-1,5-bisphosphate carboxylase. *Plant Physiol.* **85**: 152–154.

Taylor, T. and Andersson, I. (1996) Structural transitions during activation and ligand binding in hexadecameric Rubisco inferred from the crystal structure of the activated unliganded spinach enzyme. *Nature Structural Biol.* **3**: 95–101.

To, K., Suen, D., Chen, M., Chen, L. and Chen, S.C.G. (1996) *Oriza sativa* ribulose-1,5-bisphosphate carboxylase/oxygenase activase (*rca*) mRNA, complete cds. *GenBank* U74321.

van de Loo, F. and Salvucci, M. (1998) Involvement of two aspartate residues of Rubisco activase in coordination of the ATP gamma phosphate and subunit cooperativity. *Biochemistry* 37: 4621–4625.

Wang, Z.Y. and Portis, A.R. (1991) A fluorometric study with 1-analinonapthaline-8-sulfonic acid (ANS) of the interactions of ATP and ADP with Rubisco activase. *Biochim. Biophys. Acta* 1079: 263–267.

Wang, Z.Y. and Portis, A.R. (1992) Dissociation of ribulose-1,5-bisphosphate bound to ribulose-1,5-bisphosphate carboxylase/oxygenase and its enhancement by ribulose-1,5-bisphosphate carboxylase/oxygenase activase-mediated hydrolysis of ATP. *Plant Physiol.* 99: 1348–1353.

Wang, Z.Y., Snyder, G.W., Esau, B.D., Portis, A.R. and Ogren, W.L. (1992) Species-dependent variation in the interaction of substrate-bound ribulose-1,5-bisphosphate carboxylase/oxygenase (Rubisco) and Rubisco activase. *Plant Physiol.* 100: 1858–1862.

Watillon, B., Kettmann, R., Boxus, P. and Burny, A. (1993) Developmental and circadian pattern of Rubisco activase mRNA accumulation in apple plants. *Plant Mol. Biol.* 23: 501–509.

Werneke, J.M. and Ogren, W. (1989) Structure of an *Arabidopsis thaliana* cDNA encoding Rubisco activase. *Nucleic Acids Res.* 17: 2871.

Werneke, J.M., Chatfield, J.M. and Ogren, W.L. (1988a) Catalysis of ribulose bisphosphate carboxylase/oxygenase activation by the product of a Rubisco activase cDNA clone expressed in *Escherichia coli*. *Plant Physiol.* 87: 917–920.

Werneke, J.M., Zielinski, R. and Ogren, W.L. (1988b) Structure and expression of spinach leaf cDNA encoding ribulosebisphosphate caboxylase/oxygenase activase. *Proc. Natl. Acad. Sci. USA* 85: 787–791.

Wirtz, W., Stitt, M. and Heldt, H. (1982) Light activation of Calvin cycle enzymes as measured in pea leaves. *FEBS Lett.* 142: 223–226.

Yang, Z., Ramage, R. and Portis, A.R. (1993) Mg^{2+} and ATP or adenosine 5'-[thio]-triphosphate (ATP S) enhances intrinsic fluorescence and induces aggregation which increases activity of spinach Rubisco activase. *Biochim. Biophys. Acta* 1202: 47–55.

Yokota, A. and Tsujimoto, N. (1992) Characterization of ribulose-1,5-bisphosphate carboxylase/oxygenase carrying ribulose 1,5-bisphosphate at its regulatory sites and the mechanism of interaction of this form of the enzyme with ribulose-1,5-bisphosphate carboxylase/oxygenase activase. *Eur. J. Biochem.* 204: 901–909.

Zhu, Y., Lin, K., Huang, Y., Tauer, C. and Martin, B. (1997) *Lycopersicon pennelleii* Rubisco activase mRNA complete cds. *GenBank* AF037361.

11

Biochemistry of photorespiration and the consequences for plant performance

Alfred J. Keys

1. Introduction

From studies of a burst of CO_2 production by leaves in darkness immediately following steady photosynthesis in the light, the 'post-illumination burst', Decker (1955) and Decker and Tio (1959) deduced that there was a respiratory process taking place in the light that was faster than the steady rate of respiration in darkness. Photorespiration is the name for this respiratory process suggested by Decker and eventually accepted in the scientific literature (Zelitch, 1979a). The characteristics of photorespiration that distinguish it from dark respiration became evident from studies of the effect of atmospheric concentrations of CO_2 and O_2, light intensity and temperature, on the size of the post-illumination burst and on the magnitude of other parameters of gas exchange between leaves and the atmosphere surrounding them, especially the CO_2 compensation point, and net CO_2 assimilation (Canvin, 1979). These same environmental factors were observed to have related effects on the production and metabolism of glycolate, in algae and in leaves; the conclusion was that glycolate metabolism gave rise to photorespiratory CO_2. Photorespiration and glycolate production were shown to be controlled by a process having a much lower affinity for oxygen than dark respiration. Thus by decreasing the oxygen concentration in the atmosphere to 2%, photorespiration can be decreased by some 90% without decreasing dark respiration. Both photorespiration and glycolate production increase with light intensity, temperature and oxygen concentration and decrease with increases in CO_2 in the atmosphere. The controlling process, having the appropriate responses to these environmental factors, has subsequently been shown to be the oxygenation of RuBP catalysed by Rubisco (EC 4.1.1.39).

This chapter focuses on the metabolism involved in photorespiration, the energetics, the flux of carbon through the pathway, the extent to which intermediates of the

Plant Carbohydrate Biochemistry, edited by J.A. Bryant, M.M. Burrell and N.J. Kruger.
© 1999 BIOS Scientific Publishers Ltd, Oxford.

pathway may be withdrawn for growth and other plant needs, and the possibility that the process has some unique function for which there is no alternative. The aim is to identify aspects of photorespiratory metabolism needing more research, to consider current and future areas for study, and to provide basic information on photorespiration for those engaged in genetic manipulation and studies of the molecular biology of this process. Of main concern will be plants with the C_3 mechanism of photosynthesis in which photorespiratory metabolism has the greatest effect on photosynthetic carbon assimilation and which are of most importance to agriculture in temperate climates. The extensive research on photorespiration, and on the enzyme, Rubisco, that initiates it, is justified by the counter-productive effect it has on carbon assimilation under favourable growing conditions.

2. The metabolic pathway responsible for photorespiration

2.1 *Glycolate metabolism*

Intensive work on glycolate metabolism began when this acid was found to be an early product of photosynthesis in investigations using $^{14}CO_2$ (Benson and Calvin, 1950). The intramolecular labelling of glycolate produced during photosynthesis from $^{14}CO_2$ showed that it had a common origin with carbon atoms 1 and 2 of the hexose and pentose phosphates involved in the photosynthetic carbon reduction cycle (Beck, 1979). Experiments using either specifically or uniformly ^{14}C-labelled glycolate (Husic *et al.*, 1987; Tolbert, 1979), as well as identification of other early products of photosynthesis in leaves, showed that glycine and serine were intermediates in the metabolism of glycolate and that the serine carboxyl group was directly derived from a carboxyl group of glycine while carbons 2 and 3 came from carbon 2 of glycine. This was explained through the known properties of the enzyme serine hydroxymethyltransferase. Since radioactivity from [^{14}C]glycolate also appeared in phosphoglycerate and other intermediates of the photosynthetic carbon reduction cycle, it was postulated that hydroxypyruvate and glycerate were involved. Enzymes for the steps in the metabolic pathway were identified in plant material and their intracellular distribution studied by density gradient centrifugation (Bird *et al.*, 1972a; Tolbert, 1971). One of the more elusive steps in the pathway was the conversion of glycine to serine. Measured activities of serine hydroxymethyltransferase in leaf extracts made with hypotonic buffers were inadequate to account for rates of photorespiration and glycolate metabolism *in vivo*. It was found that the complete conversion of glycine to serine + CO_2 + NH_3 at appropriate rates was carried out by intact mitochondria (Bird *et al.*, 1972b, c) and that an enzyme complex, now known as the glycine decarboxylase complex (GDC) in plants (Douce *et al.*, 1994; Oliver, 1994), and as the glycine cleavage system in animals and micro-organisms, was involved. Another problem was the origin of glycolate. This was solved when it was demonstrated that the enzyme, Rubisco, catalysing the reaction of CO_2 with RuBP in the photosynthetic carbon reduction cycle also catalysed an oxygenase reaction in which phosphoglycolate was made from C-1 + C-2 of RuBP (Andrews *et al.*, 1973; Bowes *et al.*, 1971). The presence of a phosphoglycolate phosphatase in chloroplasts had already been established. As early as 1970 glycolate metabolism was conceived as a cyclic pathway (Goldsworthy, 1970) removing carbon from the photosynthetic carbon reduction cycle and subsequently returning 75% of this carbon to the cycle. The pathway has been confirmed by extensive studies of mutant plants with specific lesions in glycolate metabolism (Blackwell *et al.*, 1988a) and more recently by the production of transgenic plants (Lea and Forde, 1994; Migge *et al.*, 1997; Rawsthorne and Bauwe, 1998).

2.2 *The photorespiratory carbon oxidation cycle*

It is important to understand the stoichiometry and the energy requirements of photorespiratory metabolism. This involves integrating the photorespiratory carbon oxidation cycle and the photorespiratory nitrogen cycle with photosynthetic carbon assimilation (Keys, 1986; Keys *et al.*, 1982; Tolbert, 1997) to include the energy needed to produce the RuBP from which glycolate is made, to recycle N released as NH_3 and to recycle the PGA produced via hydroxypyruvate and glycerate from serine. The energy is required in the form of ATP and reducing equivalents. For a rate of photorespiratory CO_2 production that is 25% of net photosynthesis, and assuming net carbon assimilation results in one molecule of sucrose only, the theoretical requirement is 67 ATP and reducing equivalents equal to 42 NAD(P)H (in the recycling of the NH_3 six reduced feredoxins instead of three of the NAD(P)H); without photorespiration the theoretical requirement per sucrose is 37 ATP and 24 NAD(P)H. A simplified cycle showing the requirements for reducing equivalents (12 electrons net – two per NAD(P)H) only is shown in *Figure 1* for the release of each CO_2 by photorespiration (i.e. six electrons per oxygenation).

The pathway involves an interdependence of reactions in one part of the cycle on reactions in other parts and in other cycles. Thus the conversion of glyoxylate to glycine requires the availability of amino groups resulting from the recycling of N from the ammonia by the GOGAT cycle (Keys *et al.*, 1978; Miflin and Lea, 1976).

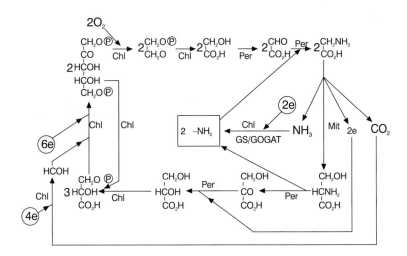

Figure 1. *Outline of photorespiratory metabolism. Chl., Per. and Mit. indicate that the reactions marked occur in the chloroplasts, peroxisomes and mitochondria, respectively. The additional reduced carbon shown as -HCOH- is shown as derived from photorespired CO_2, but can be from stored carbohydrate. The origin of amino groups (shown in the box at the centre) is not restricted to glutamate and serine although these are the main contributors (Ta and Joy, 1986). A more detailed account of the biochemistry of photorespiration can be found in Leegood et al. (1995).*

Mutants without serine glyoxylate aminotransferase, glutamate synthetase, glutamine synthase or GDC activity can function to a limited extent in conditions favourable for photorespiration if provided with a source of amino nitrogen (Blackwell *et al.*, 1988b; Somerville and Somerville, 1983). Likewise an accumulation of serine can be seen if the production of glyoxylate is decreased, for example, by decreasing atmospheric oxygen or removing the CO_2. In detached wheat leaves in CO_2-free air in the light, the production of sucrose from serine was increased if a keto acid was provided (Waidyanatha *et al.*, 1975). The reduction of hydroxypyruvate also probably depends on NADH produced in the mitochondria as a result of the oxidation step in the GDC reaction.

The involvement of chloroplast, peroxisome, mitochondrion, peroxisome and then chloroplast again, in the main flow of C through the photorespiratory pathway, means that transport through membranes is involved. Glycolate is transported from the chloroplast by a specific transporter in exchange for glycerate (Howitz and McCarty, 1985, 1991) and 2-oxoglutarate produced in the peroxisome exchanges with glutamate at the chloroplast membrane involving a specific transporter (Somerville and Ogren, 1983; Somerville and Somerville, 1985). A consequence of the transport of intermediates is that their appearance in the cytosol and vacuole of cells is not excluded.

3. Rates of photorespiration and fluxes of carbon and nitrogen in photorespiratory metabolism

Because, in the light, respiratory processes producing CO_2 and using O_2 are opposed by photosynthetic CO_2 uptake and O_2 production, accurate measurements of true rates of both processes in leaves are difficult. Thus, the size of the post-illumination burst is affected by continuing carboxylation of RuBP remaining, or synthesized, after the light is turned off. All methods of estimating photorespiration based on CO_2 exchange suffer from complications due to re-fixation of respired CO_2 within the leaf. There is also the complication of distinguishing photorespiration from 'dark' respiration continuing in the light.

3.1 *Rates of photorespiration from rates of CO_2 exchange*

Zelitch (1979b) tabulated rates of photorespiration for leaves of various species estimated by the post-illumination burst, release of CO_2 into CO_2-free air in the light, extrapolation of the rate of net photosynthesis to zero CO_2, initial rate of uptake of $^{14}CO_2$ minus net photosynthesis, and the difference between net photosynthesis in 1.5–2.5% O_2 and net photosynthesis in 21% O_2. The estimates ranged from 14 to 75% of net photosynthesis for C_3 plants at 25°C. These estimates were for leaves in various conditions, made by various researchers and subject to various errors, especially endogenous re-assimilation of photorespired CO_2. More recently, Peterson (1983), from a mathematical analysis of the post-illumination burst, reported rates of photorespiration relative to net photosynthesis of 12.5, 18.8, 45 and 85% at 20, 25, 30 and 35°C, respectively, for tobacco leaf discs in air containing 0.033% CO_2; at 32°C, rates were 25 and 100% of net photosynthesis at 0.08 and 0.013% CO_2 in air, respectively.

3.2 *Estimates from metabolic studies*

With increased knowledge of the biochemistry of photorespiration, and the availability of mutants with lesions in the pathway, other approaches have become possible. Somerville and Somerville (1983) used leaf tissue from a mutant of *Arabidopsis thaliana* without serine hydroxymethyltransferase activity in the mitochondria, supplemented with alanine and ammonia, to measure the rate of glycine accumulation on transfer of the plants from non-photorespiratory to photorespiratory conditions. From this they calculated the rate of photorespiration as 36% of net photosynthesis. In a discussion of the errors in various methods of estimating photorespiration they argue that their own method will give an overestimate whilst many of the gas-exchange measurements on whole leaves underestimate the process. Wallsgrove *et al.* (1987) estimated a rate of photorespiration of 40% of net photosynthesis from NH_3 accumulation in the leaves of a mutant of barley lacking chloroplastic glutamine synthetase. De Veau and Burris (1989) estimated, from the rate of incorporation of $^{18}O_2$ into glycine and serine, that photorespiration by wheat leaves at 25°C in 21% O_2, 0.035% CO_2 was 26.9% of net photosynthesis. Such values are consistent with the estimates of Kumarasinghe *et al.* (1977) from the decrease in glycine, increase in serine and CO_2 in darkness following steady-state photosynthesis. These measurements reflect the post-illumination burst in that the source of CO_2 for the burst is mainly from glycine in the GDC reaction (Rawsthorne and Hylton, 1991).

Biehler and Foch (1996) estimated the rate of photorespiration from the initial rates of incorporation of ^{14}C from $^{14}CO_2$ into glycolate. In unstressed wheat leaves photorespiration used 26% of the reducing equivalents assigned to photorespiration plus photosynthesis at 23°C with 0.035% CO_2 in air. Parnik and Keerberg (1995) deduced rates of photorespiration from an analysis of rates of $^{14}CO_2$ evolution from leaves previously supplied with $^{14}CO_2$ during steady-state photosynthesis at 25°C in 21% O_2, of 28, 31 and 94% of net photosynthesis for wheat, barley and tobacco leaves, respectively, after making allowance for carbon entering the photorespiratory cycle from 'end products' of photosynthesis. Essentially, this analysis distinguished CO_2 derived from photorespiration from that derived from dark respiration by its O_2 sensitivity. In an extension of the method, Parnik *et al.* (1995) showed the effect of temperature on photorespiration by barley leaves in obtaining rates that were 15, 17 and 31% of net photosynthesis at 7.5, 13.6 and 26.2°C, respectively.

3.3 *Rates from Rubisco kinetics: control of the flux of carbon to glycolate*

Reported Michaelis constants for oxygen for Rubiscos from C_3 higher plants catalysing the oxygenation of RuBP range from 200 to 800 μM; this is consistent with this reaction being that with a low affinity for oxygen which controls photorespiration and glycolate production. Because CO_2 and O_2 are competing alternative substrates for reaction with RuBP catalysed by Rubisco, the decrease in photorespiration with increased CO_2 is also explained. With increasing knowledge of the kinetics and mechanism of the oxygenation and carboxylation of RuBP catalysed by Rubisco, it has become possible to test the role of Rubisco by preparing mathematical models of whole leaf photosynthesis. The rate of oxygenation of RuBP (v_o), and hence the rate of glycolate production, can be written as follows (Farquhar *et al.*, 1980):

$$v_o = V_o \times O/[O + K_o(1 + C/K_c)] \times R/(R + K'_r) \qquad (1)$$

In this equation the concentration of RuBP (R) in a leaf depends on the use of absorbed photons and the availability of the other substrates, CO_2 and O_2. Thus at saturating light, RuBP should not be limiting [the term $R/(R + K'_r)$ approaches 1], and when CO_2 is saturating and the light is low RuBP will be limiting. V_o in the equation is the maximum velocity of oxygenation; K_c, K_o and K_r are the Michaelis constants respectively for CO_2, O_2 and RuBP. The concentrations of O_2 (O) and CO_2 (C) in the chloroplast stroma where Rubisco is located are determined by their rates of use and production, the concentrations external to the leaf, and the diffusion resistances between the stroma and the bulk atmosphere outside the leaf. Because of its high concentration in the atmosphere relative to the rates of its use and production, the concentration of oxygen in the stroma is usually taken as that determined by the solubility coefficient for the O_2 in water at the temperature of the leaf. For CO_2, the concentration is determined by the boundary layer resistance, stomatal resistance, resistance through the intercellular spaces in the leaf and a resistance due to cell wall, membranes and cytosol separating the intercellular spaces from the chloroplast stroma. Several studies have resulted in estimates of these resistances and the concentration of CO_2 in the chloroplast (Caemmerer and Evans, 1991; Harley *et al.*, 1992; Loreto *et al.*, 1992). Measurements of RuBP amounts in leaves and hence estimates of concentrations in the chloroplast stroma are possible. In theory, therefore, a calculation of the rate of glycolate production for a given situation in leaves should be possible. In practice, it is difficult to assign values for V_c, the maximum velocity of carboxylation, and V_o to Rubisco *in vivo* because of the subtle mechanisms of regulation operating on the enzyme (see Chapter 10). Photorespiratory rates have been usually expressed relative to the rate of net or gross photosynthesis and it is easier to assess the role of Rubisco in the control of photorespiratory rate through an equation derived from Equation 1 above by dividing it into the corresponding equation for the rate of carboxylation (v_c) (Farquhar *et al.*, 1980; Laing *et al.*, 1974).

$$v_c/v_o = C/O \times V_c K_o / V_o K_c \qquad (2)$$

The constant $V_c K_o / V_o K_c$ has become known as the Rubisco specificity factor (S_{rel}). Measurements of this constant are available for purified Rubisco of many species (Bainbridge *et al.*, 1995). Caemmerer and Evans (1991) concluded that at high irradiance the value of C would be about $0.5 C_a$ (the concentration in the external air). Thus with the ambient CO_2 at 0.035%, C would be about 6 µM in solution at 25°C when the oxygen in equilibrium with air is 265 µM in water. With wheat Rubisco the S_{rel} is close to 100 at 25°C, giving $v_c/v_o = 2.26$. Consideration of the stoichiometry of oxygen uptake and CO_2 production in photorespiration (*Figure 1*) shows that the rate of CO_2 production is $0.5 v_o$. Thus release of CO_2 is 22% the rate of v_c. Equating v_c with gross photosynthesis and $v_c - 0.5 v_o$ with net photosynthesis, it is evident that photorespiration calculated from Rubisco properties in this way would be 28% net assimilation at 25°C. Such considerations neglect the dark respiration component and all the complications of re-assimilation internally that confound determination of photorespiration *in vivo*. Equation 2 explains the trends in photorespiration in response to CO_2 and O_2 in leaves in terms of the properties of Rubisco at a single temperature. The trend of photorespiration to increase with temperature (e.g. Parnik *et al.*, 1995; Peterson, 1983) also finds explanation in the properties of Rubisco because the value of S_{rel} decreases dramatically with increasing temperature (Jordan and Ogren, 1984; *Figure 2*). Parnik *et al.* (1995) calculated S_{rel} values for Rubisco in leaves from their measurements of photorespiration using approximations to provide a concentration

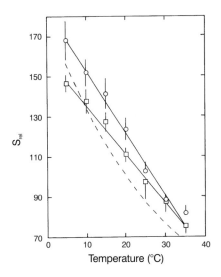

Figure 2. *Effect of temperature on the specificity factor (S_{rel}) of Rubisco. The specificity factors of purified Rubisco from wheat (○) and oil seed rape (□) were measured (unpublished work) by an oxygen electrode method depending on total consumption of the substrate RuBP (Parry et al., 1989) at the temperatures shown. Values are means ± SE of 10 measurements at each temperature. The dotted line shows the response of S_{rel} of spinach to the same temperature range (Jordan and Ogren, 1984).*

of CO_2 at the carboxylation site; the values obtained were 135, 127 and 79 for temperatures of 7.5, 13.6 and 26.2°C, respectively. Sharkey (1988) concluded that it is easier to estimate v_c and v_o from the properties of Rubisco and the internal CO_2 concentration using constants considered by Brooks and Farquhar (1985) than to persist with attempts to measure photorespiration by methods giving CO_2 release.

Brooks and Farquhar (1985) used conventional gas exchange methods and mathematical modelling of the known metabolic and leaf characteristics related to photorespiration to allow calculation of the Rubisco specificity factor at various temperatures for the enzyme in spinach leaves and showed them to correspond closely with estimates from studies of the enzyme *in vitro*. Their method was to estimate Γ^*, the hypothetical CO_2 compensation point of a leaf without dark respiration, by extrapolating plots of net photosynthesis against leaf internal CO_2 concentration at several moderate light intensities to a point of intersection. At this CO_2 concentration v_c should equal $0.5v_o$ (Equation 5, see below) whence from Equation 2: $0.5 = S_{rel}(\Gamma^*/O)$. The values of S_{rel} so determined, at different temperatures, were very close to values for the pure enzyme measured *in vitro* by Jordan and Ogren, 1984.

3.4 *Evidence from analysis of chlorophyll fluorescence quenching*

Two methods for estimating the total linear flow of electrons through photosystem II (J_1) from an analysis of chlorophyll fluorescence quenching have been devised (Genty et al., 1989; Ghashghaie and Cornic, 1994; Krall and Edwards, 1992). If, in addition, net photosynthesis (A) is measured under steady-state conditions, the reducing equivalents (as electrons) can be apportioned on the assumption that the only major consuming processes are CO_2 assimilation and photorespiration. Under defined conditions it appears that the electrons apportioned to oxygen reduction represent use in photorespiration, and that use of electrons for processes other than carbon reduction and photorespiration are minor (Cornic and Briantais, 1991). From consideration of *Figure 1* and Equation 5 below:

$$J_1 = 4(A + R_d) + 6v_o \qquad (3)$$

Electrons can be assigned to net photosynthesis as J_A:

$$J_A = 4(A + R_d) \qquad (4)$$

Since A is known from measurement, and R_d can be measured (Brooks and Farquhar, 1985), or approximated from the steady respiration in the dark by applying some arbitrary factor, for example, 0.5, to allow for inhibition by light, $J_1 - J_A$ gives $6v_o$. By substituting in Equation (5):

$$v_c = A + R_d + 0.5v_o, \qquad (5)$$

a value can be obtained for v_c. It seems to have become the custom to test the validity of such estimates by using them to compute values for S_{rel} by substitution in Equation 2 and comparing to published values obtained using purified Rubisco. The results are generally of the correct order of magnitude but the discrepancies are of much interest (Cornic and Briantais, 1991; Ghashghaie and Cornic, 1994; Habash et al., 1995, 1996; Jacob and Lawlor, 1993; Peterson, 1990). A particularly large discrepancy at low temperatures was observed by Ghashghaie and Cornic (1994) for intact leaves of *Epilobium hirsutum*.

4. Function

It can be argued that photorespiration exists because autotrophic life began when the CO_2 in the ambient atmosphere was high and that the mechanism that evolved for CO_2 fixation through Rubisco, while effective when competition with oxygen was minimal, is flawed in the modern atmosphere. Indeed the Rubisco catalytic mechanism has been said to lead inevitably to photorespiration (Morell et al., 1992). If this is so, it must be accepted that photorespiration in C_3 plants has become totally integrated into photosynthetic metabolism in which it fulfils useful functions discussed below. Whether in the absence of the flaw in Rubisco catalysis that leads to photorespiration there are other processes that could substitute in these functions without the apparent disadvantages of photorespiration is difficult to assess.

4.1 *Scavenging, idling or protecting (Lorimer and Andrews, 1981)*

The photorespiratory pathway, by its very nature, carries out all three of these functions. It recovers three carbons as PGA into the Calvin cycle for every four leaving the chloroplast as two glycolate molecules. Under water stress, where the supply of CO_2 to the chloroplasts is restricted in the light because of stomatal closure, the generation of CO_2 in the green cells by photorespiration maintains carbon in the intermediates of the Calvin cycle. The very nature of the pathway, as presented, is that it is energy consuming except for the glycine decarboxylase system in mitochondria where energy is conserved, either as ATP or as reducing power in NADH. The energy-consuming steps of reduction of hydroxypyruvate, the GOGAT cycle recovering NH_3 and especially the phosphorylative reduction of PGA in the re-generation of the RuBP used up in photorespiration, far outweigh the energy conserved. Thus it is easy to portray photorespiration as a means of dissipating energy in the leaf when the absorbed photons exceed the requirement for CO_2 assimilation and other synthetic processes coupled directly to the photochemistry. There are however other mechanisms in plants to dissipate the energy from absorbed photons, either as heat or by using excess reducing power, and also mechanisms that decrease photon absorption (Demmig-Adams and Adams, 1992; Osmond and Grace, 1995; Park *et al.*, 1996).

4.2 *Provision of intermediates for other metabolic processes*

The scheme put forward by Keys *et al.* (1982) assumed that at 25°C the rate of photorespiration by wheat leaves in the field would be close to 25% of net photosynthesis; that was 1 sucrose gained for 3 CO_2 produced. The more recent methods of estimation outlined above do not make this an unreasonable view. It is worth considering the biochemical consequences of this. Because evolution of each CO_2 by photorespiration requires that two glycolate molecules are metabolized in the photorespiratory pathway, the carbon gain by the leaf (net photosynthesis) is equal to the amount of carbon entering the glycolate pathway. At lower temperatures less carbon will be in the photorespiratory pathway relative to net assimilation, but the amount will still be considerable. In view of these high fluxes, it seems unlikely that the sole purpose of photorespiratory metabolism is to supply intermediates to other pathways. There is, however, ample evidence that carbon from the pathway is freely available within the leaf. Following photosynthesis for 15 min in air containing $^{14}CO_2$, protein in young tobacco leaves contained large amounts of radioactive glycine and serine with little radioactivity appearing in other amino acids (Ongun and Stocking, 1965). This can only be easily explained if the carbon comes from glycine and serine of the photorespiratory pathway because these were the only soluble amino acids significantly radiolabelled. It has also been shown that glycine and serine from photorespiratory metabolism appear in translocate from leaves (Madore and Grodsinski, 1984). Recently Noctor *et al.* (1998) have shown that glutathione synthesis involves glycine from the glycolate pathway. These observations are not surprising when one considers the mobility of glycine and serine in leaves. The pathway itself was investigated and characterized early on by supplying glycolate, glycine and serine in the transpiration stream. Transfer of these compounds to green cells for metabolism must have involved transfer from cell to cell as well as the crossing of membranes involved in the pathway itself. Thus, glycolate, glycine and serine within the photorespiratory pathway must be freely available in cells of leaves. The interesting question is the amounts taken out relative to what must be the major flux round the cycle. The flux as far as serine is covered by measurement (e.g. the post-illumination burst), but the further flux of carbon back into the Calvin cycle is difficult to quantify. That there is a flux to complete the cycle can be appreciated from the labelling of PGA and glycerate with ^{18}O during photosynthesis in the presence of $^{18}O_2$ (Berry *et al.*, 1978) and the greater scrambling of label among the carbons of sucrose when [3-^{14}C]serine is supplied to wheat leaves in the light at 0.015% compared to 0.099% CO_2 in air (Bird *et al.*, 1978).

Protein synthesis, lipid synthesis, nucleic acid synthesis, cell wall synthesis and other activities associated with cell division and cell growth are not major processes in mature leaf cells. Thus nitrate and sulphate reduction, processes that might be expected to compete with carbon assimilation and photorespiration for reducing power generated by the photosystems, are of more significance in young growing leaf cells (Gilbert *et al.*, 1997; Lawlor *et al.*, 1987). Studies of photosynthesis and photorespiration are often made on mature leaves in which the main metabolism is the production and export of carbohydrate. This is presumably why those studying fluorescence can make the approximation that the only major processes in leaves using reducing equivalents are carbon assimilation and photorespiration. Furthermore, only half or less of the cells in leaves contain chloro-plasts (Dean and Leech, 1982; Jellings and Leech, 1982). Cells without chloroplasts respire, store and recycle carbon and nitrogen, although photorespiratory metabolism is a major function only of green cells (Tobin *et al.*, 1988, 1989). Accepting the mobility of intermediates of the photorespiratory pathway, not necessarily being restricted to the photosynthetic cells, one can better understand such observations as those of Parnik and Keerberg (1995) of a relatively slow evolution of $^{14}CO_2$ following photosynthesis, apparently, by photorespiration. The carboxylation and decarboxylation processes not depending on RuBP, detected by Laisk and Sumberg (1994) at high CO_2 may also involve non-green cells. Parnik and Keerberg (1995) also commented on the difficulty of satu-rating pools of photorespiratory intermediates with radioactivity during steady-state photosynthesis from $^{14}CO_2$. This can be because reserve carbohydrates, and carbon returning from non-green cells of the leaf, are used to provide carbon for intermediates of the Calvin cycle. During water stress there is good evidence that reserves are pho-torespired and sugars fed to healthy leaves are used for photorespiration in competition with carbon from CO_2 (Waidyanatha *et al.*, 1975).

4.3 *Does photorespiration have a unique function?*

It is clear that in C_3 plants photorespiration uses energy and this serves to protect leaves at high photon flux densities (Osmond and Grace, 1995; Park *et al.*, 1996). However, it uses energy even at low photon flux densities and in these circumstances lowers the quantum yield of C assimilation. Also, although it can be argued that under water stress stomata will close, and because of insufficient diffusion of external CO_2 into the leaves there will be excess photons, under these conditions it has been con-firmed that photorespiration, while increased relative to carbon assimilation, is decreased in absolute terms (Biehler and Foch, 1996). There are other processes that can use excess reducing power in the leaf, and also mechanisms to dissipate excess radiation or to limit light absorption (Demmig-Adams and Adams, 1992). Thus one can have some doubts about a unique protective function for photorespiration.

5. Thoughts on current and future research trends

Upon present evidence it is reasonable to propose that it is the properties of Rubisco that determine the rate of flux of carbon into the glycolate pathway and that it is the oxygenase activity of this enzyme that is responsible for the uptake of oxygen. The release of CO_2, the other external signal identifying a respiratory process, is not directly coupled to the oxygen uptake but in the rate of post-illumination burst it is clearly a consequence of the oxygenase reaction and the formation of glycolate. Nevertheless, the factor of 0.5 (Equation 5) relating v_o to CO_2 release in photorespi-ration, often referred to as the stoichiometry, must vary if intermediates are taken from the cycle for use in other metabolic processes. Perfecting measurements of fluo-rescence and gas exchange, and further evaluation of the stoichiometry and other assumptions presently made in estimating S_{rel} values for Rubisco from whole plant studies, will allow the identification of any factors *in vivo* other than Rubisco proper-ties that determine rates of photorespiration.

Research on Rubisco, values of S_{rel}, the other kinetic constants associated with the oxygenase and carboxylase activities of the enzyme from various species, the catalytic mechanism, structure and molecular biology will continue (Cleland *et al.*, 1998; Harpel and Hartman, 1994; Spreitzer, 1993). There will continue to be attempts to decrease the oxygenase activity by *in vitro* mutagenesis (Bainbridge *et al.*, 1995; Hartman and Harpel, 1993). There is much current interest in the isolation from various algae of Rubiscos with S_{rel} values of over 100 and even over 200 (Read and Tabita, 1994; Uemura *et al.*, 1997). The prospect of a system (Svab and Maliga, 1993) that will express genes for the large and small subunit polypeptides of Rubisco in higher plants and assemble the L_8S_8 active holoenzyme presents new opportunities for making transgenic plants with altered photorespiration rates. The lack of availability of such an expression system has restricted this research mainly to studies of the L_2 enzyme of *Rhodospirillum rubrum* and the L_8S_8 enzymes of cyanobacteria and the alga, *Chlamydomonas*, all of which have lower S_{rel} values than the enzyme from higher plants.

C_4 plants and C_3/C_4 plants, which have decreased rates of photorespiration, appear to be better adapted to warm climates than C_3 species. It is not clear whether this is because they are better able to withstand water stress than C_3 plants or whether C_3 plants have advantages, or at least not major disadvantages, at lower temperatures. C_3 plants show increased rates of net photosynthesis when the oxygen concentration in the atmosphere is decreased to restrict photorespiration except at lower temperatures and with elevated CO_2 when net photosynthesis can be inhibited (i.e. an oxygen requirement is indicated). The temperature at which this oxygen requirement becomes apparent at ambient concentrations of CO_2 depends on the conditions under which plants are grown. A recent study of chlorophyll fluorescence quenching in response to temperature continues an important line of inquiry (Ghashinghaie and Cornic, 1994). Glycolate production through Rubisco oxygenation and photorespiratory metabolism are decreased relative to net photosynthesis in lower temperatures, but do not cease. The phenomenon of the oxygen requirement at low temperatures is accompanied by an increase in sugar phosphates and there is evidence that a shortage of chloroplastic orthophosphate is involved (Harley and Sharkey, 1991). Further exploration of the interaction of temperature with photorespiration and productivity by whole plants seems necessary for a proper understanding of evolution of photosynthetic mechanisms.

Genes for many of the enzymes of photorespiratory and associated metabolism have been cloned and transgenic plants with decreased and elevated activities of photorespiratory enzymes are being created. The complex effects of decreased amounts of several of the enzymes have already been described by use of selected mutants. In view of the present evidence, that it is Rubisco properties and the concentration of CO_2 and O_2 in the chloroplast stroma that determine the rate of photorespiration, we should be wary of claims that over-expression of enzymes associated with subsequent steps in the pathway can increase the rate (Brisson *et al.*, 1998; Kozaki and Takeba, 1996) except through some secondary effect on the relative concentrations of CO_2 and O_2 in the chloroplast or an effect on both photosynthesis and photorespiration by changing the activation state of Rubisco. Rather, we should encourage the use of gene technology to produce plants with decreased photorespiration, for example, by attempting (Rawsthorne and Bauwe, 1998) to follow the course of natural evolution that has produced C_3/C_4-intermediate and ultimately C_4 plants which have faster growth rates than C_3 plants in their natural habitat, or to change the properties of Rubisco. However, we should perhaps not neglect the comment of Sharkey (1988), concerning the rise in atmospheric CO_2 concentration since Decker (1955) described the post-illumination burst and the consequent 10% decrease in photorespiration in C_3 plants: 'Now if only plant physiologists could take credit for that decline'. I am sure that molecular biologists, geneticists, biochemists, crystallographers, and many others would wish to be included.

References

Andrews, T.J., Lorimer, G.H. and Tolbert, N.E. (1973) Ribulose diphosphate oxygenase. I. Synthesis of phosphoglycolate by fraction -1 protein of leaves. *Biochemistry* **12:** 11–18.

Bainbridge, G., Madgwick, P., Parmar, S., Mitchell, R., Paul, M., Pitts, J., Keys, A.J. and Parry, M.A.J. (1995) Engineering Rubisco to change its catalytic properties. *J. Exp. Bot.* **46:** 1269–1276.

Beck, E. (1979) Glycollate synthesis. In: *Encyclopedia of Plant Physiology*, New Series, Vol 6 (eds M. Gibbs and E. Latzko). Springer-Verlag, Berlin, pp. 327–325.

Benson, A.A. and Calvin, M. (1950) The path of carbon in photosynthesis VII: respiration and photosynthesis. *J. Exp. Bot.* **1:** 63–68.

Berry, J.A., Osmond, C.B. and Lorimer, G.H. (1978) Fixation of $^{18}O_2$ during photorespiration. Kinetics and steady-state studies of the photorespiratory carbon oxidation cycle with intact leaves and isolated chloroplasts of C_3 plants. *Plant Physiol.* **62:** 954–967.

Biehler, K. and Foch, H. (1996) Evidence for the contribution of the Mehler-peroxidase reaction in dissipating excess electrons in drought-stressed wheat. *Plant Physiol.* **112:** 265–272.

Bird, I.F., Keys, A.J. and Whittingham, C.P. (1972a) Intracellular localization of enzymes of the glycollate pathway. In: *Proceedings of the Second International Congress on Photosynthesis Research* (eds G. Forti, M. Avron and A. Melandri). Dr. W. Junk, The Hague, pp. 2215–2224.

Bird, I.F., Cornelius, M.J., Keys, A.J. and Whittingham, C.P. (1972b) Oxidation and phosphorylation associated with the conversion of glycine to serine. *Phytochemistry* **11:** 1587–1594.

Bird, I.F., Cornelius, M.J., Keys, A.J. and Whittingham, C.P. (1972c) Adenosine triphosphate synthesis and the natural electron acceptor for synthesis of serine from glycine in leaves. *Biochem. J.* **128:** 191–192.

Bird, I.F., Cornelius, M.J., Keys, A.J. and Whittingham, C.P. (1978) Intramolecular labelling of sucrose made by leaves from [^{14}C] carbon dioxide or [3-^{14}C] serine. *Biochem. J.* **172:** 23–27.

Blackwell, R.D., Murray, A.J.S., Lea, P.J., Kendall, A.C., Hall, N.P., Turner, J.C. and Wallsgrove, R.M. (1988a) The value of mutants unable to carry out photorespiration. *Photosynthesis Res.* **16:** 155–176.

Blackwell, R.D., Murray, A.J.S., Lea, P.J. and Joy, K.W. (1988b) Photorespiratory amino donors, sucrose synthesis and the induction of CO_2 fixation in barley deficient in glutamine synthetase and/or glutamate synthase. *J. Exp. Bot.* **39:** 845–858.

Bowes, G., Ogren, W.L. and Hageman, R.H. (1971) Phosphoglycolate production catalysed by ribulose bisphosphate carboxylase. *Biochem. Biophys. Res. Commun.* **45:** 716–712.

Brisson, L.F., Zelitch, I. and Havir, E.A. (1998) Manipulation of catalase levels produced altered photosynthesis in transgenic plants. *Plant Physiol.* **116:** 259–269.

Brooks, A. and Farquhar, G.D. (1985) Effect of temperature on the CO_2/O_2 specificity of ribulose-1,5-bisphosphate carboxylase/oxygenase and the rate of respiration in the light. *Planta* **165:** 397–406.

Caemmerer, S. von and Evans, J.R. (1991) Determination of the average partial pressure of CO_2 in chloroplasts from leaves of several C_3 plants. *Aust. J. Plant Physiol.* **18:** 287–305.

Canvin, D.T. (1979) Photorespiration: comparison between C_3 and C_4 plants. In: *Encyclopedia of Plant Physiology*, New Series, Vol 6 (eds M. Gibbs and E. Latzko). Springer-Verlag, Berlin, pp. 368–396.

Cleland, W.W., Andrews, T.J., Gutteridge, S., Hartman, F.C. and Lorimer, G.H. (1998) The mechanism of Rubisco: the carbamate as general base. *Chem. Rev.* **98**: 549–561.

Cornic, G. and Briantais, J.-M. (1991) Partitioning of photosynthetic electron flow between CO_2 and O_2 reduction in a C_3 leaf (*Phaseolus vulgaris* L.) at different CO_2 concentrations and during drought stress. *Planta* **183**: 178–184.

Dean, C. and Leech, R.M. (1982) Cellular and chloroplast numbers and DNA, RNA and protein in tissues of different ages within a 7 day old wheat leaf. *Plant Physiol.* **69**: 901–910.

Decker, J.P. (1955) A rapid post-illumination deceleration of respiration in green leaves. *Plant Physiol.* **30**: 82- 84.

Decker, J.P. and Tio, M.A. (1959) Photosynthetic surges in coffee seedlings. *J. Agric. Univ. Puerto Rico* **43**: 50–55.

Demmig-Adams, B. and Adams W.W. (1992) Photoprotection and other responses of plants to high light stress. *Annu. Rev. Plant Physiol. Plant Mol. Biol.* **43**: 599–626.

De Veau, E.J. and Burris, J.E. (1989) Photorespiratory rates in wheat and maize as determined by ^{18}O-labelling. *Plant Physiol.* **90**: 500–511.

Douce, R., Bourguignon, J., Macheral, D. and Neuberger, M. (1994) The glycine decarboxylase system in higher plant mitochondria: structure, function and biogenesis. *Biochem. Soc. Trans.* **22**: 184–188.

Farquhar, G.D., Caemmerer, S. von, and Berry, J.A. (1980) A biochemical model of photosynthetic CO_2 assimilation in leaves of C_3 species. *Planta* **149**: 78–90.

Genty, B., Briantais, J.-M. and Baker, N.R. (1989) The relationship between quantum yield of photosynthetic electron transport and quenching of chlorophyll fluorescence. *Biochim. Biophys. Acta* **990**: 87–92.

Ghashghaie, J. and Cornic, G. (1994) Effect of temperature on partitioning of photosynthetic electron flow between CO_2 assimilation and O_2 reduction and on the CO_2/O_2 specificity of Rubisco. *J. Plant Physiol.* **143**: 643–650.

Gilbert, S.M., Clarkson, D.T., Cambridge, M., Lambers, H. and Hawkesford, M.J. (1997) SO_4^{2-} deprivation has an early effect on the content of ribulose1,5-bisphosphate carboxylase/oxygenase and photosynthesis in young leaves of wheat. *Plant Physiol.* **115**: 1231–1239.

Goldsworthy, A. (1970) Photorespiration. *Bot. Rev.* **36**: 321–340.

Habash, D.Z., Paul, M.J., Parry, M.A.J., Keys, A.J. and Lawlor, D.W. (1995) Increased capacity for photosynthesis in wheat leaves grown at elevated CO_2: the relationship between electron transport and carbon metabolism. *Planta* **197**: 482–489.

Habash, D.Z., Parry, M.A.J., Parmar, S., Paul, M.J., Driscoll, S., Knight, J., Gray, J.C. and Lawlor, D.W. (1996) The regulation of component processes of photosyntheis in transgenic tobacco with decreased phosphoribulokinase activity. *Photosynthesis Res.* **49**: 159–167.

Harley, P.C. and Sharkey, T.D. (1991) An improved model of C_3 photosynthesis at high CO_2: reversed oxygen sensitivity explained by lack of glycerate re-entry into the chloroplast. *Photosynthesis Res.* **27**: 169–178.

Harley, P.C., Loreto, F., DiMarco, G. and Sharkey, T.D. (1992) Theoretical considerations when estimating the mesophyll conductance to CO_2 flux by analysis of the response of photosynthesis to CO_2. *Plant Physiol.* **98**: 1429–1443.

Harpel, M.R. and Hartman, F.C. (1994) Structure, function, regulation, and assembly of ribulose-1,5-bisphosphate carboxylase/oxygenase. *Annu. Rev. Biochem.* **63**: 197–234

Hartman, F.C. and Harpel, M.R. (1993) Chemical and genetic probes of the active site of D-ribulose-1,5-bisphosphate carboxylase/oxygenase: a retrospective based on the three-dimensional structure. *Adv. Enzymol.* **67**: 1–75.

Howitz, K.T. and McCarty, R.E. (1985) Substrate specificity of the pea chloroplast glycollate transporter. *Biochemistry* **24**: 3645–3650.

Howitz, K.T. and McCarty, R.E. (1991) Solubilization, partial purification and reconstitution of the glycollate/glycerate transporter from chloroplast inner envelope membranes. *Plant Physiol.* **96**: 1060–1069.

Husic, D.W., Husic, D.H. and Tolbert, N.E. (1987) The oxidative photosynthetic carbon cycle or C_2 cycle. *CRC Crit. Rev. Plant Sci.* **5**: 45–100.

Jacob, J. and Lawlor, D.W. (1983) Extreme phosphate deficiency decreases the *in vivo* specificity factor of ribulose 1,5-bisphosphate carboxylase-oxygenase in intact leaves of sunflower. *J. Exp. Bot.* **44**: 1635–1641.

Jellings, A.J. and Leech, R.M. (1982) The importance of quantitive anatomy in the interpretation of whole leaf biochemistry in species of *Triticum*, *Hordeum* and *Avena*. *New Phytol.* **92**: 39–48.

Jordan, D.B. and Ogren, W.L. (1984) The CO_2/O_2 specificity of ribulose 1,5-bisphosphate carboxylase/oxygenase. Dependence on ribulosebisphosphate concentration, pH and temperature. *Planta* **161**: 308–313.

Keys, A.J. (1986) Rubisco: its role in photorespiration. *Phil. Trans. Roy. Soc. Lond.* **B313**: 325–336.

Keys, A.J., Bird, I.F., Cornelius, M.J., Lea, P.J., Wallsgrove, R.M. and Miflin, B.J. (1978) Photorespiratory nitrogen cycle. *Nature* **275**: 741–743.

Keys, A.J., Bird, I.F. and Cornelius, M.J. (1982) Possible use of chemicals for the control of photorespiration. In: *Chemical Manipulation of Crop Growth and Development.* (ed. J.S. McLaren). Butterworth Scientific, London, pp. 39–53.

Kozaki, A. and Takeba, G. (1996) Photorespiration protects C_3 plants from photooxidation. *Nature* **384**: 557–560.

Krall, J.P. and Edwards, G.E. (1992) Relationship between photosystem II activity and CO_2 fixation in leaves. *Physiol. Plant.* **86**: 180–187.

Kumarasinghe, K.S., Keys, A.J. and Whittingham, C.P. (1977) The flux of carbon through the glycollate pathway during photosynthesis by wheat leaves. *J. Exp. Bot.* **28**: 1247–1257.

Laing, W.A., Ogren, W.L. and Hageman, R.H. (1974) Regulation of soybean net photosynthetic CO_2 fixation by interaction of CO_2, O_2 and ribulose 1,5-diphosphate carboxylase. *Plant Physiol.* **54**: 678–685.

Laisk, A. and Sumberg, A. (1994) Partitioning of the leaf CO_2 exchange into components using CO_2 exchange and fluorescence measurements. *Plant Physiol.* **106**: 689–695.

Lawlor, D.W., Boyle, F.A., Kendall, A.C. and Keys, A.J. (1987) Nitrate nutrition and temperature effects on wheat: enzyme composition, nitrate and total amino acid content of leaves. *J. Exp. Bot.* **38**: 378–392.

Lea, P.J. and Forde, B.G. (1994) The use of mutants and transgenic plants to study amino acid metabolism. *Plant Cell Environ.* **17**: 541–556.

Leegood, R.C., Lea, P.J., Adcock, M.D. and Hausler, R.E. (1995) The regulation and control of photorespiration. *J. Exp. Bot.* **46**: 1397–1414.

Loreto, F., Harley, P.C., DiMarco, G. and Sharkey, T.D. (1992) Estimation of mesophyll conductance to CO_2 flux by three different methods. *Plant Physiol.* **98**: 1437–1443

Lorimer, G.H. and Andrews, T.J. (1981) The C_2 chemo and photorespiratory carbon oxidation cycle. In: *The Biochemistry of Plants*, Vol. 8 (eds P.K. Stumpf and E.E. Conn). Academic Press, London, pp. 329–374.

Madore, M. and Grodsinski, B. (1984) Effect of oxygen concentration on ^{14}C-photoassimilate transport from leaves of *Salvia splendens* L. *Plant Physiol.* **76**: 782–786.

Miflin, B.J. and Lea, P.J. (1976) The pathway of nitrogen assimilation in plants. *Phytochemistry* **15**: 873–885.

Migge, A., Carrayol, E., Kunz, C., Hirel, B., Fock, H. and Becker, T. (1997) The expression of the tobacco genes encoding plastidic glutamine synthetase or ferredoxin-dependent glutamate synthase does not depend on the rate of nitrate reduction, and is unaffected by the suppression of photorespiration. *J. Exp. Bot.* **48**: 1175–1184.

Morell, M.K., Paul, K., Kane, H.J. and Andrews, T.J. (1992) Rubisco: maladapted or misunderstood. *Aust. J. Plant Physiol.* **40**: 431–441.

Noctor, G., Arisi, A.-C.M., Jouanin, L., Kunert, K.J., Rennenberg, H. and Foyer, C.H. (1998)

Glutathione: biosynthesis, metabolism and relationship to stress tolerance explored in transformed plants. *J. Exp. Bot.* **49**: 623–647.

Oliver, D.J. (1994) The glycine decarboxylase complex from plant mitochondria. *Annu. Rev. Plant Physiol. Plant Mol. Biol.* **45**: 323–337.

Ongun, A. and Stocking, C.R. (1965) Effect of light on the incorporation of serine into the carbohydrates of chloroplasts and nonchloroplast fractions of tobacco leaves. *Plant Physiol.* **40**: 819–824.

Osmond, C.B. and Grace, S.C. (1995) Perspectives on photoinhibition and photorespiration in the field: quintessential inefficiencies of the light and dark reactions of photosynthesis. *J. Exp. Bot.* **46**: 1351–1362.

Park, Y.-I., Chow, W.S., Osmond, C.B. and Anderson, J.M. (1996) Electron transport to oxygen mitigates against the photoinactivation of photosystem II *in vivo*. *Photosynthesis Res.* **50**: 23–32.

Parnik, T. and Keerberg, O. (1995) Decarboxylation of primary and end products of photosynthesis at different oxygen concentrations. *J. Exp. Bot.* **46**: 1439–1447.

Parnik, T., Talts, P., Gardestrom, P. and Keerberg, O. (1995) Influence of temperature on photosynthesis and respiration of primary leaves of barley. In: *Photosynthesis: from Light to Biosphere*, Vol. IV (ed. P. Mathis). Kluwer Academic Publishers, Netherlands, pp. 889–892.

Parry, M.A.J., Keys, A.J. and Gutteridge, S. (1989) Variation in the specificity factor of C_3 higher plant Rubiscos determined by total consumption of ribulose-P_2. *J. Exp. Bot.* **40**: 317–320.

Peterson, R.B. (1983) Estimation of photorespiration based on the initial rate of postillumination CO_2 release. *Plant Physiol.* **73**: 983–988.

Peterson, R.B. (1990) Effects of irradiance on the *in vivo* CO_2/O_2 specificity factor in tobacco using simultaneous gas exchange and fluorescence techniques. *Plant Physiol.* **94**: 892–898.

Rawsthorne, S. and Bauwe, H. (1998) C3-C4 intermediate photosynthesis. In: *Photosynthesis – A Comprehensive Treatise* (ed. A.S. Raghavendra). Cambridge University Press, Cambridge, pp. 150–162.

Rawsthorne, S. and Hylton, C.M. (1991) The post-illumination CO_2 burst and glycine metabolism in leaves of C_3 and C_3-C_4 intermediate species of Moricandia. *Planta* **186**: 122–126.

Read, B.A. and Tabita, F.R. (1994) High substrate specificity factor ribulose bisphosphate carboxylase/oxygenase from eukaryotic marine algae and properties of recombinant cyanobacterial Rubisco containing algal residue modifications. *Arch. Biochem. Biophys.* **312**: 210–218.

Sharkey, T.D. (1988) Estimating the rate of photorespiration in leaves. *Physiol. Plant.* **73**: 147–152.

Somerville, S.C. and Ogren, W.L. (1983) An *Arabidopsis thaliana* mutant defective in chloroplast dicarboxylate transport. *Proc. Natl Acad. Sci. USA* **80**: 1290–1294.

Somerville, S.C. and Somerville, C.R. (1983) Effect of oxygen and carbon dioxide on photorespiratory flux determined by glycine accumulation in a mutant of *Arabidopsis thaliana*. *J. Exp. Bot.* **34**: 415–424.

Somerville, S.C. and Somerville, C.R. (1985) A mutant of *Arabidopsis* deficient in chloroplast dicarboxylate transport is missing an envelope protein. *Plant Sci. Lett.* **37**: 217–220.

Spreitzer, R.J. (1993) Genetic dissection of Rubisco structure and function. *Annu. Rev. Plant Physiol. Plant Mol. Biol.* **44**: 411–434.

Svab, Z. and Maliga, P. (1993) High frequency plastid transformation in tobacco by selection for a chimeric aadA gene. *Proc. Natl Acad. Sci. USA* **90**: 913–917.

Ta, T.C. and Joy, K.W. (1986) Metabolism of some amino acids in relation to the photorespiratory nitrogen cycle of pea leaves. *Planta* **169**: 118–122.

Tobin, A.K., Sumar, N., Patel, M., Moore, A.L. and Stewart, G.R. (1988) Development of photorespiration during chloroplast biogenesis in wheat leaves. *J. Exp. Bot.* **39**: 833–843.

Tobin, A.K., Thorpe, J.R., Hylton, C.M. and Rawsthorne, S. (1989) Spatial and temporal influences on the cell-specific distribution of glycine decarboxylase in leaves of wheat (*Triticum aestivum* L.) and pea (*Pisum sativum* L.). *Plant Physiol.* **91**: 1219–1225.

Tolbert, N.E. (1971) Microbodies – peroxisomes and glyoxysomes. *Annu. Rev. Plant Physiol.* **22:** 45–74.

Tolbert, N.E. (1979) Glycollate metabolism by higher plants and algae. In: *Encyclopedia of Plant Physiology*, New Series, Vol 6 (eds M. Gibbs and E. Latzko). Springer-Verlag, Berlin, pp. 338–352.

Tolbert, N.E. (1997) The C_2 oxidative photosynthetic carbon cycle. *Annu. Rev. Plant Physiol. Plant Mol. Biol.* **48:** 1–25.

Uemura, K., Anwaruzzaman, Miyachi, S. and Yokota, A. (1997) Ribulose-1,5-bisphosphate carboxylase/oxygenase from thermophilic red algae with a strong specificity for CO_2 fixation. *Biochem. Biophys. Res. Commun.* **233:** 568–571.

Waidyanatha, U.P. de S., Keys, A.J. and Whittingham, C.P. (1975) Effects of carbon dioxide on metabolism by the glycollate pathway in leaves. *J. Exp. Bot.* **26:** 15–26.

Wallsgrove, R.M., Turner, J.C., Hall, N.P., Kendall, A.C. and Bright, S.W.J. (1987) Barley mutants lacking chloroplast glutamine synthetase biochemical and genetic analysis. *Plant Physiol.* **83:** 155–158.

Zelitch, I. (1979a) Photosynthesis and plant productivity. *Chem. Engng News* **5:** 28–48.

Zelitch, I. (1979b) Photorespiration: studies with whole tissues. In: *Encyclopedia of Plant Physiology*, New Series, Vol 6 (eds M. Gibbs and E. Latzko). Springer-Verlag, Berlin, pp. 353–367.

12

The use of mutants of *Amaranthus edulis* to study carbon and nitrogen metabolism in C$_4$ photosynthesis

Peter J. Lea, Louisa V. Dever, Robert J. Ireland, Karen J. Bailey and Richard C. Leegood

1. Introduction

Following on the experiments of Kortschak and Kapilov, Hatch and Slack were able to show convincingly that when sugar cane leaves were pulsed with $^{14}CO_2$, the first stable products of photosynthesis were the 4-carbon acids, malate and aspartate (see Hatch, 1987; 1997 for historical accounts of the early work). These results conflicted with the earlier $^{14}CO_2$-labelling experiments carried out by Calvin and Benson that indicated that the 3-carbon compound PGA was the first product of photosynthetic carbon fixation. It is this clear difference in $^{14}CO_2$ labelling that gave rise to the terms C$_4$ and C$_3$ plants, that will be used throughout this chapter.

In the majority of C$_4$ plants, the photosynthetic chlorophyll-containing cells are arranged in two concentric cylinders. The outer cylinder comprises thin-walled mesophyll cells, that radiate from the inner cylinder of bundle sheath cells, which have heavily thickened cell walls and large chloroplasts. This arrangement of cells within the C$_4$ plant is termed Kranz (wreath in German) anatomy. The mesophyll cells are in contact with both intercellular air spaces and the bundle sheath cells via plasmodesmata. The thickened and suberized walls of the bundle sheath cells restrict the movement of gases and, as will be discussed later, allow a high concentration of CO_2 to build up (Jenkins, 1997).

Following more detailed $^{14}CO_2$-labelling studies, the establishment of techniques to isolate pure preparations of intact mesophyll and bundle sheath cells and some painstaking enzyme activity measurements, the first steps in the pathway of C$_4$ photosynthesis became apparent (*Figure 1*).

Plant Carbohydrate Biochemistry, edited by J.A. Bryant, M.M. Burrell and N.J. Kruger.
© 1999 BIOS Scientific Publishers Ltd, Oxford.

(a)

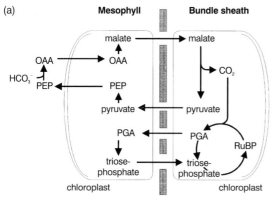

NADP-malic enzyme (NADP-ME)

(b)

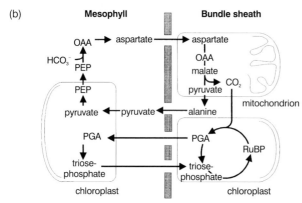

NAD-malic enzyme (NAD-ME)

(c)

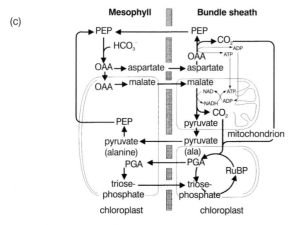

PEP carboxykinase (PCK)

Figure 1. The three biochemical pathways of C_4 metabolism, showing both the C_4 and the Benson–Calvin cycles and the shuttle of PGA and triose phosphate between the bundle sheath and mesophyll chloroplasts. Note that in the PCK types (c), NAD-ME also operates and that alanine may also return from the bundle sheath to the mesophyll in order to maintain the balance of amino groups between the two compartments. NADH generated by malate decarboxylation is used to generate the ATP required by PCK. In some NADP-ME-type plants aspartate and alanine may also move between the mesophyll and bundle sheath cells. OAA, oxaloacetate; ala, alanine. From Leegood (1997) The regulation of C_4 photosynthesis. *In:* Advances in Botanical Research, *vol. 26, pp. 252–316. Reprinted by permission of Academic Press Inc.*

(i)   The initial carboxylation reaction is catalysed by magnesium-dependent PEP car-boxylase in the cytoplasm of the mesophyll cells. The reaction utilizes bicarbon-ate (formed by carbonic anhydrase) rather than CO_2 as a substrate and is not inhibited by oxygen.

(ii) The oxaloacetate formed by the carboxylation of PEP is immediately metabo-lized. The reduction by NADP-dependent malate dehydrogenase (MDH) in the chloroplast gives rise to malate. Alternatively, the oxaloacetate is converted to aspartate in the cytoplasm by a transamination reaction using either alanine or glutamate as amino donors (Lea and Ireland, 1999). The precise distribution of carbon between malate and aspartate varies between plant species.

(iii) The malate and aspartate are transported through the plasmodesmata to the bun-dle sheath cells in a diffusion-driven process that is dependent on a large concen-tration difference between the two cell types.

Full details of these processes have been provided in recent review articles (Leegood, 1997; Leegood and Walker, 1998). The net effect of these three steps of the C_4 cycle is the transfer of a 4C acid from the mesophyll to the bundle sheath cells, at the expense of two molecules of ATP. The precise biochemical mechanisms by which the 4C acids are metabolized vary between plants. Such differences provide evidence that C_4 pho-tosynthesis evolved on more than one occasion, probably as a result of a decline in the ambient concentration of CO_2 to 200 μmol mol^{-1}, as compared to the present day value of 340 μmol mol^{-1} (Ehleringer *et al.*, 1991). C_4 plants can be classified into three sub-groups, based on the different enzymes which decarboxylate C_4 acids in the bun-dle sheath cells (*Figure 1*) as follows.

1.1 *NADP-malic enzyme type*

NADP-malic enzyme (NADP-ME) is responsible for malate decarboxylation to yield pyruvate in the bundle sheath chloroplasts of economically important grass crops such as maize, sorghum and sugar cane. The most notable feature of these plants is the pres-ence of agranal chloroplasts in the bundle sheath cells due to the absence of photosys-tem II. The dicotyledonous plant *Flaveria bidentis*, which will be discussed later with reference to the production of transgenic plants containing antisense constructs, belongs to this group. There is now evidence that this species uses aspartate as well as malate to transport the CO_2 between the mesophyll and bundle sheath cells (Meister *et al.*, 1996).

1.2 *NAD-malic enzyme type*

NAD-malic enzyme (NAD-ME) is responsible for the decarboxylation of malate in the bundle sheath mitochondria of millet (*Panicum milliaceum*) and *Amaranthus* species, which will be discussed in detail in a later section with reference to the iso-lation of mutants. These plants possess high activities of glutamate and alanine aminotransferases (Ireland and Lea, 1999), which are distributed equally between the bundle sheath and mesophyll cells. Aspartate is transported from the mesophyll cells and transaminated in the bundle sheath mitochondria to form oxalaoacetate, which is immediately reduced to malate. As a major portion of the C_4 pathway takes place in the bundle sheath mitochondria, these organelles have a highly convoluted inner membrane and are present at four times the frequency than found in NADP-ME plants. The mitochondria are often closely associated with the chloroplasts, which have normal granal stacking and photosystem II activity.

1.3 *PEP carboxykinase type*

Plants in this group (e.g. *Panicum maximum* and *Spartina anglica*) have high activities of PEP carboxykinase (PCK) in the cytoplasm of mesophyll cells. This enzyme catalyses the ATP-dependent conversion of oxaloacetate to PEP and CO$_2$ (Walker *et al.*, 1997). The ATP required for PCK activity is generated from the oxidation of malate by NAD-ME in the mitochondria (a reaction which also liberates CO$_2$) and subsequent oxidative phosphorylation.

Following the decarboxylation events described above, in the absence of carbonic anhydrase in the bundle sheath cells, CO$_2$ is assimilated by the action of Rubisco, the first step in the Benson–Calvin cycle. It has been estimated that the concentration of CO$_2$ in the bundle sheath cells can reach 70 μM (Jenkins, 1997), which is about 20-fold higher than found in the mesophyll cells and is sufficient to reduce considerably the oxygenase activity of Rubisco (see later section on photorespiration). However, somewhat surprisingly, half of the PGA formed by Rubisco is transported to the mesophyll cell chloroplasts for the subsequent reduction step and then returned as triose phosphates to the bundle sheath chloroplasts to complete the Benson–Calvin cycle and regenerate RuBP. This apparent inefficiency appears to have several benefits.

(i)   There will be a decrease in the requirement for NADPH in the bundle sheath cells, which will lower the rate of O$_2$ production from photosystem II. In NADP-ME-type plants there is sufficient NADPH generated from malate oxidation, largely to dispense with the requirement for photosystem II. As mitochondrial rates of respiration are high in the bundle sheath cells, it is possible that these two mechanisms will be sufficient to lower the concentration of O$_2$, which is unable to diffuse out through the cell walls, and thus reduce competition with CO$_2$ at the active site of Rubisco.

(ii) The PGA transported to the mesophyll cells acts as a signal of the rate of CO$_2$ assimilation in the bundle sheath cells. As will be seen in Section 2.1, this is important for the regulation of PEP carboxylase.

(iii) The shuttle is a means of maintaining the charge balance between the two cell types. The reduction of PGA to triose phosphate in the mesophyll chloroplasts consumes a proton, which counteracts the release of a proton following the hydration of CO$_2$ by carbonic anhydrase.

In order to complete the C$_4$ cycle it is essential that a 3C compound is transported back to the mesophyll cells; this takes the form of pyruvate in NADP-ME-type plants and alanine in NAD-ME-type plants. Metabolite transport has not been studied extensively in the PCK-type group and it is not clear whether pyruvate, PEP or alanine is transported back to the mesophyll cells to balance the aspartate and malate transported to the bundle sheath cells. The final step in the pathway in all C$_4$ plants is the regeneration of PEP in the mesophyll chloroplasts. The reaction is catalysed by pyruvate P$_i$ dikinase (PPDK) by a mechanism that requires P$_i$ and ATP and yields AMP and PP$_i$.

2. Regulation of C_4 photosynthesis

Sufficient space is not available to discuss all of the enzymes that are subject to regulation in the C_4 photosynthetic cycle; full details are provided in previous review articles (Leegood, 1997; Leegood and Walker, 1998). In this section the discussion centres on those enzymes that have been subject to change in activity, either by mutant selection or genetic manipulation techniques.

2.1 *PEP carboxylase*

PEP carboxylase exists as a tetramer of four identical subunits of 110 kDa, synthesis of which is encoded by a family of at least three genes (Chollet *et al.*, 1996; Westhoff *et al.*, 1997). PEP carboxylase activity in C_4 plants can be regulated by metabolites, changes in cytosolic pH and reversible phosphorylation. Metabolites that activate the enzyme include triose phosphates, hexose phosphates, serine and glycine those that inhibit include malate, aspartate and glutamate (Doncaster and Leegood, 1987; Gao and Woo, 1996). The sensitivity of PEP carboxylase to metabolites varies depending on the phosphorylation state of the enzyme. PEP carboxylase extracted from the leaves of illuminated leaves is less sensitive to inhibition by malate, than is the enzyme isolated from darkened leaves (see Section 4.1), but is activated by a lower concentration of glucose-6-phosphate. In addition, the enzyme activity is less sensitive to inhibition by malate at the pH optimum (8.0), rather than the cytosolic pH (7.3).

PEP carboxylase is phosphorylated by a highly regulated protein kinase on a serine residue near the N-terminal (Ser-8 in sorghum and Ser-15 in maize) and dephosphorylated by a mammalian-like protein phosphatase 2A. The kinase is able to operate in the absence of Ca^{2+} and is reversibly activated *in vivo* in the light (Nhiri *et al.*, 1998). It has been proposed that PGA may act as the signal molecule following transport from the Calvin cycle in the bundle sheath cells, by increasing the pH of the cytosol following uptake into the mesophyll chloroplasts. This increase in pH leads to an increase in the cytosolic Ca^{2+} concentration which may stimulate an increase in PEP carboxylase kinase activity (Chollet *et al.*, 1996; Vidal and Chollet, 1997). The rather complex system described above allows PEP carboxylase to operate in illuminated leaves when the rate of photosynthetic CO_2 assimilation (and the concentration of malate in the mesophyll cells) is high. Phosphorylation of the enzyme protein reduces the inhibitory effect of malate and increases the stimulatory effect of sugar phosphates. An increase in the pH of the cytosol leads to an increase in the catalytic rate and again modulates the response to malate and sugar phosphates.

2.2 *NADP-MDH*

NADP-MDH is located in the mesophyll chloroplasts where it reduces the oxaloacetate synthesized by PEP carboxylase to malate. Data on the molecular mass of subunits have been somewhat variable and values ranging from 38 to 60 kDa have been obtained in maize, where the enzyme can exist as a dimer or tetramer. Analysis of cDNA clones encoding the enzyme from sorghum have indicated that the molecular mass is 42 kDa (Miginiac-Maslow *et al.*, 1997). NADP-MDH is the only enzyme in the C_4 cycle that is regulated by a ferredoxin/thioredoxin reduction system, in contrast to the Benson–Calvin cycle in which four enzymes interact with thioredoxin

(Buchanan, 1991). There are eight cysteine residues in each subunit, of which four are reduced in the thioredoxin-dependent activation process, although the major activation step is thought to be via the N-terminal Cys-24 and Cys-29 residues (Miginiac-Maslow et al., 1997). Modelling of the response of the activity of the enzyme to the ratio of NADPH/NADP$^+$ indicated that at any particular ratio of oxidized to reduced thioredoxin, high proportions of active NADP-MDH (and hence high rates of oxaloacetate reduction) can only occur at high NADPH/NADP$^+$ ratios (Rebeillé and Hatch, 1986). The activation of NADPH-MDH in transgenic plants containing varying amounts of enzyme activity has been discussed by Furbank et al. (1997).

2.3 PPDK

PPDK in C$_4$ leaves is a tetramer comprising subunits of 94 kDa molecular mass, that are encoded by a gene that can target the enzyme into both the cytoplasm and chloroplast (Rosche et al., 1998). The pH optimum of PPDK is 8.3 and the enzyme requires free Mg^{2+}, suggesting that it will be maximally active in the illuminated chloroplast stroma of the mesophyll cells (see Section 5.2). PPDK is also light regulated by a process that involves phosphorylation which is totally different to that described above for PEP carboxylase. The enzyme undergoes reversible deactivation following phosphorylation of the histidine residue (His-458) at the catalytic site and of a nearby threonine residue (Thr-456). Interconversion between active and inactive forms is catalysed by a regulatory protein that has two active sites (Roeske and Chollet, 1989). ADP inactivates the enzyme by acting as a phosphoryl donor and P$_i$ activates the enzyme by phosphorylytic cleavage to yield PP$_i$ (Huber et al., 1994). The mechanism is unusual in that the catalytic reaction features in the regulation of activity. Thus if the substrates accumulate, the enzyme will be maintained in an active state.

2.4 NAD-ME

NAD-ME is normally composed of an α and β subunit with molecular masses varying between 58 and 65 kDa, dependent upon species. The subunits have 60% identity at the DNA level and 65% identity at the deduced amino acid level (Winning et al., 1994). They are however immunologically distinct, as antisera raised against the α subunit will not cross-react with the β subunit (Long et al., 1994). The distinct subunits are present in a 1:1 ratio and the native enzyme may exist as octamers, tetramers and dimers. Interconversion between the three forms has been proposed as a method of regulation, with the dimer having the lowest V_{max} and a high K_m for malate and the octamer the highest V_{max} and lowest K_m for malate. Although much of the work on the regulation of the enzyme by changes in aggregation state has been done on NAD-ME isolated from C$_3$ and CAM plants (Willeford and Wedding, 1987), the C$_4$ leaf form of the enzyme is thought to behave in a similar manner (Podesta et al., 1990).

The enzyme requires Mn^{2+} for activity and is stimulated by malate, CoA, acetyl CoA and FBPase. NAD-ME from the NAD-ME-type plants Atriplex spongiosa and Panicum miliaceum is inhibited by physiological concentrations of ATP, ADP and AMP. However, this response of enzyme activity to adenylates is unlikely to result in the regulation of the enzyme activity by energy charge (Furbank et al., 1991). In contrast, the enzyme from Urochloa panicoides (PCK-type) is activated up to 10-fold by ATP and inhibited by ADP and AMP as well as by pyruvate, oxaloacetate and 2-oxoglutarate (Furbank et al., 1991).

3. Photorespiration

As indicated previously, when Rubisco utilizes CO_2 as a substrate, the product PGA may be metabolized in the Calvin cycle through which all the CO_2 is assimilated into useful products. Rubisco is however also able to utilize O_2 as a substrate and the product glycolate-2-phosphate is metabolized via the photorespiratory carbon and nitrogen cycle in the chloroplasts, peroxisomes and mitochondria (Keys *et al.*, 1978). Following the conversion of glycine to serine, CO_2 is released at a rate of about 25% of the net rate of photosynthesis. The principal factors affecting the rate of photorespiration are the ratio of CO_2/O_2 concentrations at the active site of Rubisco and the temperature (see Chapter 11).

It is assumed that a major function of the C_4 photosynthetic cycle is to increase the concentration of CO_2 in the chloroplasts of bundle sheath cells, where the Rubisco is located, and thus to inhibit the oxygenase reaction and prevent the wasteful synthesis of glycollate-2-phosphate. If the flux of carbon through the C_4 cycle is lower than that through the Calvin cycle, then the concentration of CO_2 in the bundle sheath cells will be low. This will allow the oxygenase reaction of Rubisco and photorespiration to take place, which will cause an increase in the quantum requirement of photosynthesis. If the flux of carbon through the C_4 cycle is much greater than the action of Rubisco, then CO_2 will leak back out into the mesophyll cells with the subsequent expenditure of ATP, and again cause an increase in the quantum requirement. It is therefore very important that the rates of C_4 cycle and the Calvin cycle are co-ordinated during steady-state photosynthesis (Jenkins, 1997).

Early attempts to establish the rate of photorespiration in C_4 plants indicated that CO_2 is not released from the leaf, due to the impermeability of the bundle sheath cell wall and the presence of PEP carboxylase in the mesophyll cells (Canvin, 1979). In C_3 plants the photosynthetic CO_2 assimilation rate can be increased by 50% if the oxygen concentration is reduced from 20 to 2 kPa, due to the lower rate of RuBP oxygenation. However, the photosynthetic rate of C_4 plants is relatively insensitive to the external concentration of O_2 (Edwards and Walker, 1983). Quantum requirement measurements have also indicated that in mature leaves of C_4 plants, the rate of photorespiration is very low (Andrews and Baker, 1997), except when the internal CO_2 concentration is greatly reduced (Dai *et al.*, 1995). On the other hand, it has been clearly demonstrated that C_4 plants contain all the enzymes required for the photorespiratory nitrogen and carbon cycle in the bundle sheath cells. In addition, labelling studies with $^{14}CO_2$ and $^{18}O_2$ have also shown that the photorespiration cycle can operate in the bundle sheath cells and that it is susceptible to established inhibitors (see Dever *et al.*, 1995; Leegood, 1997 for a full description).

In order to address the difficulties of determining experimentally the rate of photorespiration in C_4 plants, and also the CO_2 concentration and leakage rate from the bundle sheath strands, Jenkins (1997) attempted to model these values at different rates of photosynthetic CO_2 fixation. At a rate of photosynthesis of 4 μmol min^{-1} mg^{-1} chlorophyll, the leakage rate from the bundle sheath cells was 16%, the bundle sheath cell CO_2 concentration was 55 μM and the photorespiration rate was 3%.

4. The isolation of mutants of the C_4 photosynthetic cycle

The method of screening for mutants was based on that originally described by Somerville and Ogren (1979), who isolated a range of mutants of *Arabidopsis thaliana*

(Somerville, 1986) that lacked key enzymes of the photorespiratory carbon and nitrogen cycle (Keys *et al.*, 1978). Somerville and Ogren (1979) argued that in C$_3$ plants grown at elevated concentrations of CO$_2$, glycollate-2-phosphate formation would be prevented due to the inhibition of the oxygenase function of Rubisco. Thus there would be no flux of carbon and nitrogen through the photorespiratory cycle and any enzyme deficiencies would not be detrimental. Mutant plants would grow normally in elevated CO$_2$ but would exhibit severe stress symptoms when exposed to ambient air. Utilizing this principle, a total of seven barley mutants (plus one double mutant) lacking enzymes of the photorespiratory cycle, have been isolated at Lancaster and Rothamsted (Blackwell *et al.*, 1988; Leegood *et al.*, 1995). More recently heterozygous barley mutants containing enzyme activity (e.g. glutamine synthetase, glutamate synthase and glycine decarboxylase) varying from 40 to 100% have been studied in detail (Häusler *et al.*, 1996; Wingler *et al.*, 1997).

Applying a similar screen to that described above, we predicted that mutants of C$_4$ plants lacking an enzyme of the C$_4$ photosynthetic cycle would only be able to grow in an atmosphere of elevated CO$_2$. This prediction was based on the work recently reviewed by Jenkins (1997), that bundle sheath cells have a low permeability to CO$_2$, but that this barrier can be overcome by high external CO$_2$ concentrations. In our initial experiments, batches of 100–200 azide-mutagenized seed (Kleinhofs *et al.*, 1978) of *Amaranthus edulis* were germinated in segmented trays and grown in a glasshouse maintained at a CO$_2$ concentration of 7000 μmol mol^{-1} for 2 weeks and plants showing any abnormal features were discarded. The seedlings were then transferred to an identical glasshouse at ambient CO$_2$ and examined carefully every day. Plants that grew slowly or showed chlorosis were immediately transferred back to the glasshouse at 7000 μmol mol^{-1} CO$_2$, and allowed to grow for a further 2–3 weeks. Those that recovered and produced new healthy green leaves were screened for enzymes of the C$_4$ photosynthesis cycle (Ashton *et al.*, 1990) and soluble amino acid content.

4.1 *A mutant lacking PEP carboxylase*

The first mutant that was positively identified was LaC$_4$ 2.16, which was shown to contain only 5% of the normal wild-type PEP carboxylase activity. Western blot analysis indicated that the mutant lacked the major leaf PEP carboxylase protein, but that a minor polypeptide of higher molecular mass was still present. Light microscope analysis of immunogold-labelled sections of leaves of the mutant LaC$_4$ 2.16 confirmed that the PEP carboxylase protein was absent from the mesophyll cells of the leaf (Dever *et al.*, 1995). Two PEP carboxylase proteins were detected in the roots, stems, petioles and flowers of the wild-type *A. edulis* which were also present at the same concentration in the mutant LaC$_4$ 2.16 (Dever *et al.*, 1996).

In order to establish the molecular basis of the mutation, poly(A$^+$)RNA was isolated from wild-type and mutant leaves and the [^{35}S]methionine-labelled translation products subjected to immunoprecipitation. Polyclonal antisera raised against full length sorghum PEP carboxylase recognized a predominant *in vitro* translation product that was 9 kDa shorter than the normal PEP carboxylase protein. This 100 kDa PEP carboxylase polypeptide was recognized by N-terminal-specific antisera, but not by C-terminal-specific antisera, suggesting that the mutant protein lacked the C-terminal domain. As the truncated form of the PEP carboxylase polypeptide was not detected in the mutant leaves, it must be assumed that it is subject to proteolysis immediately after synthesis. Sequence analysis showed that there was a point mutation at the 3' end of intron 9 of the C_4 PEP carboxylase gene, where the G of the AG splice site had been changed to A. Such a mutation prevented the normal splicing of intron 9 and induced a shift in the intron/exon boundary to the next AG sequence. If the first downstream AG sequence was used, the reading frame would be conserved but the polypeptide would lack five amino acids. If the second more distal AG sequence was used, the splicing would introduce a frame shift that would lead to the production of a truncated polypeptide of 100 kDa, due to the presence of a stop codon in the new reading frame. From the *in vitro* translation results described above, this latter inaccurate splicing system would appear to be operating in the mutant (Grisvard *et al.*, 1998).

The original LaC$_4$ 2.16 plant was grown to maturity in elevated CO_2 and both self-fertilized homozygous seed and backcrosed F$_1$ heterozygous seed were obtained. Analysis of the self-fertilized F$_1$ plants indicated that the mutation segregated in a normal Mendelian fashion in the F$_2$ generation. Homozygous mutant F$_2$ seedlings containing less than 10% of the wild-type PEP carboxylase activity grew very slowly in air and, even after 4 months, were no more than 10 cm high. The leaves, although pale, were not totally chlorotic and the plants did survive. Heterozygous F$_1$ plants containing approximately 50% of the wild-type PEP carboxylase activity appeared to grow normally in air, however a comparison of the dry weights of the root, shoot and seed indicated that there was a reduction in the biomass of approximately 15% (Dever *et al.*, 1995). When the PEP carboxylase activities in the leaves of air-viable plants derived from self-fertilized heterozygous plants were examined in detail, values from 45 to 70% of the wild-type activity were obtained. The reduction of PEP carboxylase activity had no effect on the chlorophyll and protein content of the leaf nor on the activities of other enzymes involved in C_4 photosynthesis (K.J. Bailey *et al.*, unpublished data).

The classical responses of the rate of photosynthetic CO_2 assimilation (A) to varying internal CO_2 concentration (C_i), for a wild-type C_4 plant are shown in *Figure 2*

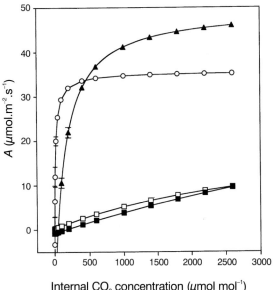

Figure 2. The response of net photosynthetic CO_2 assimilation rate (A) to varying internal CO_2 concentrations in wild-type (○), homozygous PEP carboxylase mutant LaC$_4$ 2.16 (■), barley (▲) and NAD-ME mutant LaC$_4$ 73 (□) A. edulis, determined at a photon flux density of 2000 μmol m^{-2} s^{-1}. From Dever et al. (1996) The isolation and characterization of mutants of the C$_4$ plant Amaranthus edulis. Comptes rendus de l'Académie des Sciences (Sciences de la vie), vol. 319. pp. 951–959. Reprinted by permission of the journal.

(Dever *et al.*, 1996). The rate of CO_2 assimilation by the homozygous mutant LaC$_4$ 2.16 at ambient C_i was close to zero, which is in agreement with the data obtained by Brown (1997) using an inhibitor of PEP carboxylase. However the rate displayed a linear response to increasing C_i, up to the maximum concentration tested. This response suggests that photosynthetic CO_2 assimilation in the mutant is directly dependent on the rate of CO_2 diffusion into the bundle sheath cells, and explains how the mutant is able to grow at the elevated atmospheric CO_2 concentration of 7000 μmol mol^{-1}. The heterozygous plants containing approximately 50% of PEP carboxylase activity exhibited a similar A/C_i response curve to that shown by the wild type, but the maximum rate of CO_2 assimilation was lower.

Photosynthetic CO_2 assimilation by wild-type *A. edulis* has an optimum O_2 concentration of approximately 5 kPa; below this optimum the decrease in rate is associated with lower photosystem activity, whereas above the optimum, photorespiration accounts for the inhibition of photosynthesis (Maroco *et al.*, 1997). In the homozygous mutant, the optimum O_2 concentration was reduced to 1–2 kPa, a value normally found for C$_3$ plants. Maroco *et al.* (1998) concluded that the high O_2 optimum of C$_4$ photosynthesis is linked to the O_2-dependent production of ATP by pseudocyclic/cyclic phosphorylation required for the synthesis of PEP and this requirement is absent in the LaC$_4$ 2.16 mutant.

When the A/C_i curves of the heterozygous plants containing 55–70% of the wild-type PEPcarboxylase activity were studied in more detail, it became evident that there was a decrease in the initial slope and that the C_i at which CO_2 assimilation was saturated, increased when compared to the wild type (K.J. Bailey unpublished data). This confirms the suggestion that the initial slope of an A/C_i curve of a C$_4$ plant reflects the kinetic characteristics and activity of PEP carboxylase (Collatz *et al.*, 1992). The effect of the reduction of PEP carboxylase on the CO_2 assimilation rate was also dependent upon light intensity, with differences being detected at values higher than 360 μmol m^{-2} s^{-1}, confirming that at low light intensities, C$_4$ photosyntheis is limited by RuBP or PEP regeneration (Furbank *et al.*, 1996; Trevanion *et al.*, 1997).

The degree of control exerted by PEP carboxylase on photosynthetic CO_2 assimilation was estimated using the principles of metabolic control analysis, as originally proposed by Kacser and Burns (1973) and discussed in detail with respect to plant metabolism by ap Rees and Hill (1994). At ambient C_i (130 μmol mol^{-1}), the flux-control coefficient (C^J) for PEP carboxylase increased from 0.35 in the wild type to 0.49 at 55% PEP carboxylase. At moderate C_i (60 μmol mol^{-1}), C^J increased from 0.65 in the wild type to 0.77 at 55% PEP carboxylase, whilst at low C_i (30 μmol mol^{-1}) C^J was 0.70 in the wild type increasing to 0.81 at 55% PEP carboxylase. The control exerted by PEP carboxylase on photosynthetic flux in *A. edulis* is therefore relatively high and comparable to the values obtained for PPDK and Rubisco in *F. bidentis*, as will be discussed in Section 5.

For plants containing PEP carboxylase activities below 55%, there was an upturn in the rate of CO_2 assimilation (K.J. Bailey unpublished data). This upturn was observed in each of the 3 years of study, using independently generated heterozygous plants. These findings suggest that a compensation mechanism begins to operate once PEP carboxylase activity falls below a critical amount. A similar response has been seen in the release of ammonia in barley plants which contain decreased amounts of glutamine synthetase activity (Häusler *et al.*, 1996). The possibility that the remaining PEP carboxylase protein in the heterozygous plants may become activated either via specific metabolites or phosphorylation was investigated.

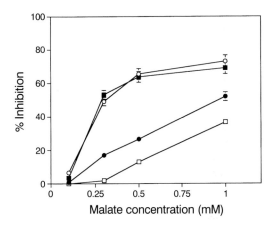

Figure 3. Malate sensitivity of PEP carboxylase isolated from illuminated (open symbols) and darkened (closed symbols) leaves of wild-type (circles) and heterozygous LaC$_4$73 (squares) A. edulis plants. Error bars represent the standard error of the mean.

The malate sensitivity of PEP carboxylase from the wild-type and heterozygous plants was compared over a physiological concentration range. *Figure 3* indicates that PEP carboxylase isolated from the leaf of a fully illuminated heterozygous plant is less sensitive to malate inhibition than the corresponding wild-type enzyme over the concentration range 0.3–1 mM. The apparent increase in phosphorylation state of the enzyme isolated from the heterozygous plant was confirmed by ^{32}P-labelling and subsequent autoradiography of the immunoprecipitated PEP carboxylase protein (Dever *et al.*, 1997). In plants containing less than 55% of the wild-type PEP carboxylase activity, there was evidence of an upturn in the concentration of compounds that would activate PEP carboxylase including triose phosphates, G6P, F6P, glycine and serine. It has been suggested that as the photorespiratory rates increase, due to reduced decarboxylation in the bundle sheath cells, the glycine and serine are able to move into the mesophyll cells and activate PEP carboxylase (K.J. Bailey unpublished data).

4.2 *A mutant lacking NAD-ME*

The mutant LaC$_4$73 was shown to contain approximately 5% of the NAD-ME activity normally found in wild-type *A. edulis* plants (Dever *et al.*, 1998). When the leaf proteins of the mutant and wild type were subject to western blot analysis following SDS–PAGE, using antisera raised against both the α and β subunits (Murata *et al.*, 1989), both subunits were clearly visible in LaC$_4$ 73 (*Figure 4*). To investigate the aggregation state of NAD-ME, western blot analysis was also carried out following

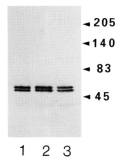

Figure 4. Western immunoblot analysis of leaf proteins of A. edulis separated by SDS–PAGE challenged with antisera raised to NAD-ME (from A. tricolor). Lane 1: wild-type, Lane 2: heterozygous and Lane 3: homozygous LaC$_4$73 mutant leaf extracts. Reprinted from Planta, The isolation and characterisation of a mutant of the C$_4$ plant Amaranthus edulis deficient in NAD-malic enzyme activity, Dever et al., vol. 206, pp. 649–656, figure 1, 1998. © Springer-Verlag.

non-denaturing gradient PAGE, utilizing the antisera described above and that raised against the α-subunit only (Long *et al.*, 1994). Both antisera recognized proteins in the octameric (480 kDa), tetrameric (240 kDa) and dimeric (120 kDa) regions of the blot (*Figure 5*). NAD-ME was present mostly in the octameric form in the wild-type leaf tissue, whether isolated from illuminated or darkened leaves. The leaf extracts of the homozygous mutant LaC$_4$ 73, although still containing predominantly octamers, exhibited a higher proportion of the dimeric NAD-ME protein. It is possible that this higher proportion is due to the low rate of photosynthetic CO$_2$ fixation. This is supported by the data in *Figure 5*, which demonstrate that when grown in elevated CO$_2$, there was little difference in the aggregation state of NAD-ME in any of the plants. It is also possible that the low activity of NAD-ME in LaC$_4$ 73 is due to a direct amino acid change in the active site of the enzyme protein or the site(s) involved in the binding of metabolite activators, for example, CoA, fumarate or fructose-1,6-bisphosphate. Such mutations must await the elucidation of the full amino acid sequence of the protein.

As described previously for the mutant lacking PEP carboxylase activity, LaC$_4$ 73 was maintained in an atmosphere of elevated CO$_2$ and back-crossed to the wild type to produce F$_1$ seed. Analysis of plants from the F$_2$ generation indicated that the mutation segregated in a normal Mendelian fashion and that leaves from heterozygous plants contained approximately 50% of the wild-type NAD-ME activity. There was no evidence of any pleiotropic effect of the deficiency on all the key enzymes of the C$_4$ pathway tested. Biomass measurements indicated that there was no difference in the growth rate in air between the wild-type and heterozygous plants, whether based on total plant dry weight or leaf area. However, the homozygous mutant grew very slowly in air and the total dry weight of the plants was only 1% and the leaf area 4% of the wild-type values (Dever *et al.*, 1998).

The A/C_i curves obtained for the homozygous LaC$_4$ 73 mutant confirmed the biomass growth data, in that the rate of photosynthetic CO$_2$ assimilation at ambient C_i was very low (Dever *et al.*, 1996). However, as was seen for the PEP carboxylase mutant, the rate increased in a linear fashion with elevated C_i (*Figure 2*). The heterozygous plants exhibited CO$_2$ assimilation rates that were 10% lower than the wild type at high C_i, however at ambient C_i values the rates were the same even up to light intensities of 1700

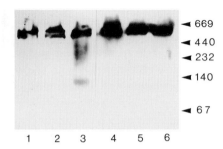

Figure 5. *Western immunoblot analysis of leaf proteins of* A. edulis *separated by native Sigma 10–17% gradient PAGE challenged with antisera raised to NAD-ME (from* A. hypochondriacus). *Lane 1: wild-type, Lane 2: heterozygous and Lane 3: homozygous mutant LaC$_4$ 73 leaf extracts from plants grown in air. Lane 4: wild-type, Lane 5: heterozygous and Lane 6: homozygous mutant LaC$_4$ 73 leaf extracts from plants grown in elevated (0.7%) CO$_2$.*
Reprinted from Planta, The isolation and characterisation of a mutant of the C$_4$ plant Amaranthus edulis deficient in NAD-malic enzyme activity, *Dever* et al., *vol. 26, pp. 649–656, figure 2, 1998.*
© *Springer-Verlag.*

µmol m^{-2} s^{-1}. These data for the heterozygous plants indicate that NAD-ME has little control over the rate of C$_4$ photosynthesis and that in the wild type the enzyme has a very low flux-control coefficient (ap Rees and Hill, 1994; Kacser and Burns, 1973).

Bundle sheath strands isolated from the wild-type *A. edulis* were able to oxidize malate in the dark and convert the pyruvate formed to alanine by the action of gluta-mate: pyruvate aminotransferase, as would be predicted from the data for other NAD-ME-type C$_4$ plants (Agostino *et al.*, 1996; Furbank *et al.*, 1990). Although bundle sheath strands isolated from the homozygous mutant LaC$_4$ 73 were able to oxidize malate at rates only slightly lower than wild type, the rate of alanine synthe-sis was less than 10% of the wild type, confirming that the NAD-ME protein was unable to convert malate to pyruvate *in vivo*. The changes in concentration of

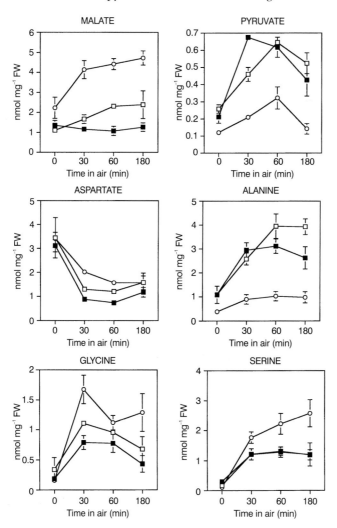

Figure 6. Changes in metabolite concentrations in the leaves of wild-type (■), heterozygous (□) and homozygous mutant LaC$_4$ 73 (○) A. edulis, when grown in elevated (0.7%) CO$_2$ and upon transfer to air for up to 180 min. Reprinted from Planta, The isolation and characterisation of a mutant of the C$_4$ plant Amaranthus edulis deficient in NAD-malic enzyme activity, Dever et al., vol. 26, pp. 649–656, figures 5 and 6, 1998. © Springer-Verlag.

metabolites in the leaves of the mutant following transfer from elevated CO_2 to air are consistent with a lack of NAD-ME activity (*Figure 6*). Firstly, the substrate malate was elevated in the mutant, whereas the products pyruvate (and alanine) were much lower, whether in high CO_2 or air. Secondly, these differences became much more pronounced upon transfer to air with a considerable accumulation of malate in the homozygous mutant and a marked deficiency in pyruvate and alanine. The rise in pyruvate and alanine in the wild-type and heterozygous plants following transfer to air is further evidence that the C_4 pathway begins to operate immediately on removal from the 7000 μmol mol^{-1} CO_2. The reason for the high concentrations of aspartate in elevated CO_2 in all plants is unclear, but might be a symptom of the inhibition of the C_4 photosynthetic pathway by CO_2 (Lal and Edwards, 1995). The accumulation of glycine and serine in the homozygous mutant is consistent with the view that the rate of photorespiration is enhanced, due to the lack of CO_2 production by NAD-ME in the bundle sheath cells.

4.3 *Mutants that accumulate glycine*

Five different mutants of *A. edulis* have been isolated that accumulate glycine following exposure to air. Of these, LaC₄ 2.11, LaC₄ 25 and LaC₄ 30 have been the most extensively characterized. Self-fertilized seed of the original mutants can be germinated in an atmosphere of 7000 μmol mol^{-1} CO_2, where the plants grow at a slower rate than the wild type, but they are not viable in air. Unlike the mutants deficient in PEP carboxylase and NAD-ME, the glycine-accumulating mutants exhibit severe symptoms of stress with a rapid loss of chlorophyll and extensive bleaching after transfer to air from elevated CO_2, the onset of which is dependent upon light intensity. The concentration of glycine in the leaves of the mutants can reach over 30 μmol g^{-1} fresh weight after 24 h and may constitute over 80% of the total soluble nitrogen, which correlates with a dramatic fall in the concentration of all other soluble amino acids. The rates of photosynthetic CO_2 assimilation of the mutants decreased dramatically in the first hour of exposure to air and fell to zero after 6 h.

The properties of the mutants are similar to those reported for photorespiratory mutants of barley that lack proteins of the glycine decarboxylase (GDC) complex (Blackwell *et al.*, 1990; Wingler *et al.*, 1997). Western blot analysis of the four proteins of the GDC (P, H, T and L) (Oliver, 1994), indicated that none of the mutants

Table 1. *Respiration rates (nmol oxygen min^{-1} mg^{-1} chlorophyll) of bundle sheath strands isolated from the leaves of wild-type and glycine-accumulating mutants of* A. edulis *in the presence of malate and glycine*

Bundle sheath source	No addition	Malate (10 mM)	Glycine (10 mM)
Wild type	206	612	389
LaC₄ 30	167	611	167
LaC₄ 25	164	492	165
LaC₄ 2.11	196	524	194

exhibited a severe loss of any of the specific proteins. The activity and protein content of serine hydroxymethyltransferase were also similar in the wild-type and mutant plants. When bundle sheath strands were isolated in the same manner as described previously for the NAD-ME mutant (Agostino *et al.*, 1996; Furbank *et al.*,1990), wild-type bundle sheath strands had the capacity to oxidize both glycine and malate (*Table 1*). However, the bundle sheath strands isolated from all the mutants were unable to oxidize glycine, whilst still maintaining the capacity to oxidize malate at similar rates to the wild type. The results confirm that the accumulation of glycine is due to a loss of glycine-metabolizing capacity in the bundle sheath cells, probably due to a lesion in the GDC complex.

As discussed previously, the magnitude of the rate of photorespiratory CO_2 release in C_4 photosynthesis, due to possible reassimilation before it is lost to the atmosphere has been the subject of considerable debate. Working on the hypothesis that the glycine accumulating in the mutants is derived from the photorespiratory carbon and nitrogen cycle, we attempted to use the mutants to establish the rate of photorespiration in *A. edulis*. Ammonia and CO_2 are released at the same rate in the conversion of glycine to serine during photorespiration (Oliver, 1994). If leaves are treated with an inhibitor of glutamine synthetase, such as phosphinothricin (PPT), there is a rapid accumulation of ammonia in a wide range of C_3 plants (Lacuesta *et al.*, 1993) and C_4 plants (González-Moro *et al.*, 1993), the ammonia may be derived from several metabolic reactions (Lea, 1991). The rate of PPT-dependent ammonia accumulation in the leaves of wild-type *A. thaliana* was reduced by 60%, either by incubation in 7000 μmol

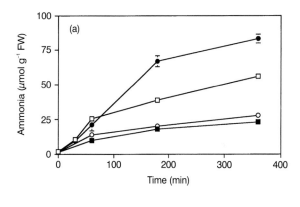

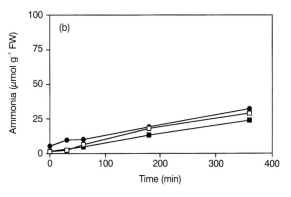

Figure 7. The increase in ammonia content of excised leaves of A. edulis *following the feeding of 1 mM PPT. (a) Wild-type (□), wild-type plus AAN (■), wild-type plus glycine (●), wild-type in 0.7% CO_2.(○). (b) LaC$_4$ 2.11 (□), LaC$_4$2.11 plus AAN (■), LaC$_4$ 2.11 plus glycine (●). From Dever et al. (1996) The isolation and characterization of mutants of the C_4 plant* Amaranthus edulis. Comptes rendus de l'Académie des Sciences (Sciences de la vie), *vol. 319, pp. 951–959. Reprinted by permission of the journal.*

mol^{-1} CO$_2$ or treatment with the GDC inhibitor aminoacetonitrile (AAN), indicating that this fraction was derived from photorespiration. The addition of glycine to wild-type leaves in the presence of PPT stimulated the accumulation of ammonia, indicating that there is spare GDC capacity within the bundle sheath mitochondria. However, when the leaves of the glycine-accumulating mutants were treated with PPT, ammonia accumulation was much lower than detected in the wild type and was not stimulated by glycine or inhibited by AAN (*Figure 7*). By comparing the rates of PPT-dependent ammonia accumulation in the wild type and mutants in the presence and absence of AAN, it proved possible to calculate a minimum value of 6% for the rate of photorespiration as expressed as a proportion of net photosynthesis (Lacuesta *et al.*, 1997). This value is somewhat higher than the rate of photorespiration calculated by Jenkins (1997) as discussed in Section 3.

5. Transgenic plants with altered photosynthetic characteristics

Following the characterization of efficient mechanisms of transforming the NADP-ME-type dicot *F. bidentis* (Chitty *et al.*, 1994), Furbank and his colleagues have used antisense techniques to study the effects of lowering the activity of three different enzymes involved in photosynthetic metabolism (Furbank *et al.*, 1997). It is of some value to discuss briefly the characteristics of these plants, in comparison with the *A. edulis* plants isolated by the screening of mutated populations.

5.1 *Rubisco*

At near-saturating illumination (2000 µmol m^{-2} s^{-1}), the rates of photosynthetic CO$_2$ assimilation declined progressively with a reduction in the amount of Rubisco activity in the leaves and C^J values of 0.5–0.7 were obtained (Furbank *et al.*, 1996). However, at low light intensity (350 µmol m^{-2} s^{-1}), the photosynthetic rate (which was much lower) was unchanged even when the Rubisco acitivity was reduced to 15% of the normal wild-type rate, indicating a C^J value of zero.

5.2 *PPDK*

PPDK has long been implicated in controlling the flux of photosynthetic CO$_2$ assimilation in C$_4$ plants, due in part to the low amount of extractable activity, and also to the complex phosphorylation mechanism of regulation, described in Section 2.3. At high light intensity, the rate of photosynthetic CO$_2$ assimilation also decreased with a reduction in PPDK activity, but the decline was not as great as that seen with Rubisco. A C^J value of 0.2–0.3 was calculated, which again reduced to zero at low light intensities (Furbank *et al.*, 1997). Interestingly, in some of the transgenic plants the measurable PPDK activity was lower than the rate of photosynthetic CO$_2$ assimilation, in some cases by as much as 50%. This cautionary tale clearly indicates that in the environment of the mesophyll chloroplast, the PPDK enzyme is in a far greater state of activation than can be achieved in the spectrophotometer cell.

5.3 *NADPH-MDH*

In these experiments, *F. bidentis* plants were transformed with sense constructs of the sorghum NADPH-MDH gene (Trevanion *et al.*, 1997). The majority of the trans-formed lines exhibited greatly decreased NADPH-MDH activity, presumably due to co-suppression (Meyer and Saedler, 1996). NADPH-MDH activity could be reduced to 10% of the wild-type activity, with little or no effect effect on the rate of photo-synthetic CO_2 assimilation at a light intensity of 1200 μmol m^{-2} s^{-1}, indicating that even in high light, the C^J for the enzyme is zero. Only at enzyme activity values below 10% of wild type was there any reduction in photosynthetic rate. Furbank *et al.* (1997) argued that the NADPH-MDH was normally present in an activated form that tracked the photosynthetic rate, so that it was present in a three-fold excess of the photosynthetic rate. In the transgenic plants with greatly reduced NADPH-MDH activity, the enzyme was always present in a 100% activated form that equalled the photosynthetic rate.

6. Conclusions

The results of the studies on mutants of *A. edulis* and the genetically engineered *F. bidentis* described in this chapter, show that for two enzymes involved in the C_4 pho-tosynthetic cycle (and also for Rubisco itself), a small reduction in enzyme activity can have a detrimental effect on the rate of photosynthetic CO_2 assimilation. It is clear from *Table 2* that the control of flux through the complete pathway of photo-synthetic CO_2 asimilation in C_4 plants does not rest with one specific enzyme-catal-ysed reaction, but is distributed between different enzymes.

Table 2. *Flux control coefficients for CO_2 assimilation in wild-type C_4 plants at high light intensities and ambient CO_2 concentrations*

Enzyme	Plants	Flux control coefficient C^J	Reference
Rubisco	*F. bidentis*	0.5–0.7	Furbank *et al.*, 1997
PEP carboxylase	*A. edulis*	0.35	Dever *et al.*, 1997
PPDK	*F. bidentis*	0.2–0.3	Furbank *et al.*, 1997
NAD-ME	*A. edulis*	zero	Dever *et al.*, 1998
NADPH-MDH	*F. bidentis*	zero	Trevanion *et al.*, 1997

It is clear that future work should be aimed at producing transgenic C_4 plants that contain more efficient forms or larger amounts of the enzymes which have high flux-control coefficients, such plants may have higher rates of photosynthesis and productivity.

Acknowledgements

The authors are grateful for the receipt of grant number 89/BR301910 from the Biotechnology and Biological Research Council for support of this work.

References

Agostino, A., Heldt, H.W. and Hatch, M.D. (1996) Mitochondrial respiration in relation to photosynthetic C_4 acid decarboxylation in C_4 species. *Aust. J. Plant Physiol.* **23**: 1–7.

Andrews, J.R. and Baker, N.R. (1997) Oxygen-sensitive differences in the relationship between photosynthetic electron transport and CO$_2$ assimilation in C$_3$ and C$_4$ plants. *Aust. J. Plant Physiol.* **24**: 495–503.

ap Rees, T. and Hill, S.A. (1994) Metabolic control analysis of plant metabolism. *Plant Cell Environ.* **17**: 587–599.

Ashton, A., Burnell, J.N., Furbank, R.T., Jenkins, C.L.D. and Hatch, M.D. (1990). Enzymes of C$_4$ photosynthesis. In: *Methods in Plant Biochemistry*, Vol. 3 (ed. P.J. Lea). Academic Press, London, pp. 39–72.

Blackwell, R.D., Murray, A.J.S., Lea, P.J., Kendall, A.C., Hall, N.P., Turner, J.C. and Wallsgrove, R.M. (1988) The value of mutants unable to carry out photorespiration. *Photosynthesis Res.* **16**: 155–176.

Blackwell, R.D., Murray, A.J.S. and Lea, P.J. (1990) Photorespiratory mutants of the mitochondrial conversion of glycine to serine. *Plant Physiol.* **94**: 1316–1322.

Brown, R.H. (1997) Analysis of bundle sheath conductance and C$_4$ photosynthesis using a PEP-carboxylase inhibitor. *Aust. J. Plant Physiol.* **24**: 549–554.

Buchanan, B.B. (1991) Regulation of CO$_2$ assimilation in oxygenic photosynthesis. *Arch. Biochem. Biophys.* **288**: 1–9.

Canvin, D.T. (1979) Photorespiration: comparison between C$_3$ and C$_4$ plants. In: *Encyclopaedia of Plant Physiology*, Vol. 6 (eds M. Gibbs and E. Latzko). Springer-Verlag, Berlin, pp. 369–96.

Chitty, J.A., Furbank, R.T, Marshall, J.S., Chen, Z. and Taylor, W.C. (1994) Genetic transformation of the C$_4$ plant *Flaveria bidentis. Plant J.* **6**: 949–956.

Chollet, R., Vidal, J. and O'Leary, M.H. (1996) Phosphoenolpyruvate carboxylase: a ubiquitous, highly regulated enzyme in plants. *Annu. Rev. Plant Physiol. Plant Mol. Biol.* **47**: 273–298.

Collatz, G.J., Ribas-Carbo, M. and Berry, J.A. (1992) Coupled photosynthesis-stomatal conductance model for leaves of C$_4$ plants. *Aust. J. Plant Physiol.* **22**: 497–509.

Dai, Z., Ku, M.S.B. and Edwards, G.E. (1995). C$_4$ photosynthesis: the effects of leaf development on the CO$_2$ concentrating mechanism and photorespiration in maize. *Plant Physiol.* **107**: 815–825.

Dever, L.V., Blackwell, R.D., Fullwood, N.J., Lacuesta, M., Leegood, R.C., Onek, L.A., Pearson, M.A. and Lea, P.J. (1995) The isolation and characterisation of mutants of the C$_4$ photosynthetic pathway. *J. Exp. Bot.* **46**: 1363–1376.

Dever, L.V., Bailey, K.J., Lacuesta, M., Leegood, R.C. and Lea, P.J. (1996) The isolation and characterisation of mutants of the C$_4$ plant *Amaranthus edulis. Comptes Rendus Acad. Sci. III. La Vie.* **319**: 951–959.

Dever, L.V., Lea, P.J., Bailey, K.J. and Leegood, R.C. (1997) Control of photosynthesis in mutants of *Amaranthus edulis. Aust. J. Plant Physiol.* **24**: 469–476.

Dever, L.V., Pearson, M., Ireland, R.J., Leegood, R.C. and Lea, P.J. (1998) The isolation and characterisation of mutants of the C$_4$ plant *Amaranthus edulis* deficient in NAD-malic enzyme activity. *Planta* **206**: 649–656.

Doncaster, H.D. and Leegood, R.C. (1987) Regulation of phospho*enol*pyruvate carboxylase in maize leaves. *Plant Physiol.* **84**: 82–87.

Edwards, G. and Walker, D.A. (1983) *C3: C4: Mechanisms and Cellular and Environmental Regulation of Photosyntheis.* Blackwells, Oxford.

Ehleringer, J.R., Sage, R.F., Flanagan, L.B. and Pearcy, R.W. (1991) Climate change and the evolution of C$_4$ photosynthesis. *Trends Ecol. Evolution* **6**: 95–99.

Furbank, R.T., Agostino, A. and Hatch, M.D. (1990) C$_4$ acid decarboxylation and photosynthesis in bundle sheath cells of NAD-malic enzyme type plants: mechanism and the role of malate and orthophosphate. *Arch. Biochem. Biophys.* **276**: 374–381.

Furbank, R.T., Agostino, A. and Hatch, M.D. (1991) Regulation of C$_4$ photosynthesis: modulation of mitochondrial NAD-malic enzyme by adenylates. *Arch. Biochem, Biophys.* **289**: 376–381.

Furbank, R.T., Chitty, J.A., von Caemerer, S. and Jenkins, C.L.D. (1996) Antisense RNA inhibition of *RbcS* gene expression reduces Rubisco level and photosynthesis in the C$_4$ plant *Flaveria bidentis*. *Plant Physiol.* **111**: 725–734.

Furbank, R.T., Chitty, J.A., Jenkins, C.L.D., Taylor, W.C., Trevanion, S.J., von Caemmerer, S. and Ashton, A.R. (1997) Genetic manipulation of key photosynthetic enzymes in the C$_4$ plant *Flaveria bidentis*. *Aust. J. Plant Physiol.* **24**: 477–485.

Gao, Y. and Woo, K.C. (1996) Regulation of phospho*enol*pyruvate carboxylase in *Zea mays* by protein phosphorylation and metabolites and their roles in photosynthesis. *Aust. J. Plant Physiol.* **23**: 25–32.

González-Moro, M.B., Lacuesta, M., Royuela, M., Muñoz-Rueda, A. and González- Murúa, C. (1993) Comparative study of the inhibition of photosynthesis caused by aminooxyacetic acid and phosphinthricin in *Zea mays*. *J. Plant Physiol.* **142**: 161–166.

Grisvard, J., Keryer, E., Takvorian, A., Dever, L.V., Lea, P.J. and Vidal, J. (1998) A spice mutation gives rise to a mutant of the C$_4$ plant *Amaranthus edulis* deficient in phosphoenolpyruvate carboxylase activity. *Gene* **213**: 31–35.

Hatch, M.D. (1987) C$_4$ photosynthesis: a unique blend of modified biochemistry, anatomy and ultrastructure. *Biochim. Biophys. Acta* **895**: 81–106.

Hatch, M.D. (1997) Resolving C$_4$ photosynthesis: trials, tribulations and other unpublished stories. *Aust. J. Plant Physiol.* **24**: 413–422.

Häusler, R.E., Bailey, K.J., Lea, P.J. and Leegood, R.C. (1996) Control of photosynthesis in barley mutants with reduced activities of glutamine synthetase and glutamate synthase. III. Aspects of glyoxylate metabolism and effects of glyoxylate on the activation state of ribulose-1,5-bisphosphate carboxylase/oxygenase. *Planta* **200**: 388–396.

Huber, S.C., Huber, J.L. and McMichael, R.W. (1994) Control of plant enzyme activity by reversible protein phosphorylation. *Int. Rev. Cytol.* **149**: 47–98.

Ireland, R.J. and Lea, P.J. (1999) The enzymes of glutamine, glutamate, asparagine and aspartate metabolism. In: *Plant Amino Acids: Biochemistry and Biotechnology* (ed. B.K.Singh). Marcel Dekker, New York, pp. 49–109.

Jenkins, C.L.D. (1997) The CO$_2$ concentrating mechanism of C$_4$ photosynthesis: bundle sheath cell CO$_2$ concentration and leakage. *Aust. J. Plant Physiol.* **24**: 543–547.

Kacser, H. and Burns, J.A. (1973) The control of flux. *Symp. Soc. Exp. Biol.* **28**: 65–104.

Keys, A.J., Bird, I.F., Cornelius, M.J., Lea, P.J., Wallsgrove, R.M. and Miflin, B.J. (1978) The photorespiratory nitrogen cycle. *Nature* **275**: 741–743.

Kleinhofs, A., Warner, R.L., Muehlbauer, F.J. and Nilan, R.A. (1978). Induction and selection of specific gene mutations in *Hordeum* and *Pisum*. *Mut. Res.* **51**: 29–35.

Lacuesta, M., González-Mono, B., González-Murúa, C. and Muñoz-Rueda, A. (1993) Time-course of the phosphinothricin effect on gas exchange and nitrate reduction in *Medicago sativa*. *Physiol. Plant.* **89**: 847–853.

Lacuesta, M., Dever, L.V., Muñoz-Rueda, A. and Lea, P.J. (1997) A study of photorespiratory ammonia production in the C$_4$ plant *Amaranthus edulis*, using mutants with altered photosynthetic capacities. *Physiol. Plant.* **99**: 447–455.

Lal, A. and Edwards, G.E. (1995) Maximum quantum yields of oxygen evolution in C$_4$ plants under high CO$_2$. *Plant Cell Physiol.* **36**: 1311–1317.

Lea, P.J. (1991) The inhibition of ammonia assimilation: a mechanism of herbicide action. In: *Topics in Photosynthesis* (eds N.R. Baker and M. Percival). Elsevier, Amsterdam, pp. 267–298.

Lea, P.J. and Ireland, R.J. (1999) Nitrogen metabolism in higher plants. In: *Plant Amino Acids: Biochemistry and Biotechnology* (ed. B.K. Singh). Marcel Dekker, New York, pp. 1–47.

Leegood, R.C. (1997) The regulation of C$_4$ photosynthesis. *Adv. Bot. Res.* **26**: 251–316.

Leegood, R.C. and Walker, R.P. (1998) Regulation of the C$_4$ pathway. In: *C$_4$ Plant Biology* (eds R.F. Sage and R.K. Monson). Academic Press, New York, pp. 85–126.

Leegood, R.C., Lea, P.J., Adcock, M.D. and Häusler, R.E. (1995) The regulation and control of photorespiration. *J. Exp. Bot.* **46**: 1397–1414.

Long, J.J., Wang, J.-L. and Berry, J.O. (1994) Cloning and analysis of the C₄ NAD-dependent malic enzyme of amaranth mitochondria. *J. Biol. Chem.* **269**: 2827–2833.

Maroco, J.P., Ku, M.S.B. and Edwards, G.E. (1997) Oxygen sensitivity of C₄ photosynthesis: evidence from gas exchange and chlorophyll fluorescence analyses with different C₄ subtypes. *Plant Cell Environ.* **20**: 1525–1533.

Maroco, J.P., Ku, M.S.B., Lea, P.J., Dever, L.V., Leegood, R.C., Furbank, R.T. and Edwards, G.E. (1998) Oxygen requirement and inhibition of photosynthesis: an analysis of C₄ plants deficient in the C₃ and C₄ cycles. *Plant Physiol.* **116**: 823–832.

Meister, M., Agostino, A. and Hatch, M.D. (1996) The roles of malate and aspartate in C₄ photosynthetic metabolism of *Flavaria bidentis* (L.). *Planta* **199**: 262–269.

Meyer, P. and Saedler, H. (1996) Homology dependent gene silencing in plants. *Annu. Rev. Plant. Physiol. Plant Mol. Biol.* **47**: 23–48.

Miginiac-Maslow, M., Issakadis, E., Lemaire, M., Ruelland, E., Jacquot, J.-P. and Decottignies, P. (1997) Light dependent activation of NADP-malate dehydrogenase: a complex process. *Aust. J. Plant Physiol.* **24**: 529–542.

Murata, T., Ikeda, J.L. and Ohsugi, R. (1989) Comparative studies of NAD malic enzyme from leaves of various C₄ plants. *Plant Cell Physiol.* **30**: 429–437.

Nhiri, M., Bakrim, N., Pacquit, V., El Hachimi-Messouak, Z., Osuna, L. and Vidal, J. (1998) Calcium dependent and independent phosphoenolpyruvate carboxylase kinase in sorghum leaves. *Plant Cell Physiol.* **39**: 241–246.

Oliver, D.J. (1994) The glycine decarboxylase complex from plant mitochondria. *Annu. Rev. Plant Physiol. Plant Mol. Biol.* **45**: 323–327.

Podesta, F., Iglesias, A. and Andreo, C. (1990) Oligomeric enzymes in the C₄ pathway of photosynthesis. *Photosynthesis Res.* **26**: 161–170.

Rebeillé, F. and Hatch, M.D. (1986) Regulation of NADP-malate dehydrogenase in C₄ plants. *Arch. Biochem Biophys.* **249**: 171–179.

Roeske, C.A. and Chollet, R. (1989) The role of metabolites in reverse light activation of pyruvate orthophosphate dikinase in *Zea mays* mesophyll cells *in vivo*. *Plant Physiol.* **90**: 330–337.

Rosche, E., Chitty, J., Westhoff, P. and Taylor, W.C. (1998) Analysis of promoter activity for the gene encoding pyruvate orthophosphate dikinase in stably transformed C₄ *Flaveria* species. *Plant Physiol.* **117**: 821–829.

Somerville, C.R. (1986) Analysis of photosynthesis with mutants of higher plants and algae. *Annu. Rev. Plant Physiol.* **37**: 467–507.

Somerville, C.R. and Ogren, W. (1979) A phosphoglycolate phosphatase deficient mutant of *Arabidopsis*. *Nature* **280**: 833–836.

Trevanion, S.J., Furbank, R.T. and Ashton, A.R. (1997) NADP-malate dehydrogenase in the C₄ plant *Flavaria bidentis*. Cosense suppression of activity in mesophyll and bundle sheath cells and consequences for photosynthesis. *Plant Physiol.* **113**: 1153–1165.

Vidal, J. and Chollet, R. (1997) Regulatory phosphorylation of C₄ PEP carboxylase. *Trends Plant Sci.* **2**: 230–237.

Walker, R.P., Acheson, R.M., Técsi, L.I. and Leegood, R.C. (1997) Phosphoenolpyruvate carboxykinase in C₄ plants: its role and regulation. *Aust. J. Plant Physiol.* **24**: 459–468.

Westhoff, P., Svensson, P., Ernst, K., Bläsing, O., Burscheidt, J. and Stockhaus, J. (1997) Molecular evolution of C₄ phosphoenolpyruvate carboxylase in the genus *Flaveria*. *Aust. J. Plant Physiol.* **24**: 429–436.

Willeford, K.O. and Wedding, R.T. (1987) Evidence for a multiple subunit composition of plant NAD malic enzyme. *J. Biol. Chem.* **262**: 8423–8429.

Wingler, A., Lea, P.J. and Leegood, R.C. (1997) Control of photosynthesis in barley plants with reduced activities of glycine decarboxylase. *Planta* **202**: 171–178.

Winning, B.M., Bourguignon, J. and Leaver, C.J. (1994) Plant mitochondrial NAD⁺-dependent malic enzyme. *J. Biol. Chem.* **269**: 4780–4786.

13

On being thick: fathoming apparently futile pathways of photosynthesis and carbohydrate metabolism in succulent CAM plants

Barry Osmond, Kate Maxwell, Marianne Popp and Sharon Robinson

1. Introduction

Tom ap Rees was a crusader of plant biochemistry, and understanding the mechanisms of coarse and fine control of carbohydrate metabolism was Tom's Holy Grail. He found particular satisfaction in the peculiarly 'plant' aspects of metabolic regulation and championed the compartmentation of plant metabolism that 'is clearly extensive, complex, universal, and, to an appreciable extent, unique' (ap Rees, 1987). More than most, he may have accepted that apparently futile metabolic pathways, pathways in which acquisition of the primary substrates of autotrophy leads simultaneously to their loss, can be a necessary inefficiency, an inherent and distinctive feature of metabolic regulation in higher plants that, by and large, are denied the benefits of homeostasis and freedom of movement in the terrestrial environment.

This chapter is about the most fundamental of carbohydrate metabolisms, the photosynthetic conversion of CO_2 and H_2O to $(CHO)_n$ in plants that spend much of their time in the light simultaneously generating CO_2 and H_2O through photorespiration. They are thick, and the notion that succulents with Crassulacean acid metabolism (CAM) actually achieve a modicum of homeostasis in stressful environments through apparently futile pathways, sustained by a large 'reciprocating carbohydrate pool' (Christopher and Holtum, 1996), must have been particularly appealing to Tom. Chris Goodsall cultivated CAM plants (pineapples and *Kalanchoë*) in the Downing Street laboratories in 1993–94, and from his thesis (Goodsall, 1995) we surmise that he and Tom had recognized the need for, and were taking steps towards, clarification

Plant Carbohydrate Biochemistry, edited by J.A. Bryant, M.M. Burrell and N.J. Kruger.
© 1999 BIOS Scientific Publishers Ltd, Oxford.

of what remains a remarkably confusing corner of plant metabolism; one in which the quality of evidence that Tom demanded still largely eludes us.

2. What is CAM?

Photosynthetic metabolism for all seasons is a reasonable synopsis of CAM, and CAM plants are testimony to totipotency of the autotrophic condition. These assessments seem to be endorsed by its presence in plant life forms ranging from submerged aquatics like *Isoetes*, to epiphytes like *Tillandsia* (Kluge *et al.*, 1973), to the most massive photosynthetic surfaces of all, the cladodes of *Carnegia*. Wisdom suggests that for an autotrophic system to avail itself of gaseous or dissolved CO_2, the carbon substrate of life, it should be thin and porous. By these standards, CAM plants have no business in photosynthesis at all, but can 33 families of vascular plants, and some 16 000 species (Winter and Smith, 1996), all be wrong? Perhaps being thick and autotrophic is not so bad after all, even though it applies constraints to metabolism that seem to be manifestly futile when measured against the rest of the plant kingdom.

The blend of taxonomy and biochemistry behind the acronym CAM signals the distinctive features of metabolism in these plants. Common among the Crassulaceae, and distinguished by a nocturnal increase and diurnal decrease in organic acidity (sometimes exceeding 1 molar protons, Popp *et al.*, 1987), CAM has been well known to botanists, at least since Grew in 1682 (Osmond, 1978). This acid metabolism is supported by, and in turn supports, a reciprocating carbohydrate pool (Christopher and Holtum, 1996) which is much less immediately evident (the sense of the sour being more acute than the sense of the sweet), and which presents a continuing challenge to plant biochemistry. Specification of the four phases of CAM some 20 years ago seems to have been a helpful step in unravelling the non-conformist metabolic capabilities of thick plants. The four phases of CAM (*Figure 1*) describe the ultimate separation of dark reactions (assimilation of external CO_2) and light reactions (photogeneration of ATP and NADPH) in photosynthesis, and identify these by their analogous carboxylation and decarboxylation reactions in C_3 and C_4 pathways of photosynthetic metabolism (Leegood *et al.*, 1997).

Thick leaves of CAM plants like *Kalanchoë* are often presented as a tightly regulated, but ponderous, CO_2-concentrating mechanism that, for part of the day at least, mitigates the inevitability of photorespiration. In CAM the CO_2-concentrating mechanism involves a huge pool of C_4 acid (several hundred μmol g^{-1} fresh weight) with a tonoplast limited turnover time of more than 10^3 s, that can be sustained as a metabolic rhythm in continuous light. In comparison, the biochemically analogous C_4 photosynthetic system is based on a symplastically separated C_4 acid pool some two orders of magnitude smaller, turning over some three orders of magnitude more rapidly. Even so, thick CAM plants sometimes show remarkably rapid metabolic transitions. Leaves of *Clusia* (tropical trees with CAM, Popp *et al.*, 1987) respond to low humidity and temperature changes by switching from normal C_3 photosynthesis to CAM in a matter of hours (Schmitt *et al*, 1988).

The extent to which the four phases of CAM are displayed depends very much on species, morphology and environment. Water supply permitting, stomata of most CAM plants open at night, allowing net CO_2 fixation in phase I, as well as recycling of respiratory CO_2 (CAM-cycling). The morphological features discussed below mean that some CAM-cycling occurs in most CAM plants most of the time. Depending on

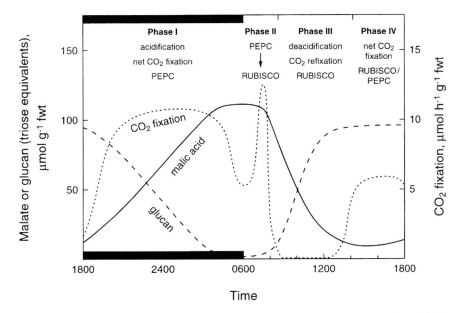

Figure 1. *The four phases of CAM in a succulent plant not exposed to water stress (modified from Osmond, 1978). PEPC, PEP carboxylase.*

morphological properties, even well-hydrated CAM plants may not engage in net CO_2 fixation in the light in phase IV. It is rare for desert stem succulents (e.g. *Opuntia*) and very thick leaf succulents (e.g. *Agave*) to do so, but occurs commonly in leaf succulents of the tropics (e.g. *Kalanchoë, Bromelia* etc.).

Morphological structures that restrain water loss from aerial photosynthetic organs inevitably keep O_2 in, and CO_2 out. When stem and leaf succulent CAM plants are exposed to severe water stress, stomata remain tightly closed throughout the 24-h cycle, preventing H_2O loss, O_2 escape, and CO_2 uptake. Recapture of respiratory CO_2 in the dark, through the closed cycle known as CAM-idling (Ting, 1985), becomes an important source of CO_2 for photosynthesis. However, for most of the daily light period the photosynthetic tissues presumably remain at a Rubisco-determined CO_2 compensation point that changes with temperature and internal O_2 concentration. Photosynthetic production of O_2, another terminal acceptor for light-driven electron transport, has potentially important ramifications, as discussed below.

In its most attenuated form, CAM-idling can be viewed as a minimalist metabolism; as one hugely futile cycle, in which phase I acidification is limited to the refixation of respiratory CO_2, and photosynthesis is limited to the machinations made possible by decarboxylation of this acid. However, survival of the hydrated photosynthetic apparatus for months on end, is as much about futile cycling through the linked photosynthetic carbon reduction (PCR) and photorespiratory carbon oxidation (PCO) cycles, and fates of electrons generated from H_2O, as it is about recycling of respiratory CO_2 in the dark. While no growth is possible under CAM-idling, when water becomes available again, leaf and stem succulents can achieve high productivity at high biomass in many habitats (Nobel, 1988), as testified by the remarkable occupation of NE Australia by *Opuntia stricta* in the century before 1930 (Osmond *et al.*, 1979).

3. Morphological matters

Being thick, being succulent, is an important but not exclusive morphological property of CAM plants. Leaves of terrestrial plants rarely exceed 1 mm in thickness, whereas those of CAM plants are rarely less than several millimetres thick (*Table 1*). Large chloroplast-containing cells, often an order of magnitude larger than cells in many other leaves, assure a high tissue water capacitance. In the massive stem succulents, chloroplast-containing tissues often extend several millimetres in depth below the epidermis. A small surface to volume ratio, low stomatal frequency, thick cuticle and epidermal waxes all potentially constrain water loss in leaf or stem succulents. In larger structures such as the leaves of *Agave* and cladodes of cacti, ridges and spines further modify water loss and energy balance (Nobel, 1988). Although most attention tends to be focused on water retention in aerial parts of CAM plants, prevention of water loss to dry soil through roots, and rapid acquisition of water by new roots when it is available in wet soil, are also fundamentally important attributes. By and large, CAM plants tend to have small root/shoot ratios, and in the case of CAM epiphytes, specialized alternatives for water uptake.

Leaves of CAM plants are therefore poor organs for gas exchange. Direct measurements of the natural abundance stable isotope composition of the C_4 carboxyl of malic acid accumulated in phase I of CAM exposed a diffusional fractionation (O'Leary and Osmond, 1980), largely because stomatal conductance of CAM plants is unusually low, and because PEP carboxylase shows little discrimination. The less negative $\delta^{13}C$ values of malic acid and dry matter deep within the thick tissues of CAM plants (Robinson *et al.*, 1993), further point to large internal gas diffusion resistances. These gradients in $\delta^{13}C$ values (*Table 1*) portray complex interactions between diffusional and biochemical isotopic fractionations, but chiefly indicate predominance of CO_2 diffusion limitations, which is not surprising because, compared with other leaves, CAM plants have very small volume gas space (Smith and Heuer, 1981).

Maxwell *et al.* (1997) recently demonstrated very low internal conductances in *Kalanchoë* leaves which, when added to low stomatal conductance, led to CO_2

Table 1. *Gradients of carbon isotopic composition (expressed as $\delta^{13}C$ values), indicating diffusional limitations during CO_2 assimilation, and possibly biochemical pathway changes, with depth in thick tissues of CAM plants (data from Robinson et al., 1993)*

Species	Thickness (mm)	$\delta^{13}C$ values (‰)	
		Outer 1 mm	Internal tissues
Ceropegia dichotoma	5	−19.9	−15.5
Crassula argentea (greenhouse)	5	−17.1	−15.4
(cabinet)	3.5	−20.5	−17.0
Echeveria colorata	–	−18.4	−15.8
Kalanchoë beharensis	4	−18.0	−14.3
Mammillaria longimamma	15	−19.5	−17.4 to −15.2
Opuntia ficus-indica (young cladode)	4	−15.8	−13.7
Opuntia stricta	11	−12.6	−11.1
Sanseveria stuckyi	–	−15.9	−13.9

concentrations of only 108 μbar at the sites of Rubisco carboxylation during photo-synthesis in phase IV (external air 380 μbar CO_2). Although these low internal CO_2 partial pressures may not become limiting during PEP carboxylase-mediated CO_2 fix-ation in phase I (indeed they are about equal to the calculated intercellular CO_2 par-tial pressures in mesophyll cells of C_4 plants) they ensure low efficiency during carboxylation by Rubisco in phase IV. These authors concluded, 'It is apparent that, in a CAM leaf, morphological constraints are in conflict for both efficient atmos-pheric CO_2 uptake during phases I, II and IV, and fixation of internally generated CO_2 during phase III'.

If sufficient malic acid can be mobilized from phase I to build high internal CO_2 concentrations during deacidification in phase III, stomata close. Cockburn *et al.* (1979) measured a record internal CO_2 concentration of 2.5% CO_2 (25000 ppm) in phase III of a CAM-cycling cladode of *O. basilaris*. Rather lower concentrations (800–8000 ppm, *Table 2*) have been measured in leaf succulents (Spalding *et al.*, 1979). Although the decarboxylation of malic acid leads to transiently high internal CO_2 concentrations (for some hours), the stoichiometry of photosynthesis suggests high internal O_2 concentrations may also build up during CO_2 fixation and carbohy-drate synthesis in phase III. Depending on decarboxylation pathway and the extent of gluconeogenesis, conversion of 100 μmol g⁻¹ fresh weight of malic acid could gen-erate up to 9 ml of O_2 in 1 ml of tissue over a few hours. The O_2 pressurization process during phase III is readily demonstrated in deacidifying tissue in a closed O_2 electrode chamber (*Figure 2*), and has been noted in a few *in vivo* measurements. Spalding *et al.* (1979) recorded 41.5% O_2 during deacidification in a *Kalanchoë* leaf (*Table 2*). Even at these high internal O_2 concentrations, calculations based on the kinetic affinities of Rubisco for O_2 and CO_2 suggest that the CO_2 concentration is such as to minimize oxygenation of RuBP and photorespiration (*Table 2*) during deacidification. Although the CO_2 concentrating function of CAM is evidently effective during phase III, it has to be remembered that deacidification is usually complete within a few hours, and CAM plants often spend much of the light period at, or near, the CO_2 compensation point.

Table 2. *Estimated ratios of carboxylase to oxygenase activities of Rubisco in CAM plants during phase III, using a gas phase CO_2/O_2 specificity factor of 2218 obtained from the kinetic properties of C_3 Rubisco (T.J. Andrews, personal communication) and using data from Spalding et al. (1979) for internal CO_2 and O_2 concentrations*

Species	Internal gas concentrations (%)			Rubisco carboxylase/ oxygenase
	$[O_2]$	$[CO_2]$	$[O_2]/[CO_2]$	
Spinach or tobacco in air	20.9	0.028	746	2.9
Ananas comosus	21.4	0.133	161	13.7
Hoya carnosa	22.7	0.080	285	7.7
Kalanchoë gastonis bonniceri	41.5	0.268	155	14.3
Kalanchoë tomentosa	30.5	0.346	88	25.2
Opuntia monocantha	24.8	0.121	205	10.8
Sedum praealtum	23.5	0.289	81	27.4

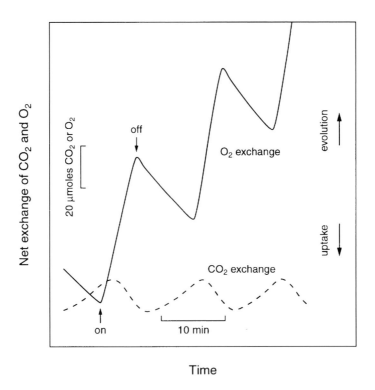

Figure 2. *Increase in O_2 pressure inside a gas exchange chamber during malate decarboxylation in* Kalanchoë *leaves (modified from Osmond et al., 1996).*

4. Apparently futile metabolic pathways in CAM

It seems unlikely that wholly futile metabolic pathways could long survive in the course of natural selection. It is much more likely that our perceptions of futility are shaped by a presently limited ability to fathom functions in particular contexts, especially when we have few clues as to selective pressures in the past. Little is known about the regulation of gene expression in CAM plants, outside of the salt-, and water stress-inducible CAM in *Mesembryanthemum crystallinum*, now recognized as a model system for stress-related gene expression (Cushman and Bohnert, 1997). Regulation at the transcriptional level seems to dominate, transcription of water stress-inducible genes increasing many-fold in response to decreased water potential. However, increased stability of mRNA may also be important. The presence, in a single chloroplast-containing cell, of high activities of all the machineries of C_3 photosynthesis and C_4 photosynthesis evidently maximizes the flexibility of metabolism, and the opportunity for apparently futile metabolic pathways, in spite of sophisticated regulatory mechanisms.

4.1 *Competing carboxylases*

Futile cycling during CO_2 fixation in CAM is strongly suggested by the fact that the expected 1:1 stoichiometry for O_2 evolution and CO_2 uptake during photosynthesis at light and CO_2 saturation is not observed in any of the near steady-state phases of CAM (*Table 3*; Osmond *et al.*, 1996). Significant CO_2 evolution and O_2 uptake evidently accompany CO_2 fixation and O_2 evolution during photosynthesis in CAM plants at all times. Some of these stoichiometric nightmares can be understood using mass spectrometry for simultaneous measurements of $^{13}CO_2$ evolution and $^{18}O_2$ uptake, and others can be resolved from ^{13}C-labelling patterns.

There was a time that the primary carboxylase of the dark reactions of photosynthesis (Rubisco, then carboxydismutase) featured as the initial carboxylase in our explanation of phase I of CAM. The notion was that two molecules of PGA from Rubisco, one C1-labelled and the other unlabelled, which when converted to PEP, would yield a mixture of doubly and singly labelled malic acid via PEP carboxylase, accounted for an apparent ratio of ^{14}C-label in malate C4:C1, of 2:1 (*Figure 3a*). In fact, only singly labelled [^{13}C]malic acid was detected by mass spectrometry (Cockburn and McAuley, 1975), and double labelling was attributed to complications arising from fumarase activity during assay. Significant progress in understanding covalent regulation of cytoplasmic PEP carboxylase (Nimmo *et al.*, 1986) now helps fathom the competition among carboxylases in phases I and III of CAM. In the dark, CAM-PEP carboxylase is phosphorylated and desensitized to feedback inhibition by malate. In the light the enzyme is dephosphorylated, and so sensitive to malate that it does not compete with Rubisco during phase III. Inactivation of PEP carboxylase in phase III (*Figure 3b*) also explains why no doubly labelled malic acid appears during deacidification (Osmond *et al.*, 1988).

Labelling studies during the transitional phases II and IV show clear evidence of parallel PEP carboxylase and Rubisco activity, linked through Rubisco-derived PEP and double carboxylation (*Figure 3a*), during the shift from PEP carboxylase-dominated to Rubisco-dominated CO_2 fixation in phase II as phosphorylation and malate control of PEP carboxylase develops. This control evidently relaxes again during phase IV, when doubly labelled malic acid appears during $^{13}CO_2$ fixation, and Rubisco initiated double carboxylation again occurs (Osmond *et al.*, 1988; Ritz *et al.*, 1986). *In vivo* assessment of these changes by natural abundance stable isotope discrimination has proved exceptionally effective (Griffiths *et al.*, 1990), with marked differences between *Kalanchoë* and *Clusia* in the rate of transition in phases II and IV (Borland and Griffiths, 1997).

4.2 *Oxygenase photorespiration and/or the Asada pathway*

Given the widely accepted notion that a principal metabolic function of CAM is to provide a CO_2-concentrating mechanism during phase III (*Table 2*), to mitigate pho-

Table 3. *Stoichiometry of net O_2 and CO_2 exchange during light- and CO_2-saturated photosynthesis in* Kalanchoë daigremontiana *(data from Osmond* et al., *1996)*

Phase of CAM	Ratio net O_2 evolution/CO_2 uptake (n)
Phase IV (C_3-like)	1.3 ± 0.2 (15)
Phase III (deacidification)	1.8 ± 0.6 (28)
Phase I (acidification)	2.7 ± 1.0 (14)

(a)

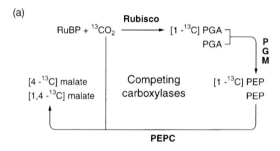

(b)

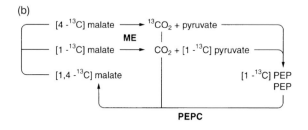

Figure 3. *Schematic outlines of apparently futile cycles of (a) double carboxylation, phases II and IV, and (b) decarboxylation–carboxylation, phase III, during photosynthesis in CAM plants. PEPC, PEP carboxylase.*

torespiration and maximize the efficiency of water use, significant light-dependent O_2 uptake during deacidification remains a matter of concern (Andre *et al.*, 1979; Cote *et al.*, 1989; Maxwell *et al.*, 1998). Time courses of photosynthetic CO_2 and O_2 exchanges, commencing at 1.5–2% CO_2 in phases III and IV, are shown in *Figure 4*. After 80 min of illumination in bright light during deacidification, CO_2 concentration in the closed gas exchange system settled to 0.3%, comparable to the internal values found for intact leaves (*Figure 4a*). Under these conditions light-dependent $^{18}O_2$ uptake remained high (21% of gross $^{16}O_2$ evolution at 600 μmol photons m^{-2} s^{-1}), and increased to 25% when light intensity was increased (if not corrected for dark $^{18}O_2$ uptake, uptake in the light was 36% of $^{16}O_2$ evolution). It is difficult to imagine that this O_2 uptake can be due to Rubisco at these CO_2 concentrations.

Much the same rates of O_2 uptake were observed during phase IV (*Figure 4b*). While CO_2 concentration declined from 1.5%, steady, light-dependent $^{18}O_2$ uptake comprised 30% of $^{16}O_2$ evolution (or 48% uncorrected). When CO_2 concentration reached the compensation point (*Figure 5a*), O_2 evolution matched O_2 uptake, as expected from the concerted interaction of PCR and PCO pathways during C_3 photosynthesis (Canvin *et al.*, 1980). Increasing the light intensity at CO_2 compensation did not accelerate $^{18}O_2$ uptake. Being thick and diffusion-limited obviously influences these measurements. Removal of the epidermis reduced $^{18}O_2$ uptake at 0.5% CO_2 in phase IV from 50% of $^{16}O_2$ evolution to about 25%, that is to not much above the dark level (Maxwell *et al.*, 1998).

These experiments show we cannot yet fully distinguish the components of O_2 exchange during photosynthesis in CAM. The only secure position is at the CO_2 compensation point in phase IV, when rates of O_2 exchange attain 50% of the maximum rate of gross O_2 evolution at CO_2 saturation. It has long been known in C_3 plants (Berry *et al.*, 1978), that the futile cycles of CO_2 fixation and O_2 evolution are precisely balanced by CO_2 evolution and O_2 uptake under these conditions. However, it is not clear how much of the O_2 uptake during CO_2-saturated photosynthesis in phase III

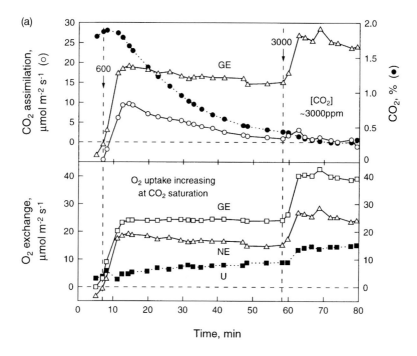

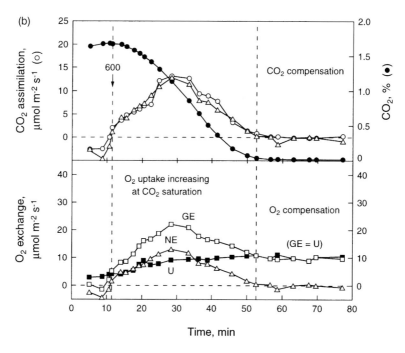

Figure 4. *Gas exchanges in* Kalanchoë pinnata *during (a) phase III and (b) phase IV of CAM, showing simultaneous* $^{18}O_2$ *uptake (U),* $^{16}O_2$ *evolution (GE) and net* O_2 *exchange (NE), as well as net* CO_2 *exchange (K. Maxwell, unpublished data; see Maxwell* et al.*, 1998). Vertical broken lines show points at which light was turned on, and then increased.*

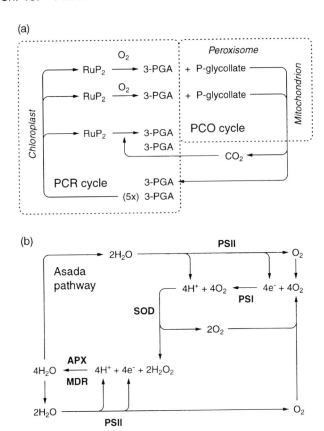

Figure 5. *Schematic outlines of futile cycles of O_2 exchange in (a) photorespiratory metabolism and (b) photosynthetic electron transport (the Asada pathway – Mehler-ascorbate peroxidase pathway of photorespiration), during photosynthesis in CAM plants. APX, ascorbate peroxidase; MDR, monodehydroascorbate reductase; PSI, photosystem I; PSII, photosystem II; SOD, superoxide dismutase.*

involves direct electron transport to O_2 in the Asada pathway (Asada, 1994; *Figure 5b*). It is difficult to imagine much Rubisco oxygenation under these conditions, yet glycine/serine-labelling has been observed (Kluge, 1969), and we are confronted again with problems of analysis of gas exchange in thick tissues.

It does not much matter, in terms of sustaining electron flow in phases III and IV, whether the O_2 uptake is due to Rubisco oxygenase, in which the PCR–PCO cycles provide a significant electron transport sink, or whether it is due to electron flow to O_2. Photosynthesis in CAM plants clearly takes place under highly oxidative conditions, given the elevated O_2 concentration that can develop when stomata are tightly closed in phase III. It is now understood that, provided adequate activities of superoxide dismutase (SOD), ascorbate peroxidase (APX) and monodehydroascorbate reductase (MDR) are present, relatively higher rates of electron flow to O_2 (generating H_2O in the process), can sustain high ΔpH and photoprotection. In this context it is not surprising that salt-induced induction of CAM in *M. crystallinum* leads to an increase in reactive O_2-scavenging system enzymes (Miszalski *et al.*, 1998; Takeda and Shigeoka, 1997).

In general, photosynthetic electron transport to O_2 has been underestimated in the past. The futile pathway of electron transport through the Asada pathway (*Figure 5b*), that involves the oxidation of water to O_2 to generate electrons that are transported to O_2 to make H_2O, may be important in the case of CAM-idling. These apparently futile metabolic pathways are likely to be the major sinks for light energy utilization and sources of ATP for maintenance processes. Both futile O_2 uptake pathways could sustain electron transport systems, resulting in high ΔpH that could assure maximum photoprotection through minimum excitation transfer to reaction centres, maintained by high zeaxanthin pools in the antennae (Demmig-Adams and Adams, 1996). Like other autotrophic systems, photosynthesis in CAM plants can be interrogated by chlorophyll fluorescence to great effect (Haag-Kerwer *et al.*, 1996; Winter and Demmig, 1987). However, being thick maximizes the probability of reabsorption of fluorescence and may constrain interpretation of fluorescence images, both at the tissue and leaf levels, that show photosynthetic efficiency is not uniform across the surface throughout the phases of CAM (B. Osmond, K. Siebke, U. Rasche and U. Lüttge, unpublished data).

One cannot escape the notion that during the minimalist metabolism of CAM-idling, the futile H_2O to H_2O cycle of the Asada pathway might also represent a biochemical water conservation mechanism, just as oxygenase photorespiration is a carbon conservation mechanism. It is not usual to take the H_2O consumption of photosynthesis into account in plant water relations (in transpiring plants they are negligible), but perhaps this should be done during CAM-idling. 'Back of the envelope' calculations show that H_2O regeneration, at the rates of electron transport to O_2 shown in *Figure 4*, could recycle 35% of tissue water over a period of 100 days in a closed CAM-idling system; in a fully hydrated and functional, illuminated photosynthetic apparatus that does not exchange gases with the atmosphere.

4.3 *Engagement of the alternative oxidase during respiration in CAM*

The alternative oxidase of plant mitochondrial respiration (Vanlerberge and McIntosh, 1997), the cyanide-insensitive, non-phosphorylating pathway that evidently reduces oxygen to water in a single four-electron step, was early identified in CAM plants (Kinraide and Marek, 1980; Rustin and Queiroz-Claret, 1985). The alternative pathway wastes two-thirds of the opportunities for oxidative phosphorylation and in this sense qualifies as a potentially futile metabolic pathway. Tom ap Rees was well acquainted with the enormous demands placed on glycolysis by the engagement of the alternative pathway in the thermogenic spadix of Aroids (ap Rees *et al.*, 1976). With him, we would agree that if an organism has to be hot and stinking to attract flies in order to reproduce, and it takes a futile respiratory pathway to achieve this condition, so be it.

It is less easy to understand why CAM plants engage peak alternative oxidase activity during phase III (Robinson *et al.*, 1992), but it is unlikely to be related to central heating. Mechanistically, it may be simply due to a stimulation of the alternative oxidase by pyruvate (Millar *et al.*, 1993), the concentration of which increases during deacidification. One role of the alternative oxidase is thought to be release of tricarboxylic acid cycle metabolism from adenylate control. This may well be important during phase III when an unspecified fraction of vacuolar malic acid may be respired to CO_2 in mitochondria. Winter and Smith (1996) proposed a role for the alternative oxidase during CAM-idling, when the carbon conservation reactions potentially generate an excess of ATP in the dark. This interpretation depends on assumption of a one-way, H^+-ATPase-driven malic acid transport system (a pump leak system would be energetically expensive) and does not allow for much ATP-dependent maintenance in the extremely stressful environments conducive to CAM-idling. On the other hand, the alternative oxidase can also be considered as an adenylate-released, H_2O recycling mechanism that might be important in a closed system exposed to extreme water deficits. If the alternative oxidase is strongly engaged, it might also represent an H_2O-generating analogue of the Asada pathway.

4.4 *Partitioning of carbohydrate metabolism in CAM*

Given we know so little about the regulation of carbohydrate metabolism in CAM plants, it would be premature to admit any sense of futile metabolic pathways, in the face of such an array of unexplained options. The core problem (*Figure 6*) is the banking of 75% of the carbohydrates synthesized during photosynthesis in phase III as a reserve sufficient to maximize CO_2 fixation in phase I, while spending the remainder on growth (Holtum and Osmond, 1995). To have found pineapples and *Kalanchoë* under cultivation in Tom ap Rees' laboratory a few years ago gave one hope that help was at hand.

There is good evidence from girdling studies that the demands of the reciprocating carbohydrate pool take precedence over export. Within 48 h of girdling that leads to accumulation of high sucrose levels in *Kalanchoë* leaves, phase IV CO_2 fixation was completely inhibited. Dark CO_2 fixation in phase I, linked to the glucan-based reciprocating carbohydrate pool, was not affected (Mayoral *et al.*, 1991). If benign environmental conditions permit a large proportion of total carbon assimilation to take place during phase IV, this carbon is preferentially used for export and growth (Mayoral and Medina, 1985). Recent data have demonstrated the increased partitioning of starch for nocturnal acidification as the induction of CAM proceeds in *M. crystallinum* (A.N. Dodd, A.M. Borland and H. Griffiths, personal communication). However, our understanding of regulated carbohydrate metabolism in CAM remains limited.

At least two C_4 acid decarboxylation pathways with different consequences for recovery of C_3 units, and three or more pathways of carbohydrate storage and mobilization are known (*Figure 6*). Substantial problems of carbohydrate analysis bedevil the last of these; the identity of the high molecular mass polyoses in *Fourcroya* remains elusive (M. Popp, unpublished data). Christopher and Holtum (1996) distinguished eight patterns of carbohydrate partitioning in 17 CAM plants. Further comparative studies show that 'although the CAM syndrome dictates that a large reciprocating carbohydrate pool must exist, it does not dictate what that carbohydrate pool should be' (Christopher and Holtum, 1998). Significantly, these authors believe that evolutionary history is a stronger determinant of the carbon-partitioning patterns in bromeliads, than decarboxylation biochemistry (NADP-ME or PCK-types) during phase III.

5. Compartmental concerns

Compartmentation in CAM is dominated by malic acid fluxes across the tonoplast, which defines the largest organelle in higher plants. Malic acid is transported into the vacuole as a divalent ion, malate^{2-}, a process serviced by two pumps, the H^+- ATPase

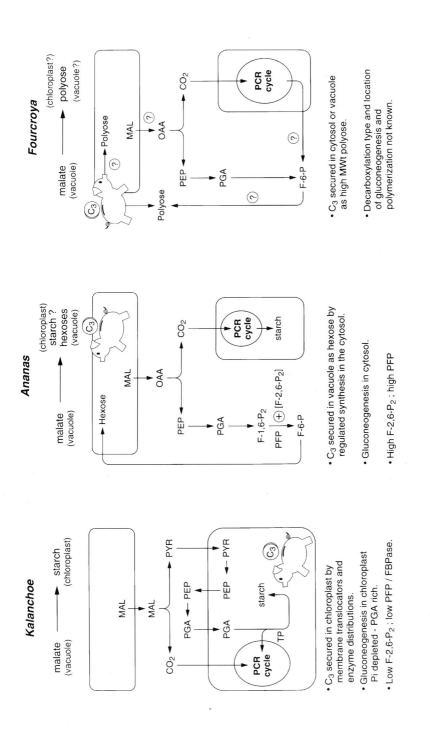

Figure 6. Some dimensions of decarboxylation and the reciprocating carbohydrate pools in different CAM plants, emphasizing the 'banking' of C_3 units for nocturnal CO_2 fixation, and summarizing potentially important regulatory steps (modified from Holtum and Osmond, 1995). $F-1,6-P_2$, fructose-1,6-bisphosphate; $F-2,6-P_2$, fructose-2,6-bisphosphate; $F-6-P$, fructose-6-phosphate; MAL, malate; OAA, oxaloacetate; PEP, phosphoenolpyruvic acid; PFP, pyrophosphate-dependent phosphofructokinase; PYR, pyruvate; TP, triose phosphate.

(Lüttge and Ratajczak, 1997), and the H^+-PP$_i$ase (Smith *et al.*, 1996). Transport of malate^{2-} is passive, down an electrochemical gradient through a malate-selective ion channel, to provide charge balance for the primary H^+- pumps. It is thought that malic acid transport slows during late phase I because the 'H^+ electrochemical potential difference against which the H^+-ATPase has to work approaches equilibrium with the free energy available from ATP hydrolysis' (Lüttge *et al.*, 1981; Smith *et al.*, 1996). However, the overall energetic demands of malic acid accumulation remain obscure, especially the cost of ATP hydrolysis in support of a pump-leak system. Compared with malic acid accumulation processes, little is known about acid release during phase II and III. Stoichiometric demands of H^+ and malate^{2-} transport implicate a proton co-transport mechanism, but in the absence of evidence, the malate influx channel is presumed held open by the diffusion gradient created by decarboxylation in the cytoplasm (Smith *et al.*, 1996).

Clearly, the decarboxylation pathways and carbohydrate recovery processes imply creative regulation of metabolite fluxes in the mitochondria and chloroplasts of CAM plants if futile metabolic pathways are to be minimized. Partitioning of malate decarboxylation in phase III between mitochondria and cytosol has important implications for overall bioenergetic costs of CAM (Winter and Smith, 1996). Chloroplasts of CAM plants evidently contain the full suite of metabolite translocators of C_3 and C_4 mesophyll chloroplasts (Leegood *et al.*, 1997) and somehow regulate these in response to the demands of overall carbon flux, in ways that are still simply mysterious.

There are striking parallels, pointed out earlier (Holtum and Osmond, 1995), between carbohydrate recovery in *Kalanchoë* chloroplasts during phase III (*Figure 6*) and glucan retention in the chloroplasts of transgenic potato plants with a 20–30% reduction in the P$_i$ translocator (Heineke *et al.*, 1994a). These transgenics show double the glucan accumulation in chloroplasts during photosynthesis and much higher glucan hydrolysis in the dark. Should we explore translocator ratios to understand the reciprocating glucan basis of CAM in *Kalanchoë*? By the same token, there are parallels between the hexose-based reciprocating carbohydrate pool of pineapples during phase III and the leaf cells of tobacco expressing yeast invertase in the vacuole (Heineke *et al.*, 1994b). Higher vacuolar hexose concentrations than wild type suggest direct vacuolar uptake of glucose and fructose, processes that could secure carbohydrate in the vacuole during the day for subsequent mobilization to the cytosol for malic acid synthesis during phase I (*Figure 6*).

6. Conclusions

That CAM plants are at times inevitably (and often monumentally) inefficient autotrophs because they are thick, is a proposition that we would dearly have loved to explore at length with Tom, in suitably leavened circumstances. It is fair to say that one of the principal consequences of being thick and autotrophic is that CAM plants present significant difficulties in deducing what is actually going on in a photosynthetic system when gas exchange is severely diffusion limited (cf. stripping the epidermis from *Kalanchoë*, Maxwell *et al.*, 1998). As Tom's first postdoctoral student in Cambridge, the senior author remembers the booming admonition, 'Don't be thick, lad!', that greeted many not clearly thought through concepts circulating in Room 19 of the Botany School in 1966–67 (Tom may not have actually spoken thus; this may be a romantic construct coloured by time!). We may have been slow to recognize these limitations, and in his more optimistic moments Tom ap Rees would have predicted that our sense of apparent futility in metabolic pathways is likely to disappear as we become more clever in dealing with what are clearly very 'clever' plants. Ultimately, perhaps, CAM is a gross compromise of most aspects of autotrophy, a compromise that tests our notions of coarse and fine control of photosynthetic metabolism. Tom would not have found this a very satisfactory conclusion. 'Back to the drawing boards!' would have been his spirited advice.

Perhaps we really are at a defining moment in plant physiology and biochemistry, one in which our problems in understanding CAM may be symbolic, even if trivial. As pointed out earlier (Holtum and Osmond, 1995), the molecular revolution already allows us to amplify the physiological and biochemical noise we cherish as plant biodiversity, well before we really understand it. As we engage in a process of 'ecotypic engineering', of designing plants for specific, often changed, habitats, progress will depend as much on past experience from plant ecophysiology and biochemistry as it does on future developments in gene technology. However, we are confronted by the fact that much of modern biochemistry and physiology has been put on hold by the diversion of what are very limited resources for research in plant molecular technology. As much as one can justify this diversion, we now have to cope with the shortfall of a generation of biochemists and physiologists when, as Tom ap Rees recognized, we can least afford 'to waste this opportunity by failing to get the practice right through ignorance of biochemical methods and physiological pitfalls' (ap Rees and Hill, 1994). We can best support Tom's crusade by stimulating interest in, and making opportunities for research in, traditional plant physiology and biochemistry. If we do not, we simply may be so thick as to lose the skills necessary to recognize and evaluate some of the most interesting transgenic plants.

Acknowledgements

Our freedom to research the magic of CAM has been supported recently by the Alexander von Humboldt Stiftung (B.O.) and the Royal Society (K.M). We are grateful to Jack Christopher and Joe Holtum for preprints, and to Tom ap Rees for setting the standards, just beyond our reach.

References

André, M., Thomas, D.A., von Willert, D.J. and Gerband, A. (1979) Oxygen and carbon dioxide exchanges in Crassulacean acid metabolism plants. *Planta* 147: 141–144.

ap Rees, T. (1987) Compartmentation of plant metabolism. In: *Physiology of Metabolism. The Biochemistry of Plants*, Vol. 12 (ed. D.D. Davies). Academic Press, San Diego, pp. 87–115.

ap Rees, T. and Hill, S.A. (1994) Metabolic control analysis of plant metabolism. *Plant Cell Environ.* 17: 587–599.

ap Rees, T., Fuller, W.A. and Wright, B.W. (1976) Measurements of glycolytic intermediates during the onset of thermogenesis in the spadix of *Arum maculatum. Biochim. Biophys. Acta* 461: 274–282.

Asada, K. (1994) Production and action of active O_2 species in photosynthetic tissues. In: *Causes of Photooxidative Stress and Amelioration by Defence Systems in Plants* (eds C.H. Foyer and P.M. Mullineaux). CRC Press, Baton Rouge, pp. 77–104.

Berry, J.A., Osmond, C.B. and Lorimer, G.H. (1978) Fixation of $^{18}O_2$ during photorespiration. Kinetic and steady state studies of the photorespiratory carbon oxidation cycle with intact leaves and isolated chloroplasts of C_3 plants. *Plant Physiol.* 62: 954–967.

Borland, A.M. and Griffiths, H. (1997) A comparative study on the regulation of C_3 and C_4 carboxylation processes in the constitutive crassulacean acid metabolism (CAM) plant *Kalanchoë daigremontiana* and the C_3-CAM intermediate *Clusia minor*. *Planta* **201**: 368–378.

Canvin, D.T., Berry, J.A., Badger, M.R., Fock, H. and Osmond, C.B. (1980) Oxygen exchange in leaves in the light. *Plant Physiol.* **66**: 302–307.

Christopher, J.T. and Holtum, J.A.M. (1996) Patterns of carbohydrate partitioning in leaves of Crassulacean acid metabolism species during deacidification. *Plant Physiol.* **112**: 393–399.

Christopher, J.T. and Holtum, J.A.M. (1998) Carbohydrate partitioning in the leaves of Bromeliaceae performing C_3 photosynthesis or Crassulacean acid metabolism. *Aust. J. Plant Physiol.* **25**: 371–376

Cockburn, W. and McAuley, A. (1975) The pathway of carbon dioxide fixation in Crassulacean plants. *Plant Physiol.* **55**: 87–89.

Cockburn, W., Ting, I.P. and Sternberg, L.O. (1979) Relationships between stomatal behavior and internal carbon dioxide concentration in Crassulacean acid metabolism plants. *Plant Physiol.* **63**: 1029–1032.

Cote, F.X., Andre, M., Folliot, M., Massimino, D., and Daguenet, A. (1989) CO_2 and O_2 exchanges in the CAM plant *Ananas comosus* (L.) Merr. *Plant Physiol.* **89**: 61–68.

Cushman, J.C. and Bohnert, H. (1997) Molecular genetics of Crassulacean acid metabolism. *Plant Physiol.* **113**: 667–676.

Demmig-Adams, B. and Adams, W.W. III (1996) The role of the xanthophyll cycle in protection of photosynthesis. *Trends Plant Sci.* **1**: 21–26.

Goodsall, C.W. (1995) The role of pyrophosphate-dependent phosphofructokinase in Crassulacean acid metabolism plants. PhD thesis, Department of Plant Sciences, University of Cambridge.

Griffiths, H., Broadmeadow, M.S.J., Borland, A.M. and Hetherington, C.S. (1990) Short-term changes in carbon-isotope discrimination identify transitions between C_3 and C_4 carboxylation during crassulacean acid metabolism. *Planta* **186**: 604–610.

Haag-Kerwer, A., Grams, T.E.E., Olivares, E., Ball, E., Arndt, S., Popp, M., Medina, E. and Lüttge, U. (1996) Comparative measurements of gas-exchange, acid accumulation and chlorophyll *a* fluorescence of different species of *Clusia* showing C_3 photosynthesis, or crassulacean acid metabolism, at the same field site in Venezuela. *New Phytol.* **134**: 215–226.

Heineke, D., Kruse, A., Flügge, I., Frommer, W.B., Riesmeier, J.W., Wilmitzer, L. and Heldt, H.W. (1994a) Effect of antisense repression on the chloroplast triose-phosphate translocator on photosynthetic metabolism in transgenic potato plants. *Planta* **193**: 174–180.

Heineke, D., Wildenberger, K., Sonnewald, U., Wilmitzer, L. and Heldt, H.W. (1994b) Accumulation of hexoses in leaf vacuoles: studies with transgenic tobacco plants expressing yeast-derived invertase in the cytosol, vacuole, or apoplasm. *Planta* **194**: 29–33.

Holtum J.A.M. and Osmond, C.B. (1995) The revolution in carbon partitioning and source-sink interactions in plants from the perspective of CAM. In: *Carbon Partitioning and Source-Sink Interactions in Plants* (eds M.A. Madore and W.J. Lucas). American Society of Plant Physiologists, Rockville, pp. 1–12.

Kinraide, T.B. and Marek, L.F. (1980) Wounding stimulates cyanide-sensitive respiration in the highly cyanide-resistant leaves of *Bryophyllum tubiflorum* Harv. *Plant Physiol.* **65**: 409–410.

Kluge, M. (1969) Veränderliche Markierungsmuster bei $^{14}CO_2$ – Fütterung von Bryophyllum tubiflorum zu verschiedenen zeitpunkten der Hell/Dunkel-Periode. I. Die CO_2 Fizierung unter Belichtung. *Planta* **88**: 113–129.

Kluge, M., Lange, O.L., von Eichmann, M. and Schmid, M. (1973) Diurnaler Säuererhythmus bei *Tillandsia usneoides*: Untersuchungen über den Weg des Kohlenstoffs sowie die Abhängigkeit des CO_2-Gaswechsels von Lichtintsnsität, Temperatur und Wassergehalt der Pflanze. *Planta* **112**: 357–372.

Lüttge, U. and Ratajczak, R. (1997) The physiology, biochemistry, and molecular biology of the plant vacuolar ATPase. *Adv. Bot. Res.* **25**: 253–296.

Lüttge, U., Smith, J.A.C., Marigo, G. and Osmond, C.B. (1981) Energetics of malate accumulation in vacuoles of *Kalanchoë tubiflora* cells. *FEBS Lett.* **126:** 81–84.

Leegood, R.C., von Caemmerer, S. and Osmond, C.B. (1997) Metabolite transport and photosynthetic regulation in C_4 and CAM plants. In: *Plant Metabolism* (eds D.T. Dennis, D.H. Turpin, D.D. Lefebvre and D.B. Layzell). Addison Wesley-Longman, Harlow, pp. 341–369.

Maxwell, K., von Caemmerer, S. and Evans, J.R. (1997) Is a low conductance to CO_2 diffusion a consequence of succulence in plants with Crassulacean Acid Metabolism? *Aust. J. Plant Physiol.* **24:** 777–786.

Maxwell, K., Badger, M.R. and Osmond, C.B. (1998) A comparison of CO_2 and O_2 exchange patterns and the relationship with chlorophyll fluorescence during photosynthesis in C_3 and CAM plants. *Aust. J. Plant Physiol.* **25:** 45–52.

Mayoral, M.L. and Medina, E. (1985) ^{14}C-translocation in *Kalanchoë pinnata* at two different stages of development. *J. Exp. Bot.* **36:** 1405–1413.

Mayoral, M.L., Medina, E. and Garcia, V. (1991) Effect of source-sink manipulations on the Crassulacean acid metabolism of *Kalanchoë pinnata*. *J. Exp. Bot* **42:** 1123–1129.

Millar, A.H., Wiskich, J.T., Whelan, J. and Day, D.A. (1993) Organic acid activation of the alternative oxidase of plant mitochondria. *FEBS Lett.* **329:** 259–262.

Miszalski, Z., Slesak, I., Niewiadomska, E., Baczek, E.R., Lüttge, U. and Ratajczak, R. (1998) Subcellular localization and stress response of superoxide dismutase isoforms from leaves in the C_3 CAM intermediate *Mesembryanthemum crystallinum* L. *Plant Cell Environ.* **21:** 169–179.

Nimmo, G.A., Nimmo, H.G., Hamilton, I.D., Fewson, C.A. and Wilkins, M.B. (1986) Purification of the phosphorylated night form and dephosphorylated day form of phosphoenolpyruvate carboxylase from *Bryophyllum fedtschenkoi*. *Biochem. J.* **239:** 213–220.

Nobel, P.S. (1988) *Environmental Biology of Agaves and Cacti*. Cambridge University Press, Cambridge.

O'Leary, M.H., and Osmond, C.B. (1980) Diffusional contribution to carbon isotope fractionation during dark CO_2 fixation in CAM plants. *Plant Physiol.* **66:** 931–934.

Osmond, C.B. (1978) Crassulacean acid metabolism: a curiosity in context. *Annu. Rev. Plant Physiol.* **29:** 379–414.

Osmond, C.B., Nott, D.L. and Firth, P.M. (1979) Carbon assimilation patterns and growth of the introduced CAM plant *Opuntia inermis* in Eastern Australia. *Oecologia* **40:** 331–350.

Osmond, C.B., Holtum, J.A.M., O'Leary, M.H., Roeske, C., Wong, O.C., Summons, R.E. and Avadhani, P.N. (1988) Regulation of malic acid metabolism in CAM plants in the dark and light: in-vivo evidence from ^{13}C-labeling patterns after $^{13}CO_2$ fixation. *Planta* **175:** 184–192.

Osmond, C.B., Popp, M. and Robinson, S.A. (1996) Stoichiometric nightmares: studies of O_2 and CO_2 exchanges in CAM plants. In: *Crassulacean Acid Metabolism: Biochemistry Ecophysiology and Evolution* (eds K. Winter and J.A.C. Smith). Ecological Studies, Vol. 114. Springer, Berlin, pp. 19–30.

Popp, M., Kramer, D., Lee, H., Diaz, M., Ziegler, H. and Lüttge, U. (1987) Crassulacean acid metabolism in tropical dicotyledonous trees of the genus *Clusia*. *Trees* **1:** 238–247.

Ritz, D., Kluge, M. and Vieth, H.J. (1986) Mass-spectrometric evidence for the double carboxylation pathway of malic acid synthesis by Crassulacean acid metabolism plants in the light. *Planta* **167:** 384–291.

Robinson, S.A., Yakir, D., Ribas-Carbo, M., Giles, L., Osmond, C.B., Siedow, J.N. and Berry, J.A. (1992) Measurements of the engagement of cyanide-resistant respiration in the Crassulacean acid metabolism plant *Kalanchoë daigremontiana* with the use of on-line oxygen isotope discrimination. *Plant Physiol.* **100:** 1087–1091.

Robinson, S.A., Osmond, C.B. and Giles, L. (1993) Interpretations of gradients in $\delta^{13}C$ value in thick photosynthetic tissues of plants with Crassulacean acid metabolism. *Planta* **190:** 271–276.

Rustin, P. and Queiroz-Claret, C. (1985) Changes in oxidative properties of *Kalanchoë blossfeldiana* leaf mitochondria during development of Crassulacean acid metabolism. *Planta* **164:** 415–422.

Schmitt, A.K., Lee, H.S.J. and Lüttge, U. (1988) Response of the C_3-CAM tree, *Clusia rosea* to light and water stress. *J. Exp. Bot.* **39**: 1581–1590.

Seravites, J.C., Parry, M.A.J., Gutteridge, S. and Keys, A.J. (1986) Species variation in the predawn inhibition of ribulose-1,5-bisphosphate carboxylase/oxygenase. *Plant Physiol.* **82**: 1161–1163.

Smith, J.A.C. and Heuer, S. (1981) Determination of the volume of intercellular spaces in leaves and some values for CAM plants. *Ann. Bot.* **48**: 915–917.

Smith, J.A.C., Ingram, J., Tsiantis, M.S., Barkela, B.J., Bartholomew, D.M., Bettey, M., Pantoja, O. and Pennington, A.J. (1996) Transport across the vacuolar membrane in CAM plants. In: *Crassulacean Acid Metabolism: Biochemistry, Ecophysiology and Evolution* (eds K. Winter and J.A.C. Smith). Ecological Studies, Vol. 114. Springer, Berlin, pp. 53–71.

Spalding, M.H., Stumpf, K.K., Ku, M.S.B., Burris, R.H. and Edwards, G.E. (1979) Crassulacean acid metabolism and diurnal variations of internal CO_2 and O_2 concentrations in *Sedum praealtum*. *Aust. J. Plant Physiol.* **6**: 557–567.

Takeda, T. and Shigeoka, S. (1997) Responses of antioxidant system to salt stress in *Mesembryanthemum crystallinum* (abstract). *Plant Physiol.* **114** (Suppl): 121 (abstract no. 547)

Ting, I.P. (1985) Crassulacean acid metabolism. *Annu. Rev. Plant Physiol.* **36**: 595–622.

Vanlerberge, G.C. and McIntosh, L. (1997) Alternative oxidase: from gene to function. *Annu. Rev. Plant Physiol. Plant Mol. Biol.* **48**: 703–734.

Vu, J.C.V., Allen, L.H. and Bowes, G. (1984) Dark/light modulation of ribulose bisphosphate carboxylase activity in plants from different photosynthetic categories. *Plant Physiol.* **76**: 843–845.

Winter, K. and Demmig, B. (1987) Reduction state of Q and nonradiative energy dissipation during photosynthesis in leaves of a Crassulacean acid metabolism plant, *Kalanchoë daigremontiana* Hamet et Perr. *Plant Physiol.* **85**: 1000–1007.

Winter, K. and Smith, J.A.C. (eds) (1996) *Crassulacean Acid Metabolism: Biochemistry, Ecophysiology and Evolution.* Ecological Studies, Vol. 114, Springer Verlag, Berlin.

Woodrow, I.E. and Berry, J.A. (1988) Enzymatic regulation of photosynthetic CO_2 fixation in C_3 plants. *Annu. Rev. Plant Physiol. Plant Mol. Biol.* **39**: 533–594

14

Phosphoenolpyruvate carboxykinase in plants: its role and regulation

Richard C. Leegood and Robert P. Walker

1. Introduction

Phosphoenolpyruvate carboxykinase (PCK; EC 4.1.1.49) in plants is a cytosolic enzyme that catalyses the reversible reaction:

$$oxaloacetate + ATP \leftrightarrow PEP + ADP + CO_2$$

although *in vivo* it probably acts as a decarboxylase because of its low affinity for CO_2 (Ray and Black, 1976; Urbina and Avilan, 1989). The enzyme shows homology to other ATP-dependent PCKs found in yeast, *Escherichia coli* and, perhaps surprisingly, *Trypanosoma cruzii* (Finnegan and Burnell, 1995; Kim and Smith, 1994; Walker and Leegood, 1995, 1996a), but is unrelated, except at the active site, to PCKs which utilize either GTP as a co-factor, with a molecular mass of about 70 kDa, that are found in mammals, birds and insects (Reymond *et al.*, 1992). In animals, PCK is principally involved in gluconeogenesis, converting fats, amino acids *et cetera* to sugars, although it may also function in metabolism of amino acids as, for example, in glutamine utilization by lymphocytes (Newsholme *et al.*, 1985).

In plants, it is well established that PCK is involved in gluconeogenesis, converting stored fats to sugars following germination in oil-storing seeds (Leegood and ap Rees, 1978), and that it is involved in CO_2-concentrating mechanisms, such as C_4 photosynthesis, in which it decarboxylates C4 acids in one sub-group of C_4 plants (Hatch and Osmond, 1976) and CAM, in which it decarboxylates C4 acids in some CAM plants (Leegood *et al.*, 1996). It also plays a role in the CO_2-concentrating mechanism of some algae (Reiskind and Bowes, 1991). However, PCK lies at an important crossroads in metabolism, where metabolism of lipids, organic acids and amino acids can give rise to oxaloacetate and sugars. Other organic and amino acids and secondary metabolites can then be formed from PEP. In keeping with this pivotal position in plant metabolism, its presence in a wide range of plant tissues is now emerging, including some C_4 plants in which it had previously been thought to be absent, in structures

Plant Carbohydrate Biochemistry, edited by J.A. Bryant, M.M. Burrell and N.J. Kruger.
© 1999 BIOS Scientific Publishers Ltd, Oxford.

involved in plant defence, such as trichomes and oil and resin ducts, in ripening fruits, in the phloem of some plants and in developing seeds.

2. Regulation of PCK

PCK catalyses a reaction which opposes the irreversible reaction catalysed by PEP carboxylase:

$$PEP + HCO_3^- \rightarrow oxaloacetate + P_i.$$

PEP carboxylase is probably ubiquitous in the cytosol of plant cells (Chollet *et al.*, 1996; Latzko and Kelly, 1983). The presence of both these enzymes in the cytosol means that they require strict regulation in order to avoid a futile cycle which leads to the hydrolysis of ATP. Recent evidence suggests that, in many plant tissues, both these enzymes are regulated by phosphorylation (Chollet *et al.*, 1996; Walker and Leegood, 1996b).

PCK has been purified from a number of plant tissues, including C_3, C_4 and CAM plants. One of these purifications (of a 62 kDa polypeptide from cucumber) resulted in the raising of antibodies which recognized a 74 kDa polypeptide in plant extracts in which proteolysis was prevented (Walker *et al.*, 1995). In addition, the gene sequence from cucumber predicted a polypeptide with a molecular mass of 74 kDa (Kim and Smith, 1994). Rapid proteolysis of PCK in crude extracts of plant tissues means that it is difficult to purify PCK in an intact form (Walker and Leegood, 1996b; Walker *et al.*, 1995, 1997). There are other examples of enzymes in plants, such as ADPglucose pyrophosphorylase and PEP carboxylase, in which proteolytic cleavage readily occurs following extraction. Although cleavage does not always affect V_{max}, it may affect regulatory properties such as sensitivity to effectors (e.g. malate sensitivity in the case of PEP carboxylase). Like these other enzymes, proteolysis of PCK does not appear to affect its V_{max} (Walker and Leegood, 1996b; Walker *et al.*, 1995).

The molecular mass of PCK varies between different plants. In gluconeogenic seedlings and most CAM plants the molecular mass is 74 kDa, but in some CAM plants (one group of *Tillandsia* spp.) the molecular mass is 78 kDa (Walker and Leegood, 1996b). In C_4 plants, PCK is often smaller and the molecular mass ranges between 68 and 74 kDa (Walker and Leegood, 1996b; Walker *et al.*, 1997). For PCK from both cucumber and *Urochloa panicoides*, the difference in size between the intact and proteolytically cleaved polypeptides is similar to the size of the N-terminal extension deduced from the cDNA sequence, and sequencing of the truncated PCK confirmed that it was cleaved at the N-terminus (Finnegan and Burnell, 1995). This observation suggests that the differences in molecular mass between PCK from different plants reflect the size of the N-terminal extension, which has presumably been modified in some plants to suit its particular role, and which is readily cleaved upon extraction of the enzyme.

Inspection of the sequence of cucumber PCK reveals two potential phosphorylation sites in the N-terminal region, one of which is a motif recognized by both cAMP-dependent protein kinases and PEP carboxylase kinase while the other is a motif recognized by SNF-1-related protein kinases. PCK is phosphorylated in gluconeogenic seedlings, such as rape, sunflower and cucumber, and in leaves of PCK-type CAM plants, such as *Tillandsia* spp. (Walker and Leegood, 1996b). Incubation of the purified intact enzyme from cucumber with mammalian cAMP-dependent protein kinase, or PEP carboxylase kinase, and [γ-^{32}P]ATP led to incorporation of ^{32}P into a part of the polypeptide which was cleaved during proteolysis and which was separate from the active site, indicating that PCK may be phosphorylated in the N-terminal region. This phosphorylation was reversed by incubation with protein phosphatase 2A. When cucumber cotyledons were supplied with $^{32}P_i$, PCK was one of five major labelled polypeptides in darkened cotyledons and this labelling was reversed by illumination (Walker and Leegood, 1995). In CAM plants, as in cucumber cotyledons, phosphorylation of PCK occurred at night and dephosphorylation occurred during the day, although phosphorylation and dephosphorylation may be controlled by a circadian rhythm, rather than by light as such (Walker and Leegood, 1996b).

In many C_4 grasses PCK is not phosphorylated. The molecular mass of PCK varies much more in C_4 plants than in C_3 and CAM plants, with a molecular mass of between 68 and 71 kDa, except in maize (Walker and Leegood, 1996b). PCK is phosphorylated during darkness in some C_4 plants with PCK of a larger molecular mass, such as *Panicum maximum* (71 kDa), but it is not phosphorylated in the PCK-type C_4 plant, *U. panicoides* (68 kDa) (Walker and Leegood, 1996b; Walker *et al.*, 1997). The sequence of PCK in *U. panicoides* lacks both putative phosphorylation sites present in cucumber (Walker and Leegood, 1996a).

The reason why PCK is not phosphorylated in these C_4 plants may be a consequence of the location of PEP carboxylase and PCK in different photosynthetic cell types, the mesophyll and bundle sheath, respectively, whereas in other plants both enzymes are located in the cytosol of the same cells. A futile cycle with PCK and PEP carboxylase is, therefore, avoided by compartmentation. Another feature is that changes in leaf metabolite concentrations between light and dark in C_4 plants are much greater than during C_3 photosynthesis and PCK may be responsive to one of these metabolites (see Walker *et al.*, 1997).

The significance of phosphorylation of PCK remains elusive. The occurrence of phosphorylation in diverse tissues makes it seem unlikely that it regulates protein turnover, for example, although PCK is inducible in CAM plants (Borland *et al.*, 1998) and in a number of other tissues (see below). There is also no change in the amount or activity of PCK between day and night in C_4 and CAM plants. By analogy with other enzymes, it seems more likely that the N-terminal extension confers regulatory properties on PCK and that these properties are modulated by phosphorylation. Certainly, CAM plants require that PCK be switched off in the dark to avoid futile cycles with PEP carboxylase and there is evidence that in gluconeogenic seedlings, gluconeogenesis is stimulated by light (Davies *et al.*, 1981) and that PCK in darkened marrow cotyledons has an unexpectedly high control coefficient, with C^l_{PEPCK} between 0.7 and 1.0 (Trevanion *et al.*, 1995), suggesting that it might be down-regulated in the dark. A number of other enzymes have regulatory properties conferred on them by phosphorylation of a residue contained within a sequence of amino acids that is absent from enzymes with otherwise very similar sequences. For example, PEP carboxylase from plants has regulatory properties conferred on it by phosphorylation/dephosphorylation of a serine residue contained in a sequence of amino acids at the N-terminus, which are lacking in the bacterial enzyme (Huber *et al.*, 1994). However, we have not yet been able to show any differential effects of metabolites on the kinetic properties of PCK isolated from illuminated or darkened cucumber cotyledons.

Of the two potential phosphorylation sites involving phosphorylation of serine or threonine residues mentioned above, the residues surrounding the first (residues 42–51; ICHDDpSPM) (Kim and Smith, 1994) form a consensus sequence for recognition by the SNF-1-related protein kinases in higher plants (Halford and Hardie, 1998). These are thought to be global regulators of carbon metabolism in plants, being implicated in the regulation of sucrose phosphate synthase, NR and 3-hydroxy-3-methylglutaryl coenzyme A (HMG-CoA) reductase, which catalyses a key step in the synthesis of iso-prenoids (Halford and Hardie, 1998). The other potential phosphorylation site (residues 64–69, KKRSTP) has homology to 14-3-3 binding sites (Muslin *et al.*, 1996). 14-3-3 proteins are involved in protein–protein interactions and participate in various kinase-related events in signal transduction pathways. They are also implicated in the regulation of plant enzymes such as NR (Huber *et al.*, 1996) and sucrose phosphate syn-thase (see Chapter 5). This raises the possibility of multi-site phosphorylation of PCK and complexities of regulation which have yet to be explored.

3. The distribution and roles of PCK in plants

Apart from its presence in C_4 and CAM plants and in gluconeogenic seedlings, PCK has been reported to be present in various other plant tissues, such as cauliflower flo-rets (Mazelis and Vennesland, 1957), turnip tap root (Cooper and Benedict, 1968) and in various fruits, such as grape (Ruffner and Kliewer, 1975), apple, kiwi fruit and aubergine (Blanke *et al.*, 1988). However, PCK activity is difficult to measure in many tissues because of its low activity and, in tissues such as fruits, the high concentrations of interfering substances, such as organic acids and phenolics. Confirmation of the presence of PCK is best made by immunoblots as well as activity measurements. For example, studies of cucumber leaves and roots have revealed that these tissues contain very low, but detectable amounts of PCK (Kim and Smith, 1994). However, PCK is not uniformly distributed in these tissues but is confined to specific cell types, such as the trichomes and the phloem.

3.1 C_4 plants

C_4 plants are traditionally classified into three groups according to the major decar-boxylase operating during photosynthesis (NADP-ME, NAD-ME and PCK/NAD-ME) (Hatch and Osmond, 1976). However, Gutierrez *et al.* (1974) reported low activities of PCK in a number of NADP-ME-type plants. A survey of a wide range of C_4 species for the presence of PCK using immunoblots showed that a 74 kDa form of PCK is present in the leaves of a number of NADP-ME plants, including maize and digit grass (*Digitaria sanguinalis*). However, PCK was not detectable in the leaves of certain other NADP-ME-type plants, such as sorghum or sugar cane or in NAD-ME-type plants (Walker *et al.*, 1997). PCK was exclusively located in the bundle-sheath cells in maize and the activity of PCK was about half that in leaves of *P. maximum*, a PCK-type C_4 plant (Walker *et al.*, 1997). PCK from maize was not phosphorylated *in vivo* (Walker *et al.*, 1997). Since it is present in the bundle sheath, PCK may play an auxiliary role in C_4 acid decarboxylation, perhaps in the decarboxylation of aspartate (Chapman and Hatch, 1981).

3.2 *Is PCK involved in the synthesis of secondary products?*

PCK is found in the trichomes of tobacco and cucumber leaves, which are known to synthesize a variety of antimicrobial secondary metabolites. It is also present in other tissues involved in plant defence, such as the resin ducts of *Clusia* spp. (Borland *et al.*, 1998) and the oil ducts in celery stems (data not shown). PCK is enriched in the pigment-containing cells of the skin of grape berries, suggesting that it might be involved in the production of pigments or secondary metabolites (Z.H. Chen *et al.*, unpublished data). PCK could act as a source of PEP for synthesis of these secondary products. Many secondary products are phenolics which derive from PEP and erythrose-4-phosphate via the shikimate pathway or terpenoids which are derived from PEP via acetyl CoA. Alternatively, it is possible that tissues such as trichomes have a carbohydrate supply which is limited by the low concentration of sugars in the epidermis and that the main carbon supply is organic or amino acids, in which case PCK would need to act in an anaplerotic capacity, as discussed in Section 3.6.

3.3 *PCK in fruits*

Recent results have shown that PCK is present in a wide range of fruits (*Table 1*) (Z.H. Chen *et al.*, unpublished data). In the climacteric fruits, tomato and blueberry, PCK was only present during ripening, and in tomato was distributed throughout the parenchyma of the pericarp, which suggests that it may play a role in the ripening process. In the non-climacteric fruits, grape and cherry, PCK was present throughout development and, in grape, PCK was present in several tissues, including the phloem, suggesting that it may have more than one function. In grape, its distribution in the pericarp was not homogeneous. In fruits, it seems likely that PCK plays a role in the dissimilation of organic acids during ripening. Many unripe fruit contain large amounts of organic acids, such as citrate in tomatoes (Hobson and Davies, 1971) and malate and tartrate in grapes (Kanellis and Roubelakis-Angelakis, 1993) which make them unpalatable. During ripening the organic acid content falls. This decline in organic acids, such as malate, may be brought about by either NADP-ME, NAD-ME or by PCK (Knee and Finger, 1992; Ruffner, 1982) followed by entry into the Krebs

Table 1. *The activity of PCK in plant tissues. Values are means symbol ± SE of three different extractions.*

Tissue	PCK activity (U g^{-1} fwt)
Leaves	
Panicum maximum (C$_4$)	8.11 ± 0.75
Cucumber (C$_3$)	0.14 ± 0.11
Barley (C$_3$)	not detectable
Tobacco (C$_3$)	0.03 ± 0.01
Germinated cucumber (cotyledon)	6.90 ± 0.82
Fruits	
Cherry flesh	0.35 ± 0.03
Tomato pericarp	0.07 ± 0.01
Seeds	
Developing pea	0.31 ± 0.06
Developing grape	0.83 ± 0.09

cycle. Early in grape berry development the parenchyma cells of the pericarp all possess crystalline inclusions (probably organic acids, such as malate, which are abundant in leaves and developing fruits) which progressively disappear as the berry develops (except below the skin). PCK and NADP-ME (R.P. Walker et al., unpublished data) appear in the parenchyma cells as the crystals disappear. However, the relative importance of these enzymes in the processes of fruit ripening are poorly understood.

If PCK is involved in the dissimilation of organic acids then carbon could either enter the Krebs cycle or be used in sugar synthesis. The latter view is supported by the observation that sucrose phosphate synthase, an enzyme essential for the synthesis of sucrose from hexose phosphates, increases during the ripening of both grape (Hawker, 1969) and tomato (Dali et al., 1992). Whether or not gluconeogenesis from organic acids occurs in ripening fruit has been investigated by feeding radiolabelled malate to slices of fruit. Evidence for gluconeogenesis has been obtained in ripening tomatoes (Farineau and Laval-Martin, 1977; Halinska and Frenkel, 1991; Paz et al., 1982), grapes (Drawert and Steffan, 1966; Hardy, 1968; Ruffner et al., 1975), in citrus and mango (Seymour et al., 1993) and in cherry. Incubation of ripe cherry fruit with [U-^{14}C]malate led to substantial incorporation of label into the neutral fraction, which comprises sugars (Table 2).

3.4 PCK in the phloem

PCK is frequently associated with the phloem. In grape berries and seeds, PCK is present in the phloem at all stages of development (Z.H. Chen et al., unpublished data). PCK is also present in the phloem of cucumber, Coleus blumei and Clusia minor (Borland et al., 1998; Leegood et al., 1999) which, like grape, load assimilates into the phloem symplastically. On the other hand, PCK was not located in the phloem of tomatoes, in common with the leaves of barley, maize and P. maximum and the stems of peas (data not shown). All of these plants are apoplastic loaders (van Bel and Gamalei, 1992). The reason for this apparent difference between symplastic and apoplastic loaders is at present unclear.

Table 2. *Metabolism of [U-^{14}C]malate by ripe cherry fruit*

Fraction	% metabolized ^{14}C	
	Control	+ 2 mM 3-mercaptopicolinic acid
Neutrals	6.4 ± 1	3.6 ± 0.8
Basics	19.3 ± 1.7	23.7 ± 0.6
Acidics	33.5 ± 5.2	30.1 ± 2.1
Insolubles	4.7 ± 0.2	4.1 ± 0.5
CO_2	36.1 ± 5.1	38.5 ± 2.2

Ripe cherries (Turkish var. Napoleon), obtained from a local supermarket, were supplied with 5 µCi [U-^{14}C]malate (Amersham, 50 mCi mmol^{-1}) in 50 µl water placed on the top of the fruit after removal of the stalk. The fruit were placed at 25°C in the dark for 4 h. Respired CO_2 was collected in KOH and the fruit extracted in boiling ethanol, then fractionated according to Leegood and ap Rees, 1978. Unmetabolized malate was removed by converting it to aspartate and removing it on ion-exchange resin. 3-Mercaptopicolinic acid, a specific inhibitor of PCK, was added to another set of samples. Values are means ± SE of three separate experiments.

In cucumber, as in other cucurbits, the situation is particularly complex because there are three types of phloem. The vascular bundles are bicollateral, with inner and outer phloem elements (Esau, 1965). In the minor veins, the inner abaxial phloem elements have symplastic connections while the outer adaxial phloem elements have apoplastic connections (Schaffer et al., 1996; Turgeon and Hepler, 1989). There is also a third type of phloem, a network of extra-fascicular phloem elements (Esau, 1965) in the leaves, petioles and stems which lie outside the vascular bundles which communicates with the vascular bundle. It has recently been shown that the abaxial phloem is involved in the synthesis and transport of sucrose and oligosaccharides (Haritatos et al., 1996), a conclusion which is supported by both histochemistry (Pristupa, 1983) and autoradiography (Schmitz et al., 1987). However, the adaxial phloem in minor veins does not appear to transport carbohydrates (Schmitz et al., 1987) and its function remains a matter for speculation. There are several features which suggest that it may be involved in the transport of amino acids. In melon (Cucumis melo), the amino acid composition of the phloem sap is markedly different from the leaf sap at certain times of the day (Mitchell et al., 1992). This contrasts with some other plants studied which have a phloem sap which is similar in composition to the mesophyll sap, as in spinach, barley and sugar beet (Lohaus et al., 1994; Riens et al., 1991; Winter et al., 1992). If the amino acid composition of the phloem sap is markedly different from the leaf sap, this requires the selective transport of amino acids, which could not occur across the symplastic boundary of the abaxial phloem. It also requires that the necessary metabolism of amino acids, such as the conversion of aspartate to glutamine, does not occur in the bulk of the leaf, but must occur within the phloem elements themselves, or in other types of cell associated with the phloem, such as companion cells or parenchyma cells, before apoplastic loading into the sieve elements. Amino acid transporters associated with the plasma membrane are known to be present in the plasma membranes of cucurbits (Hsiang and Bush, 1992). In addition, cucurbits possess a number of compounds which are specifically involved in transport, such as arginine and citrulline, which derive from glutamate and glutamine. It seems likely that, rather than being made throughout the mesophyll, these compounds may be made in or around the vasculature in cells adjacent to the site of export from the leaf. Immunolocalization of PCK in cucumber leaves and stems shows that it is associated with the adaxial phloem of minor veins, in the internal phloem of intermediate veins and throughout the extra-fascicular phloem elements (Leegood et al., 1999). Amino acids, such as glutamate and aspartate, have also been immunolocalized in leaves and petioles, showing that they are enriched in the adaxial and extra-fascicular phloem. Lastly, PCK is induced in nitrogen-starved cucumber seedlings, by compounds such as ammonia, glutamine and aspartate, suggesting that it is involved in their metabolism.

3.5 PCK in developing seeds

As well as being present following the germination of oil-storing seeds, PCK is present in developing seeds of several plants, including tomato, Iris, pea and grape (Table 1). In grape seeds, the amount of PCK is developmentally regulated and its maximal amount and activity coincides with the highest abundance of enzymes that are involved in the deposition of storage reserves such as lipids and seed storage proteins (Z.H. Chen et al., unpublished data).

Figure 1 shows the structure of a grape seed about 30 days after anthesis. Photosynthates enter the developing seed through dorsal and ventral veins. The latter terminates in the chalaza. The testa (seed coat) comprises an outer integument of three layers and an inner integument that surrounds an endosperm in which the embryo is enclosed. The inner layer of the outer integument (a layer of palisade cells) is rich in plasmodesmatal connections and overlies a layer of transfer cells in the inner integument (data not shown). It is likely that the palisade layer functions to provide an extended symplasmic route for the distribution of assimilates around the seed coat (Thorne, 1985).

Early in seed development, PCK is associated with the palisade layer of the outer integument and the inner region of the nucellus which surrounds the developing endosperm (Z.H. Chen *et al.*, unpublished data). Later in berry development (28 days after anthesis), PCK is associated with the single or double layer of palisade cells of the outer integument and in the chalaza adjacent to the developing storage tissues. In the palisade cells, PCK is also associated with enzymes involved in amino acid metabolism and with the presence of glutamate (R.P. Walker *et al.*, unpublished data). The decline in PCK activity during development coincides with a decline in its presence in the chalaza and it seems likely that, as in other seeds, such as cereals, the chalaza is important in supplying assimilates early in seed development, but its importance then declines as the cells of the chalaza lose contact with the developing endosperm and die (Felker and Shannon, 1980; Thorne, 1985). Thirty-five days after anthesis the endosperm expands at the expense of the nucellus and becomes cellular. PCK is then associated with the outer region of the endosperm. Early on in development, incubation of grape seeds for 3 days with sucrose and various amino acids leads to a very strong induction of PCK by asparagine, weak induction by glutamine and ammonia, and no induction by aspartate, or aspartate plus ammonia (the products of asparagine metabolism) (R.P. Walker *et al.*, unpublished data). These features all point to a role for PCK in the metabolism of imported assimilates in grape berries.

PCK, together with soluble acid invertase (Weber *et al.*, 1995), is also present in cells that lined the locular cavity (Z.H. Chen *et al.*, unpublished data). It is possible that

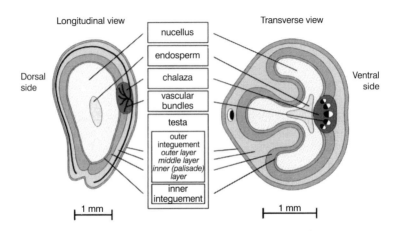

Figure 1. *The structure of a grape seed about 28 days after anthesis.*

these cells are involved in the metabolism of assimilates before delivery into the locular cavity in which the seeds are located and that, early in development, the seeds take up assimilates from the locular fluid. An analogous situation occurs in legume seeds where the inner face of the seed coat is enriched in enzymes, such as invertase (Weber *et al.*, 1995) and aminotransferases, which metabolize imported sucrose (Weber *et al.*, 1995) and amino acids (Peoples *et al.*, 1985) during passage through it. The epidermis of legume embryos is enriched in sugar transporters and takes up assimilates from the fluid-filled cavity which the seed coat encloses (Weber *et al.*, 1997). The seed coat and pods of legumes are also enriched in PCK.

3.6 *An anaplerotic function for PCK?*

The Krebs cycle not only acts to oxidize organic acids to CO_2 and reductant to power mitochondrial electron transport, but also to supply intermediates, such as isocitrate and 2-oxoglutarate for biosynthesis. This anaplerotic (Gk. αναπληρφσιζ; to fill up) function of the Krebs cycle can only be sustained if carbon from glycolysis is used to form both acetyl CoA and oxaloacetate, via PEP carboxylase, because if carbon is taken out of the cycle, the amount of oxaloacetate available to condense with acetyl CoA is diminished (*Figure 2*). However, glycolysis need not be the only source of carbon for the Krebs cycle, because carbon may also derive from organic and amino acids, which are either respired or used for biosynthesis. For example, Lohaus *et al.* (1994) have suggested that glutamate is used as a respiratory substrate in the phloem of sugar beet, because there is a net consumption of glutamate and an increase in the ratio of glutamine to glutamate from the leaves to the roots. The ATP required for phloem loading and metabolism has to be generated by the mitochondria in the companion cells of the sieve tubes and the substrates for this respiration must derive from the phloem. If glutamate were oxidized to 2-oxoglutarate which is then consumed in the

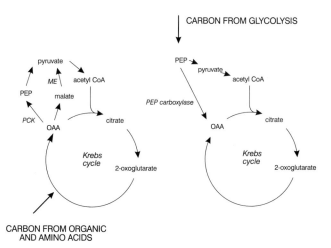

Figure 2. *The anaplerotic functions of PEP carboxylase NAD(P)-malic enzyme (ME) and PCK in relation to the Krebs cycle. Note that the Krebs cycle has been simplified to omit isocitrate, succinate, fumarate and malate as well as any co-factors and it omits the cytosolic step catalysed by NADP-isocitrate dehydrogenase (Kruse* et al., *1998).*

Krebs cycle, then either PCK or NAD(P)-ME would be required to supply PEP or pyruvate to generate acetyl CoA for the reaction catalysed by citrate synthase (*Figure 2*). This could explain why PCK is found in the phloem of some plants. Interestingly, NADP-ME is also strongly associated with the vasculature in grape (R.P. Walker *et al.*, unpublished data) and in cucumber (data not shown). On the other hand, the Krebs cycle may be involved in the interconversion of carbon skeletons of amino acids, a process that would also require the operation of PCK or NAD(P)-ME. Such inter-conversions are likely to occur in developing seeds, where the amino acids arriving in the phloem are different from those which are ultimately incorporated into storage proteins. For example, asparagine is an important transport compound in the phloem of many plants. When it arrives in seeds, it is degraded by asparaginase (Grant and Bevan, 1994) to aspartate and ammonia. Subsequent conversion of the aspartate to oxaloacetate requires transamination with 2-oxoglutarate, as does reassimilation of the ammonia by the combined actions of glutamine synthetase and glutamate syn-thase. This 2-oxoglutarate can only be generated by the anaplerotic route indicated in *Figure 2*. This may be why asparagine causes such a strong induction of PCK in devel-oping grape seeds. Such roles as these would implicate PCK in the co-ordination of carbon and nitrogen metabolism in plants in much the same manner as is already rec-ognized for PEP carboxylase (Vanleberghe *et al.*, 1990), which is also induced by amino acids and nitrate in some tissues (Sugiharto and Sugiyama, 1992) as are other enzymes of organic acid metabolism (Scheible *et al.*, 1997).

Acknowledgements

We thank Andrew Brooks for doing the feeding experiments reported in Table 2. This research was supported by research grants (PO 9495, CO5229 and RSP07804), and by a David Phillips Fellowship to R.P.W., from the Biotechnology and Biological Sciences Research Council, UK.

References

Blanke, M.M., Hucklesby, D.P. and Notton, B.A. (1988) Phosphoenolpyruvate carboxykinase in aubergine, kiwi and apple fruit. *Gartenbauwissenschaft* **53**: 65–70.

Borland, A., Tecsi, L.I., Leegood, R.C. and Walker, R.P. (1998) Inducibility of Crassulacean acid metabolism (CAM) in *Clusia*; physiological/biochemical characterisation and intercellu-lar localisation of carboxylation processes in three species which exhibit different degrees of CAM. *Planta* **205**: 342–351.

Chapman, K.S.R. and Hatch, M.D. (1981) Aspartate decarboxylation in bundle sheath cells of *Zea mays* and its possible contribution to C_4 photosynthesis. *Aust. J. Plant Physiol.* **8**: 237–248.

Chollet, R., Vidal, J. and O'Leary, M.H. (1996) Phospho*enol*pyruvate carboxylase: a ubiquitous, highly regulated enzyme in plants. *Annu. Rev. Plant Physiol. Plant Mol. Biol.* **47**: 273–298.

Cooper, T.G. and Benedict, C.R. (1968) PEP carboxykinase exchange reaction in photosyn-thetic bacteria. *Plant Physiol.* **43**: 788–792.

Dali, N., Michaud, D. and Yelle, S. (1992) Evidence for the involvement of sucrose phosphate synthase in the pathway of sugar accumulation in sucrose-accumulating tomato fruits. *Plant Physiol.* **99**: 434–438.

Davies, H.V., Gaba, V., Black, M. and Chapman, J.M. (1981) The control of food mobilisation in seeds of *Cucumis sativus* L. V. The effect of light on lipid degradation. *Planta* **152**: 70–73.

Drawert, F. and Steffan, H. (1966) Biochemisch-physiologische untersuchungen an traubenbeeren. III. Stoffwechsel von zugeführten ^{14}C-verbindungen und die bedeutung des zucker-säure-metabolismus für die reifung von traubenbeeren. *Vitis* **5**: 377–384.

Esau, K. (1965) *Plant Anatomy*. Wiley, New York, p. 271.

Farineau, J. and Laval-Martin, D. (1977) Light *versus* dark metabolism in cherry tomato fruits. II. Relationship between malate metabolism and photosynthetic activity. *Plant Physiol.* **60**: 877–880.

Felker, F.C. and Shannon, J.C. (1980) Movement of ^{14}C-labelled assimilates into kernels of *Zea mays* L. III. An anatomical examination and microautoradiographic study of assimilate transfer. *Plant Physiol.* **65**: 864–870.

Finnegan, P.M. and Burnell, J.N. (1995) Isolation and sequence analysis of cDNAs encoding phosphoenolpyruvate carboxykinase from the PCK-type C_4 grass *Urochloa panicoides*. *Plant Mol. Biol.* **27**: 365–376.

Grant, M. and Bevan, M.W. (1994) Asparaginase gene expression is regulated in a complex spatial and temporal pattern in nitrogen-sink tissues. *Plant J.* **5**: 695–704.

Gutierrez, M., Gracen, V.E. and Edwards, G.E. (1974). Biochemical and cytological relationships in C_4 plants. *Planta* **119**: 279–300.

Halford, N.G. and Hardie, D.G. (1998) SNF1-related protein kinases: global regulators of carbon metabolism in plants? *Plant Mol. Biol.* **37**: 735–748.

Halinska, A. and Frenkel, C. (1991) Acetaldehyde stimulation of net gluconeogenic carbon movement from applied malic acid in tomato fruit pericarp tissue. *Plant Physiol.* **95**: 954–960.

Hardy, P.J. (1968) Metabolism of sugars and organic acids in imature grape berries. *Plant Physiol.* **43**: 224–228

Haritatos, E., Keller, F. and Turgeon, R. (1996) Raffinose oligosaccharide concentrations measured in individual cell and tissue types in *Cucumis melo* L. leaves: implications for phloem loading. *Planta* **198**: 614–622.

Hatch, M.D. and Osmond, C.B. (1976) Compartmentation and transport in C_4 photosynthesis. In: *Transport in Plants III, Encylopedia of Plant Physiology*, Vol. 3 (eds C.R. Stocking and U. Heber). Springer-Verlag, Berlin, pp. 144–184.

Hawker, J.S. (1969) Changes in the activities of enzymes involved in sugar metabolism during the development of grape berries. *Phytochemistry* **8**: 9–17.

Hobson, G.E. and Davies, J.N. (1971) The tomato. In: *The Biochemistry of Fruits and their Products* (ed. A.C. Hulme). Academic Press, London, pp. 437–482.

Hsiang, B.C.H. and Bush, D.R. (1992). Stachyose, amino acid, and sucrose transport in plasma membrane vesicles isolated from zucchini. *Plant Physiol.* **99**: 242S.

Huber, S.C., Huber, J.L. and McMichael, R.W. (1994) Control of plant enzyme activity by reversible protein phosphorylation. *Int. Rev. Cytol.* **149**: 47–98.

Huber, S.C., Bachmann, M. and Huber, J.L. (1996) Post-translational regulation of nitrate reductase activity: a role for Ca^{2+} and 14-3-3 proteins. *Trends Plant Sci.* **1**: 432–438.

Kanellis, A.K. and Roubelakis-Angelakis, K.A. (1993) Grape. In: *Biochemistry of Fruit Ripening* (eds G. Seymour, J. Taylor and G. Tucker). Chapman and Hall, London, pp. 189–234.

Kim, D.-J. and Smith, S.M. (1994) Molecular cloning of cucumber phospho*enol*pyruvate carboxykinase and developmental regulation of gene expression. *Plant Mol. Biol.* **26**: 423–434.

Knee, M. and Finger, F.L. (1992) NADP$^+$-malic enzyme and organic acid levels in developing tomato fruits. *J. Am. Soc. Hort. Sci.* **117**: 799–801.

Kruse, A., Fieuw, S., Heineke, D. and Müller-Röber, B. (1998) Antisense inhibition of cytosolic NADP-dependent isocitrate dehydrogenase in transgenic potato plants. *Planta* **205**: 82–91.

Latzko, E. and Kelly, G.J. (1983) The many-faceted function of phosphoenolpyruvate carboxylase in C_3 plants. *Physiol. Végétale* **21**: 805–813.

Leegood, R.C. and ap Rees, T. (1978) Phosphoenolpyruvate carboxykinase and gluconeogenesis in cotyledons of *Cucurbita pepo*. *Biochim. Biophys. Acta* **524**: 207–218.

Leegood, R.C., von Caemmerer, S. and Osmond, C.B. (1996) Metabolite transport and photosynthetic regulation in C_4 and CAM plants. In: *Plant Metabolism* (eds D.T. Dennis, D.H. Turpin, D.D. Layzell and D.K. Lefebvre). Longman, London, pp. 341–369.

Leegood, R.C., Acheson, R.M., Técsi, L.I. and Walker, R.P. (1999) The many-faceted function of phosphoenolpyruvate carboxykinase in plants. In: *Regulation of Primary Metabolic Pathways in Plants* (eds N.J. Kruger, S.A. Hill and R.G. Ratcliffe). Kluwer Academic Publishers, Dordrecht, pp. 37–51.

Lohaus, G., Burba, M. and Heldt, H.W. (1994) Comparison of the contents of sucrose and amino acids in the leaves, phloem sap and taproots of high and low sugar-producing hybrids of sugar beet (*Beta vulgaris* L.). *J. Exp. Bot.* **45**: 1097–1101.

Mazelis, M. and Vennesland, B. (1957) Carbon dioxide fixation into oxaloacetate in higher plants. *Plant Physiol.* **32**: 591–600.

Mitchell, D.E., Gadus, M.V. and Madore, M.A. (1992). Patterns of assimilate production and translocation in muskmelon (*Cucumis melo* L.). *Plant Physiol.* **99**: 959–965.

Muslin, A.J., Tanner, W.J., Allen, P.M. and Shaw, A.S. (1996) Interaction of 14-3-3 with signaling proteins is mediated by the recognition of phosphoserine. *Cell* **84**: 889–897.

Newsholme, E.A., Crabtree, B. and Ardawi, M.S.M. (1985) Glutamine metabolism in lymphocytes: its biochemical, physiological and clinical importance. *Q. J. Exp. Physiol.* **70**: 473–489.

Paz, O., James, H.W., Prevost, B.A. and Frenkel, C. (1982) Enhancement of fruit sensory quality by post-harvest application of acetaldehyde and ethanol. *J. Food Sci.* **47**: 270–274.

Peoples, M.B., Atkins, C.A., Pate, J.S. and Murray, D.R. (1985) Nitrogen nutrition and metabolic interconversions of nitrogenous solutes in developing cowpea fruits. *Plant Physiol.* **77**: 382–388.

Pratt, C. (1971) Reproductive anatomy in cultivated grapes – a review. *Am. J. Enol. Vitic.* **22**: 92–109.

Pristupa, N.A. (1983) Distribution of ketosugars among cells of conducting bundles of the *Cucurbita pepo* leaf. *Fiziol. Rastenii* **30**: 492–498.

Ray, T.B. and Black, C.C. Jr (1976) Characterization of phosphoenolpyruvate carboxykinase from *Panicum maximum*. *Plant Physiol.* **58**: 603–607.

Reiskind, J.B. and Bowes, G. (1991) The role of phosphoenolpyruvate carboxykinase in a marine macroalga with C-4-like photosynthetic characteristics. *Proc. Natl Acad. Sci. USA* **88**: 2883–2887.

Reymond, P., Geourjon, C., Roux, B., Durand, D. and Feure, M. (1992) Sequence of the phosphoenolpyruvate carboxykinase-encoding cDNA from the rumen anaerobic fungus *Neocallimastix frontalis*: comparison of the amino acid sequence with animals and yeast. *Gene* **110**: 57–63.

Riens, B., Lohaus, G., Heineke, D. and Heldt, H.W. (1991) Amino acid and sucrose content determined in the cytosolic, chloroplastic and vacuolar compartments and in the phloem sap of spinach leaves. *Plant Physiol.* **97**: 227–233.

Ruffner, H.P. (1982) Metabolism of tartaric and malic acids in *Vitis*: A review – Part B. *Vitis* **21**: 346–358.

Ruffner, H.P. and Kliewer, W.M. (1975) Phosphoenolpyruvate carboxykinase activity in grape berries. *Plant Physiol.* **56**: 67–71.

Ruffner, H.P., Koblet, W. and Rast, D. (1975) Gluconeogenese in Reifenden Beeren von *Vitis vinifera*. *Vitis* **13**: 310–328.

Schaffer, A.A., Pharr, D.M. and Madore, M. (1996) Cucurbits. In: *Photoassimilate Distribution in Plants and Crops. Source–Sink Relationships* (eds E. Zamski and A.A. Schaffer). Marcel Dekker, New York, pp. 729–757.

Scheible, W.-R., González-Fontes, A., Lauerer, M., Müller-Röber, B., Caboche, M. and Stitt, M. (1997) Nitrate acts as a signal to induce organic acid metabolism and repress starch mobilisation in tobacco. *Plant Cell* **9**: 783–798.

Schmitz, K., Cuypers, B. and Moll, M. (1987) Pathway of assimilate transfer between mesophyll cells and minor veins in leaves of *Cucumis melo* L. *Planta* **171**: 19–29.

Seymour, G.B., Taylor, J.E. and Tucker, G.A. (eds) (1993) *Biochemistry of Fruit Ripening*. Chapman and Hall, London.

Sugiharto, B. and Sugiyama, T. (1992) Effects of nitrate and ammonium on gene expression of PEP carboxylase and nitrogen metabolism in maize leaf tissue during recovery from nitrogen stress. *Plant Physiol.* **98**: 1403–1408.

Thorne, J.H. (1985) Phloem unloading of C and N assimilates in developing seeds. *Annu. Rev. Plant Physiol.* **36**: 317–343.

Trevanion, S.J., Brooks, A.L. and Leegood, R.C. (1995) Control of gluconeogenesis by phosphoenolpyruvate carboxykinase in cotyledons of *Cucurbita pepo* L. *Planta* **196**: 653–658.

Turgeon, R. and Hepler, P.K. (1989) Symplastic continuity between mesophyll and companion cells in minor veins of mature *Cucurbita pepo* L. leaves. *Planta* **179**: 24–31.

Urbina, J.A. and Avilan, L. (1989) The kinetic mechanism of phosphoenolpyruvate carboxykinase from *Panicum maximum*. *Phytochemistry* **28**: 1349–1353.

van Bel, A.J.E. and Gamalei, Y.V. (1992) Ecophysiology of phloem loading in source leaves. *Plant Cell Environ.* **15**: 265–270.

Vanlerberghe, G.C., Schuller, K.A., Smith, R.G., Feil, R., Plaxton, W.C. and Turpin, D.H. (1990) Relationship between NH_4^+ assimilation rate and *in vivo* phosphoenolpyruvate carboxylase activity. Regulation of anaplerotic carbon flow in the green alga *Selenastrum minutum*. *Plant Physiol.* **94**: 284–290.

Walker, R.P. and Leegood, R.C. (1995) Purification, and phosphorylation in vivo and in vitro, of phosphoenolpyruvate carboxykinase from cucumber cotyledons. *FEBS Lett.* **362**: 70–74.

Walker, R.P. and Leegood, R.C. (1996a) Regulation of phosphoenolpyruvate carboxykinase activity in plants. In: *Current Research in Photosynthesis (Proceedings of 10th International Congress on Photosynthesis)*, Vol. 5 (ed. P. Mathis). Kluwer Academic Publishers, Dordrecht, pp. 29–34.

Walker, R.P. and Leegood, R.C. (1996b) Phosphorylation of phosphoenolpyruvate carboxykinase in plants: studies in plants with C_4 photosynthesis and Crassulacean acid metabolism and in germinating seeds. *Biochem. J.* **317**: 653–658.

Walker, R.P., Trevanion, S.J. and Leegood, R.C. (1995) Phosphoenolpyruvate carboxykinase from higher plants: purification from cucumber and evidence of rapid proteolytic cleavage in extracts from a range of plant tissues. *Planta* **195**: 58–63.

Walker, R.P., Acheson, R.M., Técsi, L.I. and Leegood, R.C. (1997) Phosphoenolpyruvate carboxykinase in C_4 plants: its role and regulation. *Aust. J. Plant Physiol.* **24**: 459–468.

Weber, H., Borisjuk, L., Heim, U., Buchner, P. and Wobus, U. (1995) Seed coat-associated invertases of Fava bean control both unloading and storage functions: cloning of cDNAs and cell type-specific expression. *Plant Cell* **7**: 1835–1846.

Weber, H., Borisjuk, L., Heim, U., Sauer, N. and Wobus, U. (1997) A role for sugar transporters during seed development: molecular characterization of a hexose and a sucrose carrier in fava bean seeds. *Plant Cell* **9**: 895–908.

Winter, H., Lohaus, G. and Heldt, H.W. (1992) Phloem transport of amino acids in relation to their cytosolic levels in barley leaves. *Plant Physiol.* **99**: 996–1004.

15

Ascorbic acid metabolism in plants

Nick Smirnoff and Glen L. Wheeler

1. Introduction

L-Ascorbic acid occurs in all metabolically active plant cells where it can comprise up to 10% of the soluble carbohydrate pool. It is found in all subcellular compartments, including the cell wall (Noctor and Foyer, 1998; Smirnoff, 1996). Despite its central role as an antioxidant, as well as other roles in metabolism, very little is known about the pathways of ascorbate metabolism or their control. Ascorbate is derived from hexoses by oxidation of C1 to an aldonic acid. This forms a relatively stable 1,4-lactone ring structure. Further oxidation at C2 forms a carbonyl group which enolizes with C3 to give rise to the enediol of ascorbate. The enediol group provides the acidic and reducing characteristics of ascorbate. The pathway by which these transformations occur in plants has remained a mystery since the first studies were carried out in the 1950s. However, a biosynthetic pathway which reconciles all the evidence has now been proposed (Wheeler et al., 1998).

The aim of this review is to discuss the ascorbate biosynthesis pathway and to consider the factors which control ascorbate pool size in plants. The use of ascorbate-deficient *Arabidopsis thaliana* mutants in studying ascorbate metabolism will be outlined. The new understanding of ascorbate metabolism which is now emerging not only fills in a major gap in our knowledge of plant carbohydrate metabolism but also has potential practical uses in increasing the nutritional quality of food plants and increasing their resistance to oxidative stress.

2. Functions of ascorbate in plants

The roles of ascorbate can be classified according to its biochemical mode of action, which can be as an antioxidant, as an electron donor/acceptor for electron transport or as a co-factor of various hydroxylase or monooxygenase reactions. This final role is important for ethylene synthesis and hydroxylation of prolyl residues to form hydroxyproline in structural proteins such as extensin (in plant cell walls) and collagen (in animal connective tissue). The symptoms of scurvy, caused by ascorbate deficiency in humans,

Plant Carbohydrate Biochemistry, edited by J.A. Bryant, M.M. Burrell and N.J. Kruger.
© 1999 BIOS Scientific Publishers Ltd, Oxford.

are caused by collagen deficiency (Davies *et al.*, 1991). While the beneficial aspects of ascorbate are usually emphasized, at high concentration ascorbate is potentially toxic because it can reduce traces of free transition metals such as Fe^{3+} and Cu^{2+}. The reduced metals then take part in the Fenton reaction in which they react with hydrogen peroxide to generate highly reactive and toxic hydroxyl radicals which can attack macromolecules (Halliwell and Gutteridge, 1989). There is evidence that high daily doses of vitamin C could cause DNA damage in humans by this mechanism (Podmore *et al.*, 1998). Exceptionally high concentrations of ascorbate occur in some fruits, for example black-currants (*Ribes* spp.), rosehips (*Rosa* spp.) and the Barbados cherry (*Malpighia glabra*) (cited in Loewus and Loewus, 1987). The significance of this is not clear; either these fruits have a large antioxidant requirement or perhaps ascorbate could act as an attrac-tant for seed dispersers.

2.1 *Antioxidant*

The antioxidant role of ascorbate is its most widely known function (Davies *et al.*, 1991; Halliwell and Guteridge, 1989; Noctor and Foyer, 1998; Smirnoff, 1996). Aerobic metabolism generates several partially reduced forms of oxygen such as superoxide radicals, hydrogen peroxide and hydroxyl radicals. Interaction of oxygen with excited pigments can excite ground state oxygen to singlet oxygen. All these oxygen species, collectively known as reactive or active oxygen species (ROS or AOS), are more reactive than ground state oxygen, and can react with DNA, proteins and membranes resulting in oxidative damage (Halliwell and Gutteridge, 1989; Smirnoff, 1995). Ascorbate can react directly with AOS thereby detoxifying them. Ascorbate is particularly important in removing hydrogen peroxide formed in vari-ous subcellular compartments and this is catalysed by ascorbate-specific peroxidase (APX). The oxidation products of ascorbate, monodehydroascorbate (MDA) radical and dehydroascorbate (DHA), are reduced back to ascorbate by MDA reductase and DHA reductase. DHA reductase uses the cysteine-containing tripeptide glutathione (GSH) as the reducing agent. Ascorbate and GSH therefore act together as antioxi-dants in the ascorbate–GSH cycle (Noctor and Foyer, 1998; Smirnoff, 1995, 1996). MDA can also be directly reduced by electrons from photosystem I and this is dis-cussed in Section 2.2 in relation to the role of ascorbate in photosynthesis. A number of environmental stresses, such as low temperature, water stress, UV-B radiation and air pollutants such as ozone, can cause damage by generating AOS and the antioxi-dant system of plants has been implicated in resistance to these (Smirnoff, 1995). Direct evidence that ascorbate is key to defence against ozone, UV-B and sulphur dioxide is provided by the *vtc1* (formerly *soz1*= sensitive to *ozone*) mutant of *A. thaliana* which is hypersensitive to all these stresses (Conklin *et al.*, 1996). The mutants contain 30% of wild-type ascorbate and grow normally until exposed to stress.

2.2 *Photosynthesis*

Measurement of ascorbate in isolated chloroplasts suggests that it can reach 25–50 mM in this compartment (Noctor and Foyer, 1998). Current evidence suggests that the last step of ascorbate synthesis is mitochondrial (see Section 4.1) therefore chloro-plasts must obtain their ascorbate from the cytosol. Chloroplasts take up ascorbate in a carrier-mediated manner but with relatively low rate and affinity (Noctor and Foyer, 1998). The ascorbate pool size in leaves is light-dependent (Section 5; Smirnoff and Pallanca, 1996). Foliar ascorbate concentration in plants grown at high altitude can be very high (Wildi and Lutz, 1996), perhaps reflecting an increased need for photoprotection under conditions of high irradiance and low temperature experienced by alpine plants. Ascorbate could play three distinct roles in chloroplasts. Firstly, it can remove hydrogen peroxide formed by oxygen photoreduction by means of the Mehler–peroxidase reaction. The resulting MDA and DHA are electron acceptors for photosystem I. Secondly, it is thought to be a co-factor for violaxanthin de-epoxidase. This enzyme, located in the intra-thylakoid space, is involved in synthesizing zeaxanthin in leaves exposed to excess excitation energy. Zeaxanthin radiates excess excitation energy in the light-harvesting complex as heat (Noctor and Foyer, 1998; Smirnoff, 1995). Thirdly, it can reduce the α-chromanoxyl radical formed by oxidation of α-tocopherol. α-Tocopherol (vitamin E) is a lipophilic antioxidant, very abundant in thylakoid membranes, which prevents lipid peroxidation and scavenges singlet oxygen produced from the interaction of oxygen and excited chlorophyll (Noctor and Foyer, 1998; Smirnoff, 1995).

It has been proposed that the Mehler–peroxidase reaction, as well as providing a high affinity mechanism of removing hydrogen peroxide (thioredoxin-regulated Calvin cycle enzymes are readily inhibited by H_2O_2), has a role in the regulation of photosynthesis (Noctor and Foyer, 1998). About 10% of electrons from photosystem I photoreduce oxygen to superoxide. This dismutes to hydrogen peroxide and the reaction is catalysed by SOD. The resulting H_2O_2 is reduced to water by ascorbate; a reaction catalysed by APX. The resulting MDA can then also act as an electron acceptor to photosystem I with regeneration of ascorbate. Thylakoid-bound isoforms of SOD and APX may facilitate this process. MDA escaping reduction here is dealt with by MDA reductase and the ascorbate–GSH cycle (Section 2.1). Overall, these reactions facilitate electron flow without CO_2 fixation and could therefore have a role in regulating the trans-thylakoid pH gradient and allowing ATP synthesis without NADPH formation. The evidence for, and significance of, the Mehler–peroxidase reaction is reviewed by Noctor and Foyer (1998).

2.3 Ascorbate in the cell wall

The possible functions of ascorbate in cell walls and in cell growth have been reviewed recently (Smirnoff, 1996). Although a number of observations support a role for ascorbate in stimulating cell expansion, or even cell division, so far no coherent hypothesis has emerged as to how this might occur. Ascorbate oxidase, a member of the blue copper oxidase family of enzymes, occurs in the cell wall. Its highest activity often coincides with the tissues undergoing the rapid expansion and there is a suggestion that the DHA/ascorbate ratio is higher in apoplastic fluid from expanding tissue. The causal relationship between this and growth is not clear, although MDA stimulates the elongation of onion roots. A low steady-state MDA concentration exists in a mixture of DHA and ascorbate (Davies et al., 1991). Alternatively, ascorbate or DHA could affect wall extensibility. Ascorbate potentially inhibits peroxidative cross-linking of cell wall polymers and lignin synthesis. Lin and Varner (1991) suggested that DHA, generated by ascorbate oxidase activity, prevents protein cross-linking in the walls by reacting with lysine and arginine residues. More recently, it has been proposed that hydroxyl radicals, generated by the interaction of ascorbate, copper ions and hydrogen peroxide in the wall could cause non-enzymatic scission of wall polysaccharides leading to wall loosening and enhanced expansion (Fry, 1998). Ascorbate crosses membranes via a carrier (Rautenkranz et al., 1994) and there is evidence that DHA, rather than ascorbate, is the species transported across the plasma membrane (Horemans et al., 1997). The plasma membrane also contains a cytochrome b system which accepts electrons from ascorbate and donates them to MDA (Horemans et al., 1994). It therefore seems likely that the redox state of wall ascorbate and ability to transport DHA across the plasma membrane has some role in cell physiology.

3. The biosynthetic pathway of ascorbate: historical perspective

Until recently the pathway of ascorbate biosynthesis in plants has been elusive. Before providing the evidence for the plant pathway an outline of the pathway in animals and some previously proposed plant pathways will be provided. The animal pathway has been known for some time (Burns, 1967). UDPglucose is oxidized at C6 to UDPglucuronic acid. UDP is cleaved and the C1 aldehyde is reduced to a hydroxymethyl function. Now, in accordance with the rules of carbohydrate nomenclature, the carboxyl group at the original C6 of the uronic acid is designated C1 and the resulting aldonic acid (L-gulonic acid) has the L-configuration. L-Gulonate exists in a stable 1,4-lactone form, L-gulono-1,4-lactone (L-GUL). L-Ascorbate is then produced by oxidation of L-GUL at C2 in a reaction catalysed by L-GUL oxidase. A number of animals, notably guinea pigs and primates (including humans), are unable to synthesize ascorbate and, in the cases studied so far, this is caused by loss of GUL oxidase activity (Nishikimi et al., 1994). The human GUL oxidase gene is heavily mutated compared with the functional version from rats and is not expressed (Nishikimi et al., 1994). An overall consequence of this pathway is that glucose labelled at C6 will produce ascorbic acid labelled at C1. Label inversion does not occur in plants, suggesting that a different type of pathway operates (Loewus, 1988).

   As a result of the first detailed investigation of ascorbate synthesis in plants, a pathway analogous to that in animals was proposed by Isherwood and Mapson (1962; Figure 1). They found that plants do not readily oxidize L-GUL although another lactone, L-galactono-1,4-lactone (L-GAL), is rapidly converted to ascorbate (Isherwood et al., 1954). They detected a mitochondrial enzyme (L-GAL dehydrogenase) which catalyses this reaction (Mapson and Breslow, 1958). However, they were not able to show how L-GAL could be synthesized. Plant extracts contained a very low affinity enzyme activity which could convert D-galacturonic acid methyl ester to L-GAL (resulting in carbon skeleton inversion) in a manner similar to the D-glucuronate to L-GUL pathway of animals (Mapson and Isherwood, 1956). Further detailed radiolabelling studies by Loewus (1963) cast doubt on the inversion route proposed by Isherwood and Mapson. Three key conclusions can be drawn from a series of detailed labelling studies carried out by Loewus using specifically labelled D-glucose as precursor (Loewus, 1963, 1988; Loewus and Loewus, 1987). Firstly, after allowing for label randomization through triose phosphate, the carbon skeleton of glucose is not inverted during ascorbate synthesis. Secondly, the hydroxymethyl group at C6 is conserved during ascorbate synthesis. Thirdly, lack of label in ascorbate synthesized from [5-³H]glucose suggests that there is an epimerization at C5 which can account for the conversion from the D to L configuration.

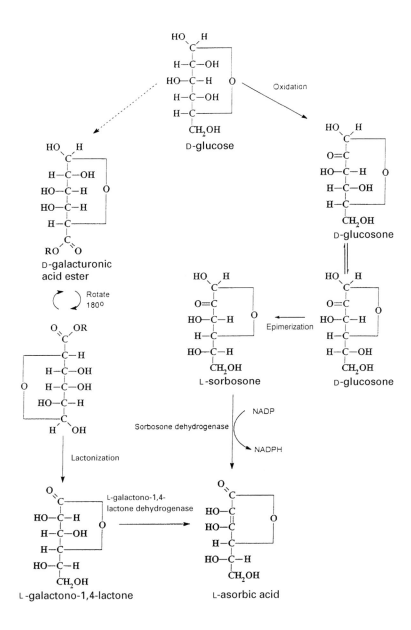

Figure 1. Outline of two previously proposed biosynthetic pathways for L-ascorbic acid in plants. Although exogenous L-galactono-1,4-lactone is rapidly converted to ascorbate, the pathway via D-galacturonate requires inversion of the original glucose carbon skeleton which is not supported by ^{14}C-labelling data. The pathway via the osones D-glucosone and L-sorbosone is supported by incorporation of ^{14}C-labelled osones into ascorbate. However, evidence that they are produced by plants is lacking. From Smirnoff (1996) The function and metabolism of ascorbic acid in plants. Annals of Botany, vol. 78, pp. 661–669. Reprinted by permission of Academic Press Inc.

Because the results of the labelling experiments could not be reconciled with the inversion pathway, an alternative pathway based on the osones as intermediates was proposed (Loewus *et al.*, 1990; Saito *et al.*, 1990). Osones are aldoses which also have a carbonyl group at C2. Some Basidiomycete fungi have pyranose-2-oxidase which forms glucosone from glucose. The oxidase generates hydrogen peroxide which is used for lignin degradation while the glucosone is converted to cortalcerone, which has antibiotic activity (Daniel *et al.*, 1994; Gabriel *et al.*, 1993). In the proposed osone pathway an aldohexose, such as glucose, is oxidized at C2 producing D-glucosone. An epimerization at C5 provides L-sorbosone which is then further oxidized at C1 to produce ascorbate (*Figure 1*). The evidence for this pathway is as follows. Firstly, [^{14}C]glucosone and [^{14}C]sorbosone, when fed to leaves, are incorporated into ascorbate and there is no carbon skeleton inversion. Secondly, isotope dilution experiments show that unlabelled glucosone and sorbosone compete with [^{14}C]glucose for incorporation into ascorbate. Thirdly, NADP-sorbosone dehydrogenase activity, able to oxidize L-sorbosone to ascorbate was partially purified from bean leaves. However, this enzyme has a very high K_m for sorbosone and a very high pH optimum. Evidence against the pathway includes lack of evidence that plants can synthesize osones and the inability of exogenous glucosone or sorbosone to increase the ascorbate pool (Conklin *et al.*, 1997).

4. The biosynthetic pathway of ascorbate

The evidence for either of the two pathways described in Section 3 is not strong. A new pathway for ascorbate synthesis in plants which reconciles the previous proposals has now been proposed (Wheeler *et al.*, 1998). The pathway is shown in *Figure 2*. In this pathway ascorbate is derived from the hexose phosphate pool by way of mannose-1-phosphate and GDPmannose. This part of the pathway is also used for synthesis of mannose containing polysaccharides, lipids and glycoproteins. GDPmannose is converted to GDP-L-galactose after which free L-galactose is released. The L-galactose is then oxidized to L-GAL by a novel L-galactose dehydrogenase. Finally, L-GAL is oxidized to ascorbate by mitochondrial L-GAL dehydrogenase. Aspects of this pathway are discussed in more detail in Sections 4.1 and 4.2.

4.1 *L-Galactose and L-GAL dehydrogenases*

Because L-GAL is an effective precursor for ascorbate (Section 3), a pathway in which it could be formed from a hexose precursor without carbon skeleton inversion would be possible. A key observation is that exogenous L-galactose is rapidly converted to ascorbate (Wheeler *et al.*, 1998; *Figure 2*). This conversion is as rapid as that previously found for L-GAL. L-GAL must be formed from L-galactose by oxidation of the C1 aldehyde and a novel enzyme catalysing this reaction, named L-galactose dehydrogenase (L-galactose:NAD 1-oxidoreductase) has been detected in pea seedlings (Wheeler *et al.*, 1998). It has a preference for NAD$^+$ over NADP$^+$ and is specific for L-galactose ($K_m = 0.4$ mM at pH 7.5), not oxidizing any other commonly occurring aldohexoses or D-arabinose. However, it does oxidize L-sorbosone with very low affinity. This reaction would produce L-ascorbate (Loewus *et al.*, 1990), which could explain why [^{14}C]sorbosone is incorporated into ascorbate (Wheeler *et al.*, 1998). L-Galactose dehydrogenase differs from G6P dehydrogenase, another enzyme which

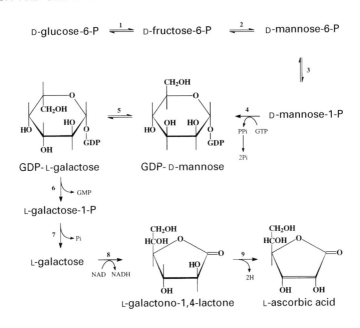

Figure 2. *A proposed biosynthetic pathway for L-ascorbic acid in plants. The enzymes (1–5) which convert D-glucose-6-phosphate to GDP-D-mannose and GDP-L-galactose have been previously identified and provide GDP-sugar precursors for polysaccharide synthesis. L-Galactose, the first dedicated intermediate, is provided by hydrolysis of GDP-L-galactose. A two step hydrolysis (steps 6 and 7) via L-galactose-1-phosphate is shown although this is speculative. L-Galactose is oxidized at C1 by L-galactose dehydrogenase (8, L-galactose:NAD⁺ oxidoreductase) forming L-galactono-1,4-lactone. This is oxidized by mitochondrial L-galactono-1,4-dehydrogenase (9) to L-ascorbic acid. Enzymes: 1, hexose phosphate isomerase; 2, phosphomannose isomerase; 3, phosphomannose mutase; 4, GDP-D-mannose pyrophosphorylase; 5, GDP-D-mannose-3,5-epimerase; 8, L-galactose dehydrogenase; 9, L-galactono-1,4-lactone dehydrogenase. Reprinted with permission from* Nature *(Vol. 393, Wheeler* et al., *The biosynthetic pathway of vitamin C in higher plants, pp. 365–369). Copyright 1998 Macmillan Magazines Limited.*

oxidizes the aldose C1, in using a non-phosphorylated substrate but is similar to bacterial D-galactose dehydrogenase (Maier and Kurtz, 1982), mammalian L-fucose dehydrogenase (Schacter *et al.*, 1969) and yeast D-arabinose dehydrogenase (Kim *et al.*, 1996). The fucose and arabinose dehydrogenases have a wider substrate specificity than plant L-galactose dehydrogenase; they can also oxidize L-galactose. However, both these enzymes have very low affinity for their substrates compared to plant L-galactose dehydrogenase. Yeast D-arabinose dehydrogenase is used for D-erythroascorbate synthesis, the fungal analogue of L-ascorbate (see Section 4.1).

L-GAL is then oxidized to L-ascorbate by L-GAL dehydrogenase (L-GALDH). L-GALDH has been purified from potato (Ôba *et al.*, 1994), sweet potato (Ôba *et al.*, 1995), spinach (Mutsuda *et al.*, 1995) and cauliflower (Østergaard *et al.*, 1997). The K_m for L-GAL varies from 0.12 to 3 mM and the pH optimum between 8.0 and 8.5 (Mapson and Breslow, 1958; Ôba *et al.*, 1994; Østergaard *et al.*, 1997). Cauliflower L-GALDH is specific for L-GAL and does not oxidize L-GUL or a range of other lactones (Østergaard *et al.*, 1997). It has a subunit molecular mass of 56 kDa (Østergaard *et al.*, 1997). In all cases it co-sediments with mitochondria (Mapson and Breslow,

1958; Mutsuda *et al.*, 1995; Ôba *et al.*, 1994, 1995; Østergaard *et al.*, 1997) and it appears to be membrane-bound (Mutsuda *et al.*, 1995). The enzyme has no oxidase activity, it is generally assayed with cytochrome *c* as electron acceptor and has been suggested to contain a flavin adenine dinucleotide (FAD) co-factor (Ôba *et al.*, 1995). However, on the basis of lack of inhibition by riboflavin and inability to detect it spectrophotometrically, Østergaard *et al.* (1997) suggest the cauliflower enzyme does not contain FAD. There is no strong evidence at present to support the view that cytochrome *c* is the immediate physiological electron acceptor. The exact location of L-GALDH in the mitochondria is not known, and this will have implications for control of ascorbate synthesis. L-Galactose dehydrogenase is probably not mitochondrial (G.L. Wheeler and N. Smirnoff, unpublished), so if L-GALDH was inside the inner membrane there would need to be a carrier for L-GAL on this membrane. Alternatively, L-GALDH may be attached to the outer surface of the inner membrane. L-GALDH is analogous to mammalian L-GUL oxidase which is also used for L-ascorbate synthesis (Chatterjee, 1970; Koshizaka *et al.*, 1988). In contrast to L-GALDH, the mammalian enzyme can donate electrons directly to oxygen (oxidase activity) or to phenazine methosulphate (PMS) (dehydrogenase activity) (Chatterjee, 1970). The mammalian enzyme contains a bound flavin (Kiuchi *et al.*, 1982). It is possible that the GUL oxidase activity results from autoxidation of the bound flavin. Absence of flavin in the plant enzyme could therefore explain its lack of oxidase activity. GUL oxidase is microsomal (endoplasmic reticulum)-bound and hydrogen peroxide is a product of its oxidase activity. In assays carried out on microsomal membranes, lipid peroxidation occurs as a consequence of the reaction (Chatterjee, 1970). The alkaloid lycorine has been suggested to inhibit ascorbate biosynthesis in plants by inhibiting L-GALDH activity (Arrigoni *et al.*, 1997). This needs further investigation since other reports show that lycorine does not inhibit L-GALDH (Østergaard *et al.*, 1997).

Yeasts (*Saccharomyces cerevisiae* and *Candida albicans*) can oxidize L-GAL to ascorbate and D-arabinono-1,4-lactone to D-erythroascorbate (Huh *et al.*, 1994; C.M. Spickett and N. Smirnoff, unpublished). These reactions are probably catalysed by the same enzyme; two independently cloned genes from yeast (GenBank accession numbers AB009401 and P54783) catalysing each of these reactions have identical amino acid sequences. Both should probably be called D-arabinono-1,4-lactone oxidase since, in the absence of exogenous precursors, yeast contains only erythroascorbate (Nick *et al.*, 1986). The plant, mammalian and fungal genes have considerable homology, being 23–28% identical in amino acid sequence. The properties reported for yeast aldono-1,4-lactone oxidase enzymes vary (Bleeg and Christensen, 1982; Huh *et al.*, 1994; Niishikimi *et al.*, 1978). There is still much to learn about the plant, mammalian and yeast aldono-1,4-lactone oxidizing enzymes and their properties are compared in *Table 1*.

4.2 L-*Galactose synthesis*

For L-galactose to be a physiological precursor for ascorbate, it must be present in plants. It is a minor constituent of cell wall polysaccharides and the suggested scheme for its synthesis is via GDPmannose, itself a well-known precursor of wall polysaccharides and mannose in glycoproteins (Herold and Lewis, 1977). GDPmannose-3,5-epimerase has been detected in *Chlorella*, the product being GDP-L-galactose (Barber,

Table 1. A comparison of the properties of aldono-1,4-lactone oxidase and dehydrogenase enzymes which produce ascorbate by oxidizing C2 of aldono-1,4-lactones in various groups of organisms. After oxidation at C2 there is an enolization between C2 and C3 to give the enediol group of L-ascorbate (in the case of L-GAL and L-GUL) or D-erythroascorbate (in the case of D-ARAL)

	L-Galactono-1,4-dehydrogenase (higher plant)[a]		L-Arabinono-1,4-lactone oxidase (yeast)[b]		L-Gulono-1,4-lactone oxidase/ dehydrogenase (mammal)[c]	
	Relative activity (%)	K_m (mM)	Relative activity (%)	K_m (mM)	Relative activity (%)	K_m (mM)
Substrate						
L-GAL	100	0.2–3.0	100	0.30	68–87	–
L-GUL	0	–	0–32	–	100	0.12
D-ARAL	–	–	97	0.16–44	–	–
Electron acceptor						
Oxygen	No		Yes		Yes	
Cytochrome c or PMS	Yes		?		Yes	
Bound flavin co-factor	None detected		Yes		Yes	
Molecular mass (kDa)	56 (native and denatured)		70–290 (native) 18–67 (denatured)		51 (denatured)	
Subcellular location	Mitochondrial		Mitochondrial		Microsomal[d]	
GenBank accession no. and amino acid identity with plant sequence (%)	Z97060 100		AB009401 and P54783 28		D12754 28	

[a] Mutsuda *et al.*, 1995; Ôba *et al.*, 1994, 1995; Ostergaard *et al.*, 1997.
[b] Bleeg and Christensen, 1982; Huh *et al.*, 1994; Nishikimi *et al.*, 1978.
[c] Chatterjee, 1970; Kiuchi *et al.*, 1982; Koshizaka *et al.*, 1988.
[d] Small proportion may be mitochondrial.
–, not determined.
L-GAL, L-galactono-1,4-lactone; L-GUL, L-gulono-1,4-lactone; D-ARAL, D-arabinono-1,4-lactone; PMS, phenazine methosulphate.

1979). The results from *Chlorella* also suggest there is an enzyme which cleaves GDP-L-galactose to release galactose-1-phosphate (Barber, 1971). Pea seedling extracts convert GDPmannose to GDP-L-galactose/L-galactose-1-phosphate and to free L-galactose (Wheeler *et al.*, 1998). This suggests that there is a GDPmannose-3,5-epimerase and enzymes which subsequently release L-galactose. The latter enzymes have not yet been characterized. GDPmannose is formed from mannose-1-phosphate and GTP by GDPmannose pyrophosphorylase (Feingold, 1982). Mannose-1-phosphate is formed from G6P via F6P and mannose-6-phosphate (Herold and Lewis, 1977). If ascorbate is synthesized from mannose, it would be predicted that [14C]mannose is effectively incorporated into ascorbate. This is indeed the case; after short labelling times, 10% of the labelled mannose is converted to ascorbate by leaves

(Wheeler *et al.*, 1998). Exogenous mannose has very little effect on the ascorbate pool size. Since exogenous mannose is readily phosphorylated (Herold and Lewis, 1977) this suggests that the reactions subsequent to mannose-6-phosphate are potentially limiting the supply of L-galactose. As noted above, L-galactose feeding rapidly elevates the ascorbate pool. Mannose is toxic at high concentration, since it sequesters inorganic phosphate as slowly metabolized mannose phosphate (Herold and Lewis, 1977), so it is possible that ascorbate synthesis could be inhibited by a limited GTP supply. A possible explanation of the ability of glucosone to act as an ascorbate precursor could be its reduction to mannose or fructose (Wheeler *et al.*, 1998). A mixture of mannose and glucose would result from reduction at C2, while fructose could be formed by reduction at C1.

5. The control of ascorbate synthesis

The first committed step for ascorbate synthesis, according to the proposed pathway, is formation of L-galactose from GDP-L-galactose. Prior to this step, GDP-D-mannose and GDP-L-galactose are used as precursors for the synthesis of mannose, L-fucose (6-deoxy-L-galactose) and L-galactose residues in cell wall polysaccharides and mannose in glycoproteins (Baydoun and Fry, 1988; Roberts, 1971). L-Fucose is synthesized from GDP-D-mannose using two enzymes (GDPmannose-4,6-dehydratase and GDP-4-keto-6-deoxy-D-mannose 3,5-epimerase-4-reductase). An *Arabidopsis* mutant deficient in cell wall fucose (*mur1*) has been isolated. The *MUR1* gene encodes GDPmannose-4,6-dehydratase (Bonin *et al.*, 1997). The mutant is not affected in ascorbate concentration (G.L. Wheeler and N. Smirnoff, unpublished), perhaps because fucose is a relatively minor wall constituent and therefore presents only a small sink for GDPmannose. The exact steps between GDP-L-galactose and L-galactose have not yet been characterized, but they are likely to have an important role in determining how rapidly carbon skeletons are diverted from the sugar nucleotide pool. There is more likely to be competition in rapidly growing tissues between cell wall and ascorbate synthesis than in mature tissue where wall synthesis is much slower. In fully expanded tissue it would seem that the major uses for mannose are ascorbate synthesis and protein glycosylation.

Neither L-galactose nor L-GAL have easily measurable pools. L-Galactose can be detected while feeding an exogenous supply but it disappears very rapidly when the supply is removed (G.L. Wheeler and N. Smirnoff, unpublished results). This implies that it is metabolized very effectively by L-galactose dehydrogenase. A compound with the same retention time as L-GAL in several HPLC systems, and which serves as a substrate for purified L-GALDH, has been detected in cauliflower extracts (Østergaard *et al.*, 1997). This suggests that low concentrations of L-GAL exist. The very rapid elevation of the ascorbate pool in tissues fed with L-galactose and L-GAL suggests that L-galactose dehydrogenase and L-GALDH capacities are high. Considering the very small pool sizes of the substrates and the relatively high K_m of both enzymes, it seems likely that this high capacity is required to allow ascorbate synthesis to continue at an appreciable rate.

The pool size of ascorbate depends on the rate of turnover as well as on biosynthesis. Ascorbate is degraded and is a precursor for oxalate and tartrate (see Section 7). Exogenous labelled ascorbate is readily metabolized (Loewus, 1988) while barley and *Arabidopsis* leaves placed in the dark lose ascorbate (Smirnoff and Pallanca, 1996). These experiments suggest that about 50% of the leaf ascorbate pool turns over in 24 h (Conklin *et al.*, 1997). Turnover could result from loss of unstable DHA, since severe oxidative stress causes loss of the total ascorbate pool (Smirnoff, 1995), or from synthesis of other compounds. In germinating pea seedlings, increasing the ascorbate pool by ascorbate feeding both decreases the rate of ascorbate synthesis and accelerates the rate of [^{14}C]ascorbate metabolism (J.E. Pallanca and N. Smirnoff, unpublished data). These results imply that ascorbate synthesis could be controlled by feedback inhibition or repression of expression of biosynthesis genes by ascorbate. Further detailed investigation will be possible now that the pathway has been identified.

It has been realized for some time that the ascorbate pool size in leaves is dependent on light (Smirnoff, 1995; Smirnoff and Pallanca, 1996). The central role of ascorbate in photoprotection and photosynthesis was described in Section 2.2. As might be predicted from this role, ascorbate concentration in leaves is higher in plants grown at high light intensity. In barley leaves the concentration of ascorbate increases from the base to the tip of the primary leaf of germinating seedlings (Smirnoff, 1995; Smirnoff and Pallanca, 1996) in parallel with photosynthetic capacity. Because cereal leaves grow from a basal meristem, photosynthetic activity increases from base to tip as the chloroplasts develop (Rogers *et al.*, 1991). It is clear that ascorbate is intimately associated with photosynthesis and photoprotection but the important question of how light influences ascorbate synthesis and turnover remains to be answered. The loss of ascorbate in barley leaves in the dark can be partially reversed by feeding exogenous sucrose or glucose (Smirnoff and Pallanca, 1996) which suggests that hexose phosphate supply is limiting in the absence of photosynthesis. Sugar supply could therefore influence ascorbate synthesis in the light. Additionally, the expression of genes encoding the ascorbate biosynthesis enzymes could be influenced via mechanisms similar to those used to sense light intensity (Karpinski *et al.*, 1997) or photosynthetic products (Koch, 1996) in other metabolic contexts.

6. Ascorbate-deficient (*vtc*) *Arabidopsis thaliana* mutants

Four *A. thaliana vtc* mutants which are deficient in ascorbate have been isolated (P.L. Conklin and R.L. Last, unpublished). These will be important aids in confirming the biosynthetic pathway and understanding the functions of ascorbate. The first of these mutants to be characterized is *vtc1* (formerly known as *soz1*). It has 30% of wild-type ascorbate (Conklin *et al.*, 1996). Further biochemical characterization of *vtc1* suggests that the low ascorbate pool is caused by impairment of its biosynthesis (Conklin *et al.*, 1997). The rate of [^{14}C]glucose incorporation into ascorbate is reduced in the mutant, while the rate of [^{14}C]ascorbate metabolism is slower. [^{14}C]Mannose is also incorporated into ascorbate more slowly in *vtc1* (G.L. Wheeler and N. Smirnoff, unpublished data). Activities of ascorbate oxidase and ascorbate–GSH cycle enzymes are unaffected in the mutant. Ascorbate peroxidase activity is lower which could be attributed to the requirement of ascorbate to stabilize the chloroplast isoform (Conklin *et al.*, 1997).

7. The products of ascorbate metabolism

Ascorbate is metabolized to a range of products; oxalate and tartrate have been identified as the major products in a number of species. These compounds are labelled in plants supplied with [^{14}C]ascorbate. The following information is summarized from the review by Loewus (1988). The carbon skeleton of (dehydro)ascorbate is cleaved at the C2/C3 bond, producing oxalic acid and L-threonic acid. L-Threonic acid is then either oxidized to L-tartaric acid or decarboxylated to provide a C3 fragment which can be recycled into carbohydrate metabolism. These reactions occur in a variety of species (e.g. *Pelargonium* and *Rumex*). In contrast, members of the Vitaceae, such as grapes (*Vitis vinifera*), synthesize L-tartrate by cleaving ascorbate at the C4/C5 bond to form tartrate plus a C2 fragment, possibly glycoaldehyde.

Enzymes which catalyse ascorbate catabolism have not yet been identified and further information would be useful since tartrate contributes to wine quality and, more importantly, binding of aluminium by oxalate has been implicated as a mechanism of aluminium tolerance in buckwheat (Ma *et al.*, 1998). However, although oxalate accumulates to high concentration in species such as spinach (*Spinacia oleracea*) and buckwheat (*Fagopyron esculentum*), ascorbate may not be the major source. Leaves can also convert glycolate to oxalate via glyoxylate at a faster rate than ascorbate (Fujii *et al.*, 1993). Both reactions are catalysed by glycolate oxidase in the peroxisomes. Lactate dehydrogenase can also catalyse oxidation of glyoxylate to oxalate, although the physiological relevance of this reaction in plants is questioned (Sugiyama and Tamaguchi, 1997). Glycolate metabolism is part of the photorespiratory cycle, thus in photorespiring leaves it is possible that glycolate, not ascorbate, is the main source of oxalate.

8. Comparative biochemistry of ascorbate

Ascorbate is not synthesized by all organisms. Those mammals which lack capacity for synthesis have lost the ability because L-GUL oxidase is not expressed (Nishikimi and Udenfriend, 1976). The human gene has accumulated many mutations compared to the rat gene (Nishikimi *et al.*, 1994). Why this gene should have become inactive in a range of unrelated species is not known; perhaps it has an unrecognized deleterious effect or perhaps these animals were able to obtain sufficient ascorbate in their diets so that selective pressure to retain the capacity for ascorbate synthesis was absent. Other interesting features are the diversity of biosynthetic pathways for L-ascorbate synthesis and the occurrence of D-erythroascorbate in fungi. D-Erythroascorbate has an identical structure to L-ascorbate from carbon atoms one to four and can presumably fulfil the same biochemical functions as L-ascorbate. Bacteria apparently lack ascorbate. All this suggests a late and separate evolutionary origin for ascorbate and its analogues in plants, animals and fungi, in each case the last step in the pathway co-opting a similar gene having aldono-1,4-lactone oxidase/dehydrogenase activity.

9. Conclusions

The difficulties, controversies and rivalries which resulted in the discovery, isolation and structure determination of ascorbic acid make interesting and illuminating reading (Davies *et al.*, 1991). The compound we now call ascorbic acid was isolated from cabbage by Albert Szent-Györgyi in 1928; his originally proposed names 'ignose' and 'Godnose' being rejected by the *Biochemical Journal*. Instead, it was inappropriately named hexuronic acid. Later it was determined to be identical to the anti-scorbutic factor in fruit and vegetables and to vitamin C. Vitamin C was independently isolated from animal tissue as a compound able to prevent browning (oxidation). After the first detailed investigations of the biosynthesis of ascorbate in plants in the 1950s, it has taken a further 40 years for a pathway to be elucidated. Now we are in a position to understand the factors controlling the ascorbate content and to manipulate its levels in crop plants with potential benefits for human nutrition and crop stress resistance. Investigations of this newly recognized product of sugar nucleotide metabolism should also produce further insights into plant carbohydrate metabolism.

References

Arrigoni, O., De Gara, L., Paciolla, C., Evidente, A., de Pinto, M.C. and Liso, R. (1997) Lycorine: a powerful inhibitor of L-galactono-γ-lactone dehydrogenase activity. *J. Plant Physiol.* 150: 362–364.

Barber, G.A. (1971) The synthesis of L-galactose by plant enzyme systems. *Arch. Biochem. Biophys.* 147: 619–623.

Barber, G.A. (1979) Observations on the mechanism of the reversible epimerization of GDPmannose to GDP-L-galactose by an enzyme from *Chlorella pyrenoidosa*. *J. Biol. Chem.* 254: 7600–7603.

Baydoun, E.A.-H. and Fry, S.C. (1988) [2-^{3}H]Mannose incorporation in cultured plant cells: investigation of L-galactose residues of the primary wall. *J. Plant Physiol.* 132: 484–490.

Bleeg, H.S. and Christensen, F. (1982) Biosynthesis of ascorbate in yeast. Purification of L-galactono-1,4-lactone oxidase with properties different from mammalian L-gulonolactone oxidase. *Eur. J. Biochem.* 127: 391–396.

Bonin, C.P., Potter, I., Vanzin, G.F. and Reiter, D.-W. (1997) The *MUR1* gene of *Arabidopsis thaliana* encodes an isoform of GDP-mannose-4,6-dehydratase, catalysing the first step in the *de novo* synthesis of GDP-L-fucose. *Proc. Natl Acad. Sci. USA* 94: 2085–2090.

Burns, J.J. (1967) Ascorbic acid. In: *Metabolic Pathways*, 3rd edn, Vol. 1 (ed. D.M. Greenberg). Academic Press, New York, pp. 394–411.

Chatterjee, I.B. (1970) Biosynthesis of L-ascorbate in animals. In: *Methods in Enzymology*, Vol. XVIII. *Vitamins and Coenzymes* (eds D.B. McCormick and L.D. Wright). Academic Press, New York, pp. 28–34.

Conklin, P.L., Williams, E.H. and Last, R.L. (1996) Environmental stress tolerance of an ascorbic acid-deficient *Arabidopsis* mutant. *Proc. Natl Acad. Sci. USA* 93: 9970–9974.

Conklin, P.L., Pallanca, J.E., Last, R.L. and Smirnoff, N. (1997) L-Ascorbic acid metabolism in the ascorbate-deficient *Arabidopsis* mutant *vtc1*. *Plant Physiol.* 115: 1277–1285.

Daniel, G., Volc, J. and Kubatova, E. (1994) Pyranose oxidase, a major source of H$_2$O$_2$ during wood degradation by *Phanerochaete chrysosporium*, *Trametes versicolor* and *Oudemansiella mucida*. *Appl. Environ. Microbiol.* 60: 2524–2532.

Davies, M.B., Austin, J. and Partridge, D.A. (1991) *Vitamin C: its Chemistry and Biochemistry*. Royal Society of Chemistry, Cambridge.

Feingold, D.S. (1982) Aldo (and keto) hexoses and uronic acids. In: *Encyclopedia of Plant Physiology*, Vol. 13A (eds F.A. Loewus and W. Tanner). Springer, Berlin, pp. 3–76.

Fry, S.C. (1998) Oxidative scission of plant cell wall polysaccharides by ascorbate-induced hydroxyl radicals. *Biochem. J.* 332: 507–515.

Fujii, N., Watanabe, M., Watanabe, Y. and Shimada, N. (1993) Rate of oxalate synthesis from glycolate and ascorbic acid in spinach leaves. *Soil. Sci. Plant Nutr.* 39: 627–634.

Gabriel, J., Volc, J., Sedmera, P., Daniel, G. and Kubátová, E. (1993) Pyranosone dehydratase from the basidiomycete *Phanerochaete chrysosporium*: improved purification and identification of 6-deoxy-D-glucosone and D-xylosone reaction products. *Arch. Microbiol.* 160: 27–34.

Halliwell, B. and Gutteridge, J.M.C. (1989) *Free Radicals in Biology and Medicine*, 2nd edn. Clarendon Press, Oxford.

Herold, A. and Lewis, D.H. (1977) Mannose and green plants: occurrence, physiology and metabolism, and use as a tool to study the role of orthophosphate. *New Phytol.* 79: 1–40.

Horemans, N., Asard, H. and Caubergs, R.J. (1994) The role of ascorbate free radical as an electron acceptor to cytochrome *b*-mediated trans-plasma membrane electron transport in higher plants. *Plant Physiol.* 104: 1455–1458.

Horemans, N., Asard, H. and Caubergs, R.J. (1997) The ascorbate carrier of higher plant plasma membranes preferentially translocates the fully oxidized (dehydroascorbate) molecule. *Plant Physiol.* **114**: 1247–1253.

Huh, W.K., Kim, S.T., Yang, K.S., Seok, Y.J., Hah, Y.C. and Kang, S.O. (1994) Characterization of D-arabinono-1,4-lactone dehydrogenase oxidase from *Candida albicans* ATCC-10231. *Eur. J. Biochem.* **225**: 1073–1079.

Isherwood, F.A. and Mapson, L.W. (1962) Ascorbic acid metabolism in plants. Part II. Metabolism. *Annu. Rev. Plant Physiol.* **13**: 329–350.

Isherwood, F.A., Chen, Y.-T. and Mapson, L.W. (1954) Synthesis of L-ascorbic acid in plants and animals. *Biochem. J.* **56**: 1–14.

Karpinski, S., Escobar, C., Karpinska, B., Creissen, G. and Mullineaux, P.M. (1997) Photosynthetic electron transport regulates the expression of cytosolic ascorbate peroxidase genes in *Arabidopsis* during excess light. *Plant Cell* **9**: 627–640.

Kim, S.-T., Huh, W.-K., Kim, J.-Y., Hwang, S.-W. and Kang, S.-O. (1996) D-Arabinose dehydrogenase and biosynthesis of erythroascorbate in *Candida albicans*. *Biochim. Biophys. Acta* **1297**: 1–8.

Kiuchi, K., Nishikimi, M. and Yagi, K. (1982) Purification and characterization of L-gulono-γ-oxidase from chicken liver microsomes. *Biochemistry* **21**: 5076–5082.

Koch, K.E. (1996) Carbohydrate-modulated gene expression in plants. *Annu. Rev. Plant Physiol. Plant Mol. Biol.* **47**: 509–540.

Koshizaka, T., Nishikimi, M., Ozawa, T. and Yagi, K. (1988) Isolation and sequence analysis of a complementary DNA encoding rat liver L-gulono-γ-oxidase, a key enzyme for ascorbic acid biosynthesis. *J. Biol. Chem.* **263**: 1619–1621.

Lin, L.-S. and Varner, J.E. (1991) Expression of ascorbate oxidase in zucchini squash (*Cucurbita pepo* L.). *Plant Physiol.* **96**: 159–165.

Loewus, F.A. (1963) Tracer studies of ascorbic acid formation in plants. *Phytochemistry* **2**: 109–128.

Loewus, F.A. (1988) Ascorbic acid and its metabolic products. In: *The Biochemistry of Plants*, Vol. 14 (ed. J. Preiss). Academic Press, New York, pp. 85–107.

Loewus, F.A. and Loewus, M.W. (1987) Biosynthesis and metabolism of ascorbate in plants. *Crit. Rev. Plant Sci.* **5**: 101–119.

Loewus, M.W., Bedgar, D.L., Saito, K. and Loewus, F.A. (1990) Conversion of L-sorbosone to L-ascorbic acid by a NADP-dependent dehydrogenase in bean and spinach leaf. *Plant Physiol.* **94**: 1492–1495.

Ma, J.F., Hiradate, S. and Matsumoto, H. (1998) High aluminium resistance in buckwheat. II. Oxalic acid detoxifies aluminium internally. *Plant Physiol.* **117**: 753–759.

Maier, E. and Kurtz, G. (1982) D-Galactose dehydrogenase from *Pseudomonas fluorescens*. *Methods Enzymol.* **89**: 176–181

Mapson, L.W. and Breslow, E. (1958) Biological synthesis of L-ascorbic acid: L-galactono-γ-lactone dehydrogenase. *Biochem. J.* **68**: 395–406.

Mapson, L.W. and Isherwood, F.A. (1956) Biological synthesis of L-ascorbic acid: the conversion of derivatives of D-galacturonic acid to L-ascorbate in plant extracts. *Biochem. J.* **64**: 13–22

Mutsuda, M., Ishikawa, T., Takeda, T. and Shigeoka, S. (1995) Subcellular localization and properties of L-galactono-γ-lactone dehydrogenase in spinach leaves. *Biosci. Biotechnol. Biochem.* **59**: 1983–1984.

Nick, J.A., Leung, C.T. and Loewus, F.A. (1986) Isolation and identification of erythroascorbic acid in *Saccahromyces cerevisiae* and *Lymphomyces starkeyi*. *Plant Sci.* **46**: 181–187.

Nishikimi, M. and Udenfriend, S. (1976) Immunologic evidence that the gene for L-gulono-γ-lactone oxidase is not expressed in animals subject to scurvy. *Proc. Natl Acad. Sci. USA* **73**: 2066–2068.

Nishikimi, M., Noguchi, E. and Yagi, K. (1978) Occurrence in yeast of L-galactonolactone oxidase which is similar to a key enzyme for ascorbic acid biosynthesis in animals, L-gulonolactone oxidase. *Arch. Biochem. Biophys.* **191**: 479–486.

Nishikimi, M., Fukuyama, R., Minoshima, S., Shimizu, N. and Yagi, K. (1994) Cloning and chromosomal mapping of the human nonfunctional gene for gulono-γ-lactone oxidase, the key enzyme for L-ascorbic acid biosynthesis missing in man. *J. Biol. Chem.* **269**: 13685–13688.

Noctor, G. and Foyer, C.H. (1998). Ascorbate and glutathione: keeping active oxygen under control. *Annu. Rev. Plant Physiol. Plant Mol. Biol.* **49**: 249–279.

Ôba, K., Fukui, M., Imai, Y., Iriyama, S. and Nogami, K. (1994) L-Galactono-γ-lactone dehydrogenase: partial characterization, induction of activity and role in synthesis of ascorbic acid in wounded white potato tuber tissue. *Plant Cell Physiol.* **35**: 473–478.

Ôba, K., Ishikawa, S., Nishikawa, M., Mizuno, H. and Yamamoto, T. (1995) Purification and properties of L-galactono-γ-lactone dehydrogenase, a key enzyme for ascorbic acid biosynthesis, from sweet potato roots. *J. Biochem.* **117**: 120–124.

Østergaard, J., Persiau, G., Davey, M.W., Bauw, G. and Van Montagu, M. (1997) Isolation and cDNA cloning for L-galactono-γ-lactone dehydrogenase, an enzyme involved in the biosynthesis of ascorbic acid in plants. *J. Biol. Chem.* **272**: 30009–30016.

Podmore, I.D., Griffiths, H.R., Herbert, K.E., Mistri, N., Mistri, P. and Lunec, J. (1998) Vitamin C exhibits pro-oxidant properties. *Nature* **392**: 559.

Rautenkranz, A.A.F., Li, L., Machler, F., Martinoia, E. and Oertli, J.J. (1994) Transport of ascorbic and dehydroascorbic acids across protoplast and vacuole membranes isolated from barley (*Hordeum vulgare* cv. Gerbil) leaves. *Plant Physiol.* **106**: 187–193.

Roberts, R.M. (1971) The metabolism of D-mannose-^{14}C to polysaccharide in corn roots. Specific labelling of L-galactose, D-mannose, and L-fucose. *Arch. Biochem. Biophys.* **145**: 685–692.

Rogers, W.J., Jordan, B.R., Rawsthorne, S. and Tobin, A.K. (1991) Changes to the stoichiometry of glycine decarboxylase subunits during wheat (*Triticum aestivum* L.) and pea (*Pisum sativum* L.) leaf development. *Plant Physiol.* **96**: 952–956.

Saito, K., Nick, J.A. and Loewus, F.A. (1990) D-Glucosone and L-sorbosone, putative intermediates of L-ascorbic acid biosynthesis in detached bean and spinach leaves. *Plant Physiol.* **94**: 1496–1500.

Schachter, H., Sarney, J., McGuire, E.J. and Roseman, S. (1969) Isolation of diphosphopyridine nucleotide-dependent L-fucose dehydrogenase from pork liver. *J. Biol. Chem.* **244**: 4785–4792.

Smirnoff, N. (1995) Antioxidant systems and plant response to the environment. In: *Environment and Plant Metabolism: Flexibility and Acclimation* (ed. N. Smirnoff). BIOS Scientific Publishers, Oxford, pp. 217–243.

Smirnoff, N. (1996) The function and metabolism of ascorbic acid in plants. *Ann. Bot.* **78**: 661–669.

Smirnoff, N. and Pallanca, J.E. (1996) Ascorbate metabolism in relation to oxidative stress. *Biochem. Soc. Trans.* **24**: 472–478.

Sugiyama, N. and Tamaguchi, N. (1997) Evaluation of the role of lactate dehydrogenase in oxalate synthesis. *Phytochemistry* **44**: 571–574.

Wheeler, G.L., Jones, M.A. and Smirnoff, N. (1998) The biosynthetic pathway of vitamin C in higher plants. *Nature* **393**: 365–369.

Wildi, B. and Lutz, C. (1996) Antioxidant composition of selected high alpine plant species from different altitudes. *Plant Cell Environ.* **19**: 138–146.

16

Implications of inter- and intracellular compartmentation for the movement of metabolites in plant cells

Michael J. Emes, Caroline G. Bowsher, Phillip M. Debnam, David T. Dennis, Guy Hanke, Steven Rawsthorne and Ian J. Tetlow

1. Introduction

Metabolism in plants is highly compartmentalized (Emes and Dennis, 1997). This is true in terms of the distribution of enzyme activities between cells which have different biochemical and physiological functions within the same tissue, and the sequestration of pathways into distinct subcellular compartments within an individual cell. Nowhere is this better exemplified than in the area of carbohydrate metabolism.

The regulation of gene expression within cells provides the mechanism that elicits both inter- and intracellular compartmentation. Further, although an enzyme activity may be present in adjacent cells of the same tissue, it may be that distinctive isozymes are being expressed and that their particular biochemical properties or localization within a cell gives rise to a distinctive function. This segregation of events into different cell types and/or different subcellular compartments has profound implications for the movement of small molecules which form the metabolic currency of an organism and which may also signal the physiological state of a cell or organelle.

It is becoming increasingly clear that the concentration of particular intermediates may serve to regulate the expression of genes (Koch, 1996) and consequently may provide either a feedback loop within the cells in which they originate, sometimes regulating their own synthesis, or, a feedforward mechanism to adjacent cells which will produce an activation of genes, the products of which may utilize the substrate concerned. Since this may involve detection of intermediates which are not necessarily synthesized in the cytoplasm, the movement of such compounds will also be a function of

Plant Carbohydrate Biochemistry, edited by J.A. Bryant, M.M. Burrell and N.J. Kruger.
© 1999 BIOS Scientific Publishers Ltd, Oxford.

the capacity of the subcellular compartment, in which they are contained, to transport them across lipid bilayers via specific metabolite transporters. Clearly, therefore, an understanding of which metabolites are transported is central to an understanding of the growth and development of plants.

Whilst in some cases the nature of the transported compound and the route via which it is transported is well established and directly measurable, for example sucrose (see Chapter 4), in many instances it is inferred from the localization of the pathway(s) from which it is produced. As such, a major tool in developing our understanding of these events has been the ability to determine where a particular protein or pathway is located: either within a cell by using techniques of subcellular fractionation; or between cells, which often involves some form of histochemical staining of either the protein itself or the mRNA which encodes it, or alternatively, by observing the localization of the products of reporter-genes which are linked to the promotor for the gene of interest. The purpose of this chapter is to use a few examples of how these approaches have influenced our thinking and, often, undermined much-loved dogma. It is important not to lose sight of the fact that, whilst we may be interested in, for example, photosynthesis, starch or lipid metabolism, as far as the plant is concerned these are only the means to an end, and that its prime purpose is to reproduce itself and ensure the success of the species. Consequently, we should not be surprised at the diversity of strategies which plants apply to ensure their continuation. Nor should we be disturbed to find that different cells, and compartments within cells, show an apparent duplication of activity which, at first glance, may seem inefficient. One of the most obvious features of plants is that they are, literally, firmly rooted in their surroundings and cannot escape the changes in environment to which they are continually exposed. It has been a striking feature of the study of transgenic plants that enzymes which had been thought to be central to, and essential for, the metabolism, growth and well-being of a plant can be happily dispensed with or reduced to a few per cent of the normal wild-type activity, without detriment (Stitt and Sonnewald, 1995). In some cases this can be explained by the fact that their role may be taken up by what had been considered minor isozymes, possibly in a different subcellular compartment or, alternatively, novel mechanisms may be induced to compensate for the loss of the activity, allowing the plant to by-pass the metabolic block.

The examples which follow are intended to illustrate how the delineation of metabolic compartmentation has influenced our understanding of both intercellular and intracellular transport, and the extent to which our knowledge of some of the transporters themselves has developed. Excellent specific examples are illustrated by experts throughout this book and it is not the intention to reproduce these in detail, although in some cases there will, inevitably, be some overlap.

2. Intercellular compartmentation

One of the best understood examples of the implications of intercellular compartmentation for metabolite transport is, of course, the comparison between C_3 and C_4 photosynthesis. In C_3 plants carbon dioxide is fixed into 3-phosphoglycerate by the Calvin cycle where it is reduced within the same organelle to triose phosphate. As a result of the oxygenase reaction of Rubisco, significant amounts of carbon are diverted away from this route into the pathway of photorespiration, which substantially decreases the yield of photosynthesis. Photorespiration involves transport of carbon through two other organelles, the peroxisome and the mitochondrion, in addition to the chloroplast. In C_4 plants, the consequences of photorespiration are minimized by the spatial separation of events. The initial fixation of CO_2 gives rise to oxaloacetate which is reduced to malate in the mesophyll cells. Malate (or the transamination product of oxaloacetate, aspartate) is then transported into the bundle sheath layer of cells where it is decarboxylated and the CO_2 released is fixed by the Calvin cycle. All this is well established and the implications for the movement of metabolites within and between cells is well understood (Leegood et al., 1997). However, there is a third group of plants which do not fall into either of these categories, the so-called C_3–C_4 intermediates. Such plants can be recognized by their CO_2 compensation point (T) and their unusual micro-anatomy. As will become apparent, an understanding of the biochemistry which underpins this phenomenon opens up a whole new set of questions relating to the movement of metabolites between cells, which are distinct from those which have to be addressed in C_3 and C_4 photosynthesis.

As indicated by the name applied to this group of plants, the value of T is intermediate between the values observed for C_3 species (typically 45–50 µl l^{-1}) and those of C_4 plants (5 µl l^{-1}) lying within the range of 10–30 µl l^{-1}. These values are indicative of the fact that plants which belong to this category show a reduction in the loss of photorespiratory CO_2 compared to C_3 species. However, unlike C_4 species, C_3–C_4 intermediates do not fix CO_2 into a C_4 acid. Members of this group include species of *Flaveria* and *Moricandia*, which have been studied in detail by Rawsthorne and his colleagues (Rawsthorne et al., 1992). In addition to the reduction in T, examination of the leaves of *Moricandia* and *Flaveria* species reveals an unusual anatomy. The arrangement of cells is similar to that of C_4 species in that the vascular bundles are surrounded by a sheath of large cells adjacent to a mesophyll layer. However, unlike C_4, where photosynthetic metabolism is confined to adjacent cells, in C_3–C_4 leaves there are several cell distances between the outer mesophyll layers and the bundle sheath. Further there is a fascinating micro-anatomy which provides the basis for understanding the reduction in T and which, in one sense, arises from the intercellular localization of a single enzyme, glycine decarboxylase.

In the bundle sheath cells, the organelles involved in photorepiration are arranged in a distinctive orientation. The mitochondria are located largely along the cell wall immediately adjacent to the vascular cells and are overlain by the chloroplasts. As a consequence of this arrangement, the possibility of the CO_2, which arises from photorespiratory glycine decarboxylation, being recaptured by the chloroplast is enhanced (*Figure 1*). Immuno-electron microscopy has revealed that enzymes involved in chloroplastic and peroxisomal photosynthetic carbon metabolism are found in both mesophyll and bundle sheath cells (Rawsthorne et al., 1988a,b) of intermediate species of *Moricandia*. The major distinction is that a functional form of the glycine decarboxylase is absent from the mitochondria of the mesophyll cells. This complex enzyme consists of four sub-units termed P, H, L, and T (Oliver et al., 1990) and accounts for 50% of the soluble protein content of the mitochondria in a fully expanded leaf of a C_3 plant. What is particularly fascinating about the situation in C_3–C_4 intermediates is that whilst all four types of subunit are expressed in the bundle sheath cells, only three are found in the mesophyll and the pyridoxal phosphate-containing P sub-unit is absent (Hylton et al., 1988). As a result of this, glycine can only be oxidized in the mitochondria of bundle sheath cells, thus enhancing the possibility of recapturing the CO_2 released during photorespiration.

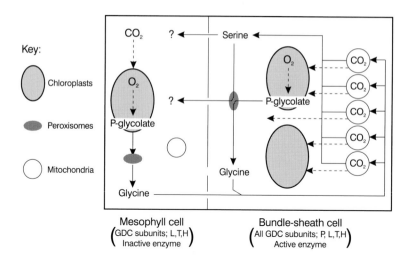

Figure 1. *The pathway of photorespiratory meatbolism in leaves of* Moricandia arvensis. *The expression of subunit proteins of the glycine decarboxylase complex in the bundle sheath and mesophyll cells in this C_3–C_4 intermediate species is indicated. Solid and dashed lines indicate movement of metabolites and gases, respectively. Redrawn from* Planta, Photorespiratory metabolism and immunogold localization of photorespiratory enzymes in leaves of C3 and C3–C4 intermediate species of Moricandia, Rawsthorne *et al., vol. 173, pp. 298–308, fig. 7. Copyright Springer-Verlag 1988.*

Implicit in this is that glycine must be transported from the mesophyll to the bundle-sheath layer of cells and that an equivalent amount of serine must be returned in the other direction (*Figure 1*). In other words, as a consequence of the absence of a single sub-unit of just one of the enzymes involved in photorespiratory carbon metabolism in mesophyll cells, there must be considerable traffic of amino acids between the two cell types which does not occur in either C_3 or C_4 plants. This is only apparent to us because of an understanding of where different enzymic reactions are compartmentalized within a leaf. Whether this movement of metabolites occurs via specific carriers in the plasma membrane or via diffusion between adjacent cells via plasmodesmata is unknown. Estimates of the pool sizes of these two amino acids has led to the suggestion that this shuttling could occur down a concentration gradient (Leegood and von Caemmerer, 1994). However, unlike the transport of organic acids which takes place in C_4 species between adjacent cells, in the case of C_3–C_4 intermediates much longer path-lengths may be involved since there are several cell layers between the outermost mesophyll and the bundle sheath. It is therefore not clear whether diffusion of glycine and serine in opposite directions can account for this movement alone or whether there are specific carriers which facilitate this process. Whatever the mechanism, given that the flux of carbon involved in photorespiration may be 15–30% of the rate of CO_2 fixation, this exchange represents one of the largest components of carbon and nitrogen transport between these cell types and underlines the importance of knowing where different reactions are located.

3. Intracellular compartmentation

Our understanding of the location of metabolic pathways within distinct subcellular compartments has increased considerably in recent years and has served to indicate the forms of carbon which cross lipid bilayers, usually by means of a specific transport protein embedded in the membrane surrounding the compartment. Two examples will be used to illustrate how, even now, our understanding of these events is changing as new data require a re-appraisal of what had become the new accepted dogma. The first of these is of pertinence to both green and non-green tissue and concerns the oxidative pentose phosphate pathway (OPPP). The second concerns starch synthesis in storage organs such as developing seeds and tubers.

3.1 *The oxidative pentose phosphate pathway*

This example has been chosen because it illustrates how the pendulum has swung from one position to almost the complete opposite. For many years the pathways of carbohydrate oxidation, glycolysis and the OPPP, were regarded as being localized solely in the cytoplasm. As a result of the studies of several groups, that view changed to a realization that many of these reactions are duplicated in plastids of plant cells where distinct isozymes are often found (Emes and Neuhaus, 1997; Emes and Tobin, 1993). In particular it was demonstrated that the complete sequence of OPPP reactions could be found within plastids extracted from either photosynthetic or non-photosynthetic tissue. The pathway can be thought of as having two components (*Figure 2*). The first two reactions are catalysed by dehydrogenases which generate NADPH for biosynthetic and assimilatory pathways, such as fatty acid synthesis and nitrogen assimilation. The second stage consists of a series of non-oxidative interconversions of sugar phosphates, eventually giving rise to fructose-6-phosphate and triose phosphate, which, in principle, could enter glycolysis. Three of these non-oxidative reactions (the exception being transaldolase) are common to the Calvin cycle and in chloroplasts futile cycling is avoided by the light-dependent inactivation of the first enzyme of the pathway, glucose-6-phosphate dehydrogenase (Lendzian and Zeigler, 1970). The demonstration that glucose-6-phosphate (G6P) could support nitrite reduction when supplied to intact plastids (Emes and Fowler, 1983) opened up the possibility that hexose phosphates might be able to traverse the plastid envelope, ultimately overturning the view that all plastids transported triose phosphates (see below). Recently, Flügge and co-workers have cloned the G6P transporter from pea root plastids (Kammerer *et al.*, 1998) and shown that it catalyses a counter-exchange with inorganic phosphate (P_i) or triose phosphate as had been implicated from kinetic studies of transport in root plastids (Borchert *et al.*, 1989).

Based on this kind of analysis it has become accepted that the OPPP operates in both cytoplasm and plastids. Interestingly, our view of this is now being tempered further with the suggestion that whilst the dehydrogenases are found in both compartments, the non-oxidative reactions may be confined to plastids and are not found in the cytoplasm at all (Schnarrenberger *et al.*, 1995). This view has been propagated as the result of subcellular fractionation of spinach leaf extracts on sucrose gradients and the apparent inability to demonstrate isozymes which could be separated chromatographically. The latter is a negative result and does not constitute proof that such isozymes do not exist. However, it does at least support the fractionation studies and

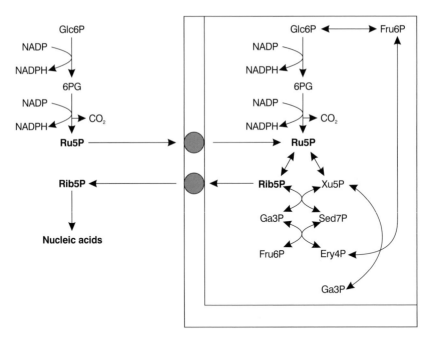

Figure 2. The oxidative pentose phosphate pathway in higher plants. The complete sequence of reactions is shown inside the plastid, whereas only the dehydrogenases are shown as present in the cytoplasm. As a consequence of this, pentose phosphates must be transported across the plastid envelope. Ery4P, erythrose-4-phosphate; Fru6P, fructose-6-phosphate; Ga3P, glyceraldehyde-3-phosphate; Glc6P, glucose-6-phosphate; 6PG, 6-phosphogluconate; Rib5P, ribose-5-phosphate; Ru5P, ribulose-5-phosphate; Sed7P, seduheptulose-7-phosphate; Xu5P, xylulose-5-phosphate.

lends weight to this point of view. It also raises the question as to whether this is true of all plants or just some, and one must be wary of extrapolating too far from studies of a single species.

We have therefore re-examined the localization of all the OPPP reactions in roots and leaves of a number of species including spinach. Plastids were rapidly purified from roots and leaves of spinach, pea, maize and tobacco and all the enzymes of the OPPP assayed alongside marker enzymes (specific for different subcellular compartments including the cytosol). A summary of these data is presented in *Table 1* (P.M. Debnam and M.J.Emes, unpublished data). As a result of this we have concluded that the non-oxidative reactions are lacking in the cytoplasm of some tissues and species, but not all. Tobacco is the exception amongst the four species examined in that between 30 and 50% of the capacity of each of the non-oxidative reactions (data not shown) appears to be outside organelles and therefore, presumably, cytosolic.

The consequences which flow from this are, at the very least for spinach and pea, considerable. First, the absence of the non-oxidative reactions of the OPPP in the cytosol implies that the pentose phosphates which are precursors for nucleic acid synthesis in the cytosol must be exported from the plastid, since they cannot be synthesized *in situ*. Study of the transport of metabolites across plastid envelopes has been largely confined to consideration of triose phosphates, hexose phosphates and P_i but there is some evidence for believing that pentose phosphates at least may be able to

Table 1. Localization of enzymes of the oxidative pentose phosphate pathway in plants. The distribution of the test enzyme was compared with the distribution of a plastid marker enzyme. If there is a significant difference between the distribution of a plastid marker enzyme and the test enzyme (indicated by +), the test enzyme is determined to be localized in both the cytosolic and plastidic fraction (p < 0.05 as determined by a paired one-tailed t-test). – indicates the enzyme is present only in the plastid fraction (3–7 replicates obtained for each tissue)

Enzyme	Tobacco leaf	Tobacco root	Maize root	Pea leaf	Pea root	Spinach leaf
Glucose-6-phosphate dehydrogenase	+	+	+	+	+	+
6-Phosphogluconate dehydrogenase	+	+	+	+	+	+
Ribulose-5-phosphate 3-epimerase	+	+	–	–	–	–
Ribose-5-phosphate isomerase	+	+	–	–	–	–
Transketolase	+	+	–	–	–	–
Transaldolase	+	+	–	–	–	–

cross the plastid envelope. In pea root plastids the reduction of nitrite is sustained by the oxidation of G6P by the OPPP (Bowsher *et al.*, 1989). The NADPH produced by the latter is used to reduce root ferredoxin, the immediate electron donor for nitrite reductase. Hartwell *et al.* (1996) investigated the extent to which recycling of carbon occurs within the plastidial OPPP and as part of that approach demonstrated that exogenous ribose-5-phosphate is able to sustain nitrite reduction as effectively as G6P. The conclusion from this is that ribose-5-phosphate enters the plastid and is rapidly converted by the non-oxidative reactions of the OPPP to fructose-6-phosphate (F6P), and then to G6P by hexose phosphate isomerase. The concentration dependence of this process is shown in *Figure 3* and one must conclude from this that ribose-5-phosphate at least is able to pass from the outside to the inside of plastids, and therefore, presumably, also in the other direction in order to sustain those cytosolic processes which require pentose sugars. Since, in the case of spinach and pea, the cytosolic OPPP stops after the 6-phosphogluconate dehydrogenase reaction, the pentose which would have to enter the plastid *in vivo* would be ribulose-5-phosphate. It is therefore a reasonable proposition that, as a result of the distribution of the non-oxidative reactions, ribose-5-phosphate must be able to leave plastids to support nucleic acid synthesis, and that ribulose-5-phosphate produced in the cytoplasm from the oxidative steps of the OPPP must enter the plastid for further metabolism. The mechanism by which the transfer of either pentose occurs is entirely a matter for speculation. Whether the triose phosphate translocator of leaves, or the hexose phosphate translocator of roots is able to carry out this exchange should be readily testable given the ability to follow counter-exchange with radiolabelled P_i. However, we should not necessarily give in to this temptation too easily, since Flügge and co-workers have recently demonstrated the existence of a phosphoenolpyruvate (PEP) transporter which also counter-exchanges with P_i, and which is distinct from both of the other two transporters (Fischer *et al.*, 1997). Again, it appears that a more precise understanding of where a metabolic pathway is located has opened up new questions as to the nature of the metabolites which are transported intracellularly. There is also some

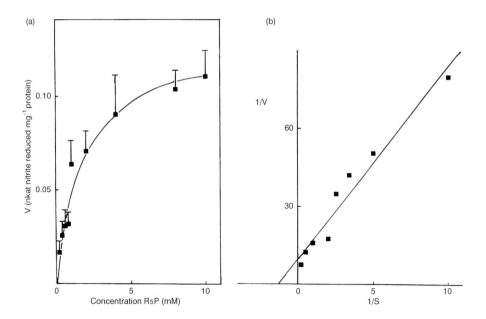

Figure 3. (a) *Ribose-5-phosphate (R5P) dependent reduction of nitrite in intact pea root plastids. Purified organelles were supplied with 1 mM NaNO$_2$ in the presence of varied concentrations of the pentose phosphate for 1 h and reduction determined by measurement of nitrite remaining in the assay.* (b) *Results as a Lineweaver–Burke plot giving a K$_m$ of 1.6 mM for ribose-5-phosphate.*

irony in having to address the question as to whether there is any cytosolic activity of several of the OPPP reactions, when for many years it had been assumed to be exclusively located in the cytosol.

The significance of the above observations should not be overlooked with respect to our understanding of the regulation of the other major pathway of carbohydrate oxidation, glycolysis. Whilst there is some doubt as to whether all of the reactions of glycolysis are located in plastids (Trimming and Emes, 1993) there is no doubt that they are all present in the cytosol. The consequences of the presence or absence of the non-oxidative reactions of the OPPP on the regulation of glycolysis in the cytosol are highly significant. Phosphofructokinase (PFK) is often regarded as an important regulatory enzyme in glycolysis, because of its sensitivity to PEP (inhibitor) and P$_i$ (reversal of PEP inhibition). However, the presence of the non-oxidative reactions of the OPPP in the cytosol would allow oxidation of G6P as far as triose phosphate, thus by-passing this regulatory step. The absence of these reactions would place greater emphasis on PFK in controlling cytosolic carbohydrate oxidation, an issue already complicated by the presence of a pyrophosphate-dependent PFK which is regulated by the metabolism of fructose-2,6-bisphosphate (Stitt, 1990). Given that non-photosynthetic tissues lack a cytosolic fructose-1,6-bisphosphatase (Entwistle and ap Rees, 1990), this would mean that the pyrophosphate- and ATP-dependent PFKs constitute the only means of regulating the interconversion of hexose phosphate and triose phosphate. By contrast, in a tissue such as tobacco leaf, the present evidence implies that the cytosol would additionally contain both a fructose bisphosphatase and the wherewithal, via a complete

OPPP, to by-pass all three reactions involved in the interconversion of F6P and fructose-1,6-bisphosphate. This would suggest that in this instance the ATP–PFK is likely to exert less control over flux between hexose phosphates and triose phosphates.

3.2 *Starch biosynthesis and metabolite transport into amyloplasts*

Starch is the major storage compound in graminaceous seeds and tuberous crops. Its significance both to the plant and mankind has made the understanding of its biosynthesis a priority and the subject is covered in detail in Chapters 7 and 9. Photosynthate delivered to the cytosol of heterotrophic tissue is converted to a form of carbon which can then enter the amyloplast for the synthesis of starch. Our understanding of the form of carbohydrate which crosses the amyloplast membrane has undergone considerable modification in recent years, and, as a consequence, our view of the regulation of starch synthesis has also changed. Much of this has arisen from studies of the localization of particular enzymes and as a result has led to important discoveries about the nature of the metabolites which can move in and out of the amyloplast. It should be emphasized that these conclusions could not have been reached by a study of chloroplast starch metabolism as that is quite a distinct process, occurring in the light and exploiting reduced carbon which has been assimilated *in situ*.

It had been assumed, by analogy with the chloroplast, that triose phosphates would be imported into amyloplasts, from the cytosol, and converted to ADPglucose (the substrate for the starch synthases) within the organelle. This assumption was dispelled by the demonstration that (i) the re-arrangement of carbon atoms in glucose fed to whole grains of wheat was not consistent with the formation of triose phosphates during starch synthesis (Keeling *et al.*, 1988); (ii) many heterotrophic tissues lack fructose-1,6-bisphosphatase and therefore the means to convert triose phosphate to hexose phosphate within the amyloplast (Entwistle and ap Rees, 1990); and (iii) hexose phosphates could be transported into amyloplasts and the glucose moiety incorporated into starch (Hill and Smith, 1991; Tetlow *et al.*, 1994).

The nature of the hexose phosphate which enters amyloplasts seems to vary between species. In some cases G6P is the form which enters the amyloplasts, as in the case of pea root plastids where a K_m for G6P transport of 0.54 mM was found (Borchert *et al.*, 1989). This value is similar to the value observed for other plastids such as those from cauliflower buds (Möhlmann *et al.*, 1995), and in all cases studied so far G6P transport involves a counter-exchange with either P_i or triose phosphate as confirmed by the recent cloning and expression of the cDNA for the pea root G6P translocator (Kammerer *et al.*, 1998). There is still debate as to the extent to which some amyloplasts are able to import glucose-1-phosphate (G1P). Möhlmann *et al.* (1995) demonstrated that cauliflower bud amyloplasts are able to import G1P, but this activity was not affected by G6P or triose phosphate at concentrations that might have been expected to cause inhibition, and did not depend on counter-exchange with P_i. However, there are examples where G1P is clearly imported in exchange for other phosphorylated intermediates. When the envelope proteins of highly purified wheat endosperm amyloplasts were solubilized and incorporated into liposomes, G1P was shown to counter-exchange effectively with G6P, dihydroxyacetone phosphate, 3-phosphoglycerate or P_i at physiologically relevant concentrations and with a K_m for G1P of 0.4 mM (Tetlow *et al.*, 1996). Significantly, the counter-exchange between G1P and P_i had a stoichiometry of 1:1 and the same relationship was subsequently found for the G6P transporter for the protein from pea roots (Kammerer *et al.*, 1998). G1P/P$_i$ counter-exchange has also been observed for tomato fruit chromoplasts and chloroplasts (Schünemann and Borchert, 1994).

There is conflicting evidence over the form of carbon which enters potato amyloplasts during starch synthesis. Schott *et al.* (1995) reconstituted envelope proteins made from amyloplasts of transgenic tubers, possessing a reduced starch content, and observed that G6P was transported in exchange for P$_i$, but that G1P was not. Expression of a homologue of the pea root G6P transporter has been observed in potato tubers (and not found in any other tissues) reinforcing the view that such a translocator is present. However Naeem *et al.* (1997) have shown that G1P, but not G6P, could support starch synthesis in potato amyloplasts, offering indirect evidence that G1P is transported. It is therefore possible that both hexose phosphates are transported across potato amyloplast membranes, though it is not yet clear whether more than one transporter is involved.

Irrespective of the form of hexose phosphate which might permeate the amyloplast envelope, its subsequent conversion to ADPglucose within the organelle requires ATP. This could be generated by oxidative metabolism within the organelle (Kleppinger-Sparace *et al.*, 1992) or by import from the cytosol. Schünemann *et al.* (1993) kinetically characterized an adenylate translocator of pea root plastids. The plastid protein is distinct from its mitochondrial counterpart as emphasized by the work of Neuhaus and co-workers (Kampfenkel *et al.*, 1995) who obtained a full-length cDNA clone for a plastidic adenylate translocator termed AATP1. This protein is closely related to a homologue found in *Rickettsia prowazekii*. The translated protein has a molecular mass of 68 kDa, which is reduced to 62 kDa after incorporation into spinach leaf chloroplasts following the removal of a transit peptide (Neuhaus *et al.*, 1997). By comparison, the mitochondrial translocator has a mass of only half this and shows limited sequence similarity. Northern hybridization of a potato homologue indicated that the mRNA encoding for the plastidic adenylate translocator is expressed mainly in heterotrophic tissue (J. Tjaden and H.E. Neuhaus, pers. comm.).

However, a further option for the movement of carbon into amyloplasts has arisen with the discovery that in developing maize and barley endosperm (see Chapter 18) the majority of the ADPglucose pyrophosphorylase (AGPase) is located outside the organelle. The cDNAs which have been cloned for the small AGPase subunit (Bt2) from these species lack a transit peptide (Giroux and Hannah, 1994; Villand and Kleczkowski, 1994), though a minor form is found within amyloplasts (Thorbjørnsen *et al.*, 1996). To underline this point it has recently been found that ADPglucose is the most effective precursor for starch synthesis in isolated amyloplasts (Möhlmann *et al.*, 1997).

Consistent with this proposal is the observation that the Bt1 mutant of maize lacks a 44 kDa protein in the amyloplast envelope which brings about an increase in ADPglucose concentration (Shannon *et al.*, 1996). It has therefore been proposed that Bt1 encodes an ADPglucose transporter, though direct evidence that this is the function of the protein is still lacking.

In our earlier studies of starch biosynthesis in purified wheat endosperm amyloplasts (Tetlow *et al.*, 1994) we also observed that the highest rates of starch synthesis were observed when ADPglucose was supplied to intact organelles, some 20-fold in excess of the rate observed with G1P. At the time this was largely discounted since there was no clear evidence that ADPglucose could be synthesized in the cytoplasm. We have recently re-examined the subcellular distribution of AGPase in this organ,

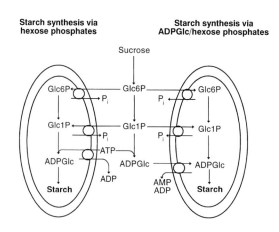

Starch synthesis via hexose phosphates

Starch synthesis via ADPGlc/hexose phosphates

Figure 4. Alternative routes of starch synthesis in non-photosynthetic amyloplasts. Amyloplasts appear capable of importing glucose-6-phosphate (Glc6P), glucose-1-phosphate (Glc1P), ATP or ADPglucose (ADPGlc) depending on the species in which they are found. From Emes and Neuhaus (1997) Metabolism and transport in non-photosynthetic plastids. Journal of Experimental Botany, *vol. 48, pp. 1995–2005. Reprinted with permission of Oxford University Press.*

and found that activity is located in the amyloplast early in development, but that it is predominantly cytosolic during the most active period of grain-filling from around 10 days post-anthesis, although an amyloplastidic form is still present. The implication of this is that ADPglucose import is the major route of carbon entry into the amyloplast during the most active period of starch synthesis in wheat. We have subsequently purified an ADPglucose transporter to homogeneity from wheat amyloplasts. This transporter is able to counter-exchange AMP, ADP, ATP and ADPglucose and is therefore kinetically distinct from the adenylate translocator cloned from *Arabidopsis* which counter-exchanges ADP and ATP (Neuhaus *et al.*, 1997). The ADPglucose transporter has a molecular mass of 38 kDa, though the peptide sequences which have so far been obtained (Tetlow *et al.*, unpublished data) do not show any homology to Bt1. Whether it is related to Bt1 will be determined when the complete sequence is known. Antibodies have been raised to the wheat protein and shown to cross-react specifically with the ADPglucose translocator. Western blots of wheat amyloplast envelope proteins using this antibody and the antibody to the plastidic adenylate transporter of *Arabidopsis* have indicated the presence of two immunologically distinct proteins of 38 and 62 kDa, respectively. This implies that both types of adenylate translocator are present within amyloplast preparations and it will be interesting to see how their expression is regulated during development.

Again, an awareness of where an enzyme is located in the cell, and how this may alter during development has led to a re-appraisal of a significant pathway of plant metabolism. It lays open the possibility that *in vivo* starch synthesis may occur as a result of ADPglucose synthesis in both amyloplasts and cytoplasm (*Figure 4*). The extent to which these occur simultaneously is an open question, though there is evidence of developmental regulation, and an important question that will have to be addressed is whether the two translocators are expressed in the same cell or whether there is spatial separation of the two different pathways of starch synthesis.

4. Conclusions

In 1985, Tom ap Rees wrote 'The first requirement for the understanding of a metabolic pathway is a knowledge of the reactions involved and the way in which they are organised in the cell'. In this chapter we have used a few examples to indicate how our comprehension of metabolism has changed through a knowledge of the compartmentation of events at both a cellular and subcellular level. Further, we have tried to illustrate that there is seldom just one way of doing things and that species have evolved a variety of mechanisms which may lead to the same outcome, be it the fixation of carbon dioxide, the oxidation of carbohydrate or the synthesis of starch, but the detail of which often makes it necessary to invoke previously unconsidered transport processes and the discovery of new proteins.

References

ap Rees, T. (1985) The organisation of glycolysis and the oxidative pentose phosphate pathway in plants. In: *Encyclopaedia of Plant Physiology*, Vol. 18 (eds R. Douce and D.A. Day). Springer, Berlin, pp. 391–417.

Borchert, S., Grosse, H. and Heldt, H.W. (1989) Specific transport of inorganic phosphate, glucose 6-phosphate, dihydroxyacetone phosphate and 3-phosphoglycerate into amyloplasts from pea roots. *FEBS Lett.* **253**: 183–186.

Bowsher, C.G., Hucklesby, D.P. and Emes, M.J. (1989) Nitrite reduction and carbohydrate-metabolism in plastids purified from roots of *Pisum sativum* L. *Planta* **177**: 359–366.

Emes, M J. and Dennis, D.T. (1997) Regulation by compartmentation. In: *Plant Metabolism* (eds D.T. Dennis, D.H. Turpin, D.D. Lefebvre and D.B. Layzell). Longman, pp. 69–80.

Emes, M.J. and Fowler, M.W. (1983) The supply of reducing power for nitrite reduction in plastids of seedling pea roots (*Pisum sativum* L.). *Planta* **158**: 97–102.

Emes, M.J. and Neuhaus, H.E. (1997) Metabolism and transport in non-photosynthetic plastids. *J. Exp. Bot.* **48**: 1995–2005.

Emes, M.J. and Tobin, A.K. (1993) Control of metabolism and development in higher plant plastids. *Int. Rev. Cytol.* **145**: 149–216.

Entwistle, G. and ap Rees, T. (1990) Lack of fructose 1,6-bisphosphatase in a range of higher plants that store starch. *Biochem. J.* **271**: 467–472.

Fischer, K., Kammerer, B., Gutensohn, M., Arbinger, B., Weber, A., Hausler, R. and Flügge, U.-I. (1997) A new class of plastidic phosphate translocator: a putative link between primary and secondary metabolism by the phosphoenolpyruvate/phosphate antiporter. *Plant Cell* **9**: 453–462.

Giroux, M.J. and Hannah, L.C. (1994) ADP-glucose pyrophosphorylase in shrunken-2 and brittle-2 mutants of maize. *Mol. Gen. Genet.* **243**: 400–408.

Hartwell, J., Bowsher, C.G. and Emes, M.J. (1996) Recycling of carbon in the oxidative pentose phosphate pathway in non-photosynthetic plastids. *Planta* **200**: 107–11.

Hill, L.M. and Smith, A.M. (1991) Evidence that glucose 6-phosphate is imported as the substrate for starch synthesis by the plastids of develping pea embryos. *Planta* **185**: 91–96.

Hylton, C.M., Rawsthorne S., Smith, A.M., Jones, D.A. and Woolhouse, H.W. (1988) Glycine decarboxylase is confined to the bundle-sheath cells of leaves of C3–C4 intermediate species. *Planta* **175**: 452–459.

Kammerer, B., Fischer, K., Hilpert, B., Schubert, S., Gutensohn, M., Weber, A. and Flügge, U.-I. (1998) Molecular characterisation of a carbon transporter in plastids from heterotrophic tissues : the glucose 6-phosphate anitporter. *Plant Cell* **10**: 105–117.

Kampfenkel, K., Möhlmann, T., Batz, O., van Montagu, M., Inze, D. and Neuhaus, H.E. (1995) Molecular characterization of an *Arabidopsis thaliana* cDNA encoding a novel putative adenylate translocator of higher plants. *FEBS Lett.* **374**: 351–355.

Keeling, P.L., Wood, J.R., Tyson, R.H. and Bridges, I.J. (1988) Starch biosynthesis in developing wheat grain. Evidence against the direct involvement of triose phosphates in the metabolic pathway. *Plant Physiol.* **87**: 311–319.

Kleppinger-Sparace, K.F., Stahl, R.J. and Sparace, S.A. (1992) Energy requirements for fatty acid and glycerolipid biosynthesis from acetate by isolated pea root plastids. *Plant Physiol.* **98**: 723–727.

Koch, K.E. (1996) Carbohydrate-modulated gene expression in plants. *Annu. Rev. Plant Physiol. Plant Mol. Biol.* **47**: 509–540.

Leegood, R.C. and von Caemmerer, S. (1994) Regulation of photosynthetic carbon assimilation in leaves of C3–C4 intermediate species of *Moricandia* and *Flaveriea. Planta* **192**: 232–238.

Leegood, R.C., von Caemmerer, S. and Osmond, C.B. (1997) Metabolite transport and photosynthetic regulation in C4 and CAM plants. In: *Plant Metabolism* (eds D.T. Dennis, D.H. Turpin, D.D. Lefebvre and D.B. Layzell). Longman, pp. 341–369.

Lendzian, K. and Zeigler, H. (1970) Regulation of glucose 6-phosphate dehydrogenase in spinach chloroplasts by light. *Planta* **94**: 27–36.

Möhlmann, T., Batz, O., Maaß, U. and Neuhaus, H.E. (1995) Analysis of carbohydrate transport across the envelope of isolated cauliflower-bud amyloplasts. *Biochem. J.* **307**: 521–526.

Möhlmann, T., Tjaden, J., Henrichs, G., Quick, W.P., Hausler, R. and Neuhaus, H.E. (1997) ADPglucose drives starch synthesis in isolated maize endosperm amyloplasts: characterisation of starch synthesis and transport properties across the amyloplast envelope. *Biochem. J.* **324**: 503–509.

Naeem, M., Tetlow, I.J. and Emes, M.J. (1997) Starch synthesis in amyloplasts purified from developing potato tubers. *Plant J.* **11**: 1095–1103.

Neuhaus, H.E., Thom, E., Möhlmann, T., Steup, M. and Kampfenkel, K. (1997) Characterisation of a novel eukaryotic ATP/ADP translocator located in the plastid envelope of *Arabidopsis thaliana* L. *Plant J.* **11**: 73–82.

Oliver, D.J., Neuberger, M., Bourguignon, J. and Douce, R. (1990) Interaction between the component enzymes of the glycine decarboxylase multienzyme complex. *Plant Physiol.* **94**: 834–839.

Rawsthorne, S., Hylton, C.M., Smith, A.M. and Woolhouse, H.W. (1988a) Photorespiratory metabolism and immunogold localization of photorespiratory enzymes in leaves of C-3 and C-3–C-4 intermediate species of *Moricandia. Planta* **173**: 298–308.

Rawsthorne, S., Hylton, C.M., Smith, A.M. and Woolhouse, H.W. (1988b) Distribution of photorespiratory enzymes between bundle-sheath and mesophyll-cells in leaves of the C3–C4 intermediate species *Moricandia arvensis* (L) DC. *Planta* **176**: 527–532.

Rawsthorne, S., von Caemmerer, S., Brooks, A. and Leegood, R.C. (1992) Metabolic interactions in leaves of C3 + C4. In: *Plant Organelles* SEB Seminar Series 50, pp. 113–139.

Schnarrenberger, C., Flechner, A. and Martin, W. (1995) Enzymatic evidence for a complete oxidative pentose phosphate pathway in chloroplasts and an incomplete pathway in the cytosol of spinach leaves. *Plant Physiol.* **108**: 609–614.

Schott, K., Borchert, S., Müller-Rober, B. and Heldt, H.W. (1995) Transport of inorganic phosphate and C3- and C6-sugar phosphates across the envelope membranes of potato tuber amyloplasts. *Planta* **196**: 647–652.

Schünemann, D. and Borchert, S. (1994) Specific transport of inorganic phosphate and C3- and C6-sugar phosphates across the envelope membranes of tomato (*Lycopersicon esculentum*) leaf-chloroplasts, tomato fruit-chloroplasts and fruit-chromoplasts. *Bot. Acta* **107**: 461–467.

Schünemann, D., Borchert, S., Flügge, U.I. and Heldt, H.W. (1993) ATP/ADP translocator from pea-rot plastids. Comparisons with translocators from spinach chloroplasts and pea leaf mitochondria. *Plant Physiol.* **103**: 131–137.

Shannon, J.C., Pien, F.-M. and Lui, K.C. (1996) Nucleotides and nucleotide sugars in developing maize endosperms. Synthesis of ADPglucose in brittle-1. *Plant Physiol.* **110**: 835–843.

Stitt, M. (1990) Fructose-2,6-bisphosphate as a regulatory molecule in plants. *Annu. Rev. Plant Physiol. Plant Mol. Biol.* **41**: 153–185.

Stitt, M. and Sonnewald, U. (1995) Regulation of metabolism in transgenic plants. *Annu. Rev. Plant Physiol. Plant Mol. Biol.* **46**: 341–368.

Tetlow, I.J., Blissett, K.J. and Emes, M.J. (1994) Starch synthesis and carbohydrate oxidation in amyloplasts from developing wheat endosperm. *Planta* **194**: 454–460.

Tetlow, I.J., Bowsher, C.G. and Emes, M.J. (1996) Reconstitution of the hexose phosphate translocator from the envelope membranes of wheat endosperm amyloplasts. *Biochem. J.* **319:** 717–723.

Thorbjørnsen, T., Villand, P., Denyer, K., Olsen, O.A. and Smith, A.M. (1996) Distinct iso-forms of ADPglucose pyrophosphorylase occur inside and outside the amyloplasts in barley endosperm. *Plant J.* **10:** 243–250.

Trimming, B.A. and Emes, M.J. (1993) Glycolytic enzymes in non-photosynthetic plastids of pea (*Pisum sativum* L.). *Planta* **190:** 439–445.

Villand, P. and Kleczkowski, L.A. (1994) Is there an alternative pathway for starch biosynthe-sis in cereal seeds? *Zeitschr. Naturfors.* **49:** 215–219.

17

Metabolite transporters in plastids and their role in photoassimilate allocation

Ulf-Ingo Flügge

1. Introduction

Like other eukaryotic cells, plant cells contain different organelles, amongst them plastids, compartments that are specific for plants. Chloroplasts are the site of photosynthesis in which atmospheric CO_2 is assimilated into intermediates that are used within as well as outside the chloroplasts for a variety of metabolic pathways. The daily fixed carbon is exported from the chloroplasts as triose phosphates in exchange with inorganic phosphate. Chloroplasts are also involved in nitrogen assimilation, the synthesis of amino acids, transitory starch and a series of secondary compounds. Plastids of non-photosynthetic tissues have to import carbon, mainly hexose phosphates, as a source of energy and biosynthetic precursors for the synthesis of, for example, fatty acids, amino acids as well as starch. All plastids are double membrane-bounded organelles. The inner envelope membrane is the actual permeability barrier of the plastid and contains a variety of different metabolite translocators that co-ordinate the metabolism in both compartments. The molecular characterization of transporters that transport carbon in exchange with inorganic phosphate and their role in plant metabolism will be described.

2. The triose phosphate/phosphate translocator

The triose phosphate/phosphate translocator (TPT) mediates the export of fixed carbon in the form of triose phosphates or PGA from the chloroplast into the cytosol (Fliege *et al.*, 1978). In the mature leaves of most plants, the photosynthates exported by the TPT are used in the formation of sucrose and amino acids which are the main products allocated to heterotrophic plant organs (*Figure 1*). Photosynthates can also be stored during the day in the form of transitory starch which is mobilized during the following night period, exported from the chloroplast via a glucose translocator, and transformed to sucrose. Sucrose and amino acids are actively loaded into the sieve element/companion cell complex by specific H^+/symporters (*Figure 1*).

Plant Carbohydrate Biochemistry, edited by J.A. Bryant, M.M. Burrell and N.J. Kruger.
© 1999 BIOS Scientific Publishers Ltd, Oxford.

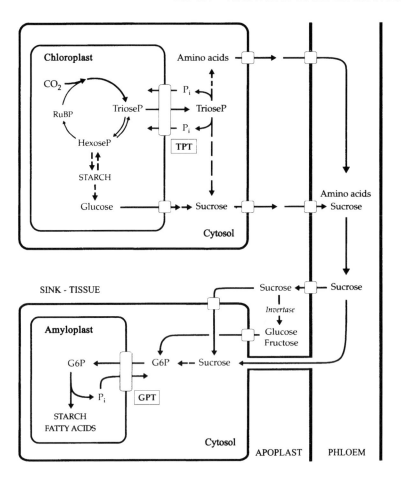

Figure 1. Processes involved in the transport of photoassimilates in leaves and non-green tissues. G6P, glucose-6-phosphate; GPT, glucose-6-phosphate/phosphate translocator; HexoseP, hexose phosphates; P$_i$, inorganic phosphate; RuBP, ribulose-1,5-bisphosphate; TPT, triose phosphate/phosphate translocator; TrioseP, triose phosphates. For details, see text.

The TPT antiport system accepts as substrates inorganic phosphate, triose phosphates and PGA. These substrates are transported in a 1:1 counter-exchange via a ping-pong reaction mechanism (Fliege *et al.*, 1978; Flügge, 1992).

The spinach TPT was the first plant membrane transport system for which the primary sequence was determined (Flügge *et al.*, 1989). At present, TPT sequences that share high similarities are available from various plants (Fischer *et al.*, 1994, 1997). All TPTs are nuclear-encoded and possess N-terminal transit peptides that direct the protein to the chloroplasts. In its functional state, the TPT forms a homodimer and belongs to the group of translocators with a 6+6 helix folding pattern, similar to mitochondrial carrier proteins (Flügge, 1985; Walker and Runswick, 1993).

The final proof for the identity of a transporter cDNA is the expression of the functional protein in heterologous systems, for example, yeast cells, bacteria or oocytes. The recombinant and histidine-tagged TPT protein, produced in yeast cells, was purified to apparent homogeneity by affinity chromatography and reconstituted into artificial membranes. As shown in *Table 1*, the recombinant TPT protein exhibited transport characteristics almost identical to those of the authentic chloroplast protein: it accepts as counter-substrates only triose phosphates and PGA, but not PEP or hexose phosphates (Loddenkötter *et al.*, 1993).

The TPT is almost exclusively present in photosynthetically active tissues (Fischer *et al.*, 1997; Flügge, 1995). To assess the role of the TPT on photosynthetic metabolism, transgenic antisense potato plants were created in which both the amount and the activity of the TPT were reduced to about 70% of the controls (Riesmeier *et al.*, 1993). In ambient CO_2 and intermediate light, there was no significant effect on photosynthetic rates, growth and tuber development. However, the starch content in the leaves of the transformants was much higher than in wild-type plants. At the expense of sucrose biosynthesis, the transformants retain the daily assimilated carbon mainly within the plastids and direct it into the accumulation of starch (*Figure 2*). In contrast to wild-type plants, which show an almost continuous carbon supply to the heterotrophic tissues even in the absence of photosynthesis, the transformants have to compensate their deficiency in the TPT activity by mobilization of the accumulated starch during the following night period (Heineke *et al.*, 1994). The products of starch breakdown are then exported from the chloroplasts as hexoses via the glucose translocator. Bypassing the TPT, the transformants thus export the major part of the carbon during the night period and possess, compared to wild-type plants, an altered day–night rhythm of carbon allocation (*Figure 2*).

Interestingly, antisense TPT tobacco plants accumulate starch as potato plants do, but begin to mobilize the accumulated starch in the light while photosynthesis is still occurring. Hexoses are exported via a GT, the activity of which is two to three times higher in the transformants compared to the wild type (Häusler *et al.*, 1998). The importance of this translocator in starch mobilization and export is underlined by the observation that a mutant of *Arabidopsis thaliana*, TC265, that is able to degrade starch but obviously unable to export the products of starch degradation via a GT, accumulate large amounts of starch (Caspar *et al.*, 1991; Trethewey and ap Rees, 1994).

Table 1. Substrate specificities of recombinant phosphate translocators isolated from yeast cells. The phosphate transport activities were reconstituted into liposomes that had been preloaded with the indicated substrates. They are given as a percentage of the activity measured for proteoliposomes preloaded with inorganic phosphate. Data from Kammerer et al. (1998)

| | Recombinant histidine-tagged translocator proteins, purified from *S. pombe* cells | | |
	TPT (%)	GPT (%)	PEPPT (%)
Liposomes loaded with			
Phosphate	(= 100)	(= 100)	(= 100)
Triose phosphate	92	112	22
PGA	90	50	16
PEP	5	20	72
G6P	5	92	2
G1P	4	< 1	4

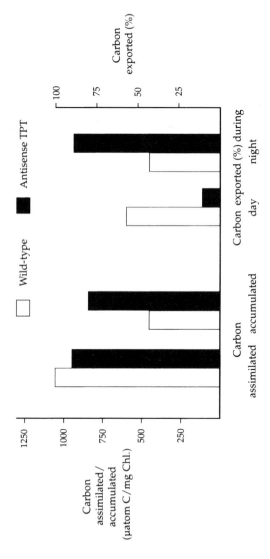

Figure 2. *Effect of antisense repression of the TPT in potato plants on carbon assimilation and assimilate utilization. Data from Heineke et al. (1994).*

3. The hexose phosphate/phosphate translocator

In sink tissues, sucrose is unloaded from the phloem either via symplasmic connections or via the apoplast (*Figure 1*). Sucrose is cleaved by either invertase or sucrose synthase and the resulting products are converted to hexose phosphates. In non-green tissues from most plants, G6P is the preferred hexose phosphate taken up by amyloplasts and subsequently used in biosynthetic processes such as starch formation or fatty acid biosynthesis (Borchert *et al.*, 1993; Hill and Smith, 1991; Kang and Rawsthorne, 1996; Neuhaus *et al.*, 1993). However, in amyloplasts from some tissues, both G6P and/or G1P can be transported and used for biosynthetic processes (Schünemann and Borchert, 1994; Tetlow *et al.*, 1996).

The import of G6P into the plastids is mediated by a recently identified G6P/phosphate translocator, the GPT (Kammerer *et al.*, 1998). Corresponding cDNA clones for GPTs of non-green tissues from various plants have been isolated and the analysis of the deduced GPT protein sequences revealed that the GPT proteins share only about 38% identical amino acids with members of the TPT family. Thus, the GPTs represent a second group of plastidic phosphate translocators. The expression of the corresponding cDNA in yeast cells and the subsequent reconstitution of the produced protein in artificial membranes revealed that inorganic phosphate, triose phosphates and G6P are about equally well accepted as counter-substrates by the GPT, whereas PEP is only poorly transported (*Table 1*). Other hexose phosphates, such as G1P and F6P, are virtually not transported by the GPT.

Inside the plastids, G6P is the precursor for starch biosynthesis or fatty acid biosynthesis during which processes inorganic phosphate is released. In addition, G6P can be transformed to triose phosphates via the oxidative pentose phosphate pathway in which redox equivalents are delivered for nitrogen metabolism and fatty acid biosynthesis. Evidently, both substrates, triose phosphates and inorganic phosphate, can be used as counter-substrates by the GPT in exchange with G6P.

A G6P transport activity is also present in chloroplasts from guard cells (Overlach *et al.*, 1993). Like non-green plastids, these chloroplasts are devoid of FBPase activity (Hedrich *et al.*, 1985), the key enzyme for the conversion of triose phosphates into hexose phosphates, and therefore rely on the provision of hexose phosphates for starch biosynthesis. Starch is mobilized during stomatal opening and converted to malate that is then used as a counter-ion for potassium.

Transcripts of the GPT gene are abundant in heterotrophic tissues such as roots, developing maize kernels, potato tubers or reproductive organs (Kammerer *et al.*, 1998) which is in line with the proposed function of the GPT protein in these tissues that utilize G6P as a precursor for starch biosynthesis.

It has been shown recently that in the seed endosperm of some cereals ADPglucose can be taken up by the plastids and directly used for starch biosynthesis. These tissues contain a cytosolic ADPglucose pyrophosphorylase (Denyer *et al.*, 1996; Thorbjørnsen *et al.*, 1996) and the ADPglucose formed in the cytosol is presumably transported into the plastids for starch biosynthesis via an ADPglucose/adenylate antiporter. This transporter is different from the recently identified ADP/ATP translocator (Kampfenkel *et al.*, 1995; Schünemann *et al.*, 1993). It is assumed, but not yet proven, that the Brittle1 protein serves as an ADPglucose/adenylate transporter which would thus represent an alternative route to provide the plastids with a precursor for starch biosynthesis (Sullivan *et al.*, 1991; Sullivan and Kaneko, 1995).

4. The phosphoenolpyruvate/phosphate translocator

Except plastids of lipid-storing tissues, chloroplasts and non-green plastids from most plants lack the complete set of glycolytic enzymes for the conversion of hexose phosphates and/or triose phosphates into PEP and, therefore, rely on the supply of PEP from the cytosol (Miernyk and Dennis, 1992; Stitt and ap Rees, 1979). In plastids, PEP can be used for the biosynthesis of PEP- and pyruvate-derived amino acids, for fatty acids, or as an immediate precursor for the plastid-localized shikimate pathway (*Figure 3*). Erythrose-4-phosphate, another precursor for the shikimate pathway, can be provided either by the reductive or the oxidative pentose phosphate pathway inside the plastids. The aromatic amino acids synthesized from PEP via the shikimate pathway are not only constituents of proteins but are also utilized as precursors for the biosynthesis of a large number of secondary metabolites (e.g. alkaloids, flavonoids and lignin) which are important in plant defence mechanisms and stress responses (for review, see Herrmann, 1995).

A transporter that enables the transport of PEP into plastids has been recently identified and corresponding cDNAs have been isolated from non-green and photosynthetic tissues of various plants (Fischer *et al.*, 1997). All of these clones were highly homologous to each other but the identities with members of the TPT- and GPT-families were only about 35%. Thus, these membrane transporters represent a third class of phosphate translocators that are different from the TPTs and GPTs. Expression of functional protein in yeast cells revealed that this translocator transports inorganic phosphate in exchange with PEP (*Table 1*). Triose phosphates and PGA, which are the only counter-substrates for the TPT, are only poorly transported

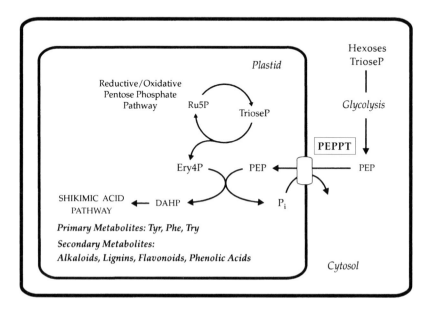

Figure 3. *Proposed function of the PEPPT. Ery4P, erythrose-4-phosphate; PEP, phosphoenolpyruvate; P$_i$, inorganic phosphate; Phe, phenylalanine; PPT, phosphoenolpyruvate/phosphate translocator; Ru5P, ribulose-5-phosphate; TrioseP, triose phosphates; Tyr, tyrosine; Try, tryptophan. For details, see text.*

and hexose phosphates are virtually not transported by this transporter. Thus, the translocator represents a PEP/phosphate translocator (PEPPT) which is able to supply the plastids with PEP even in the presence of triose phosphates and PGA. It is expected that mesophyll chloroplasts of C_4 plants possess a PEPPT-like protein which mediates the export of PEP to provide the substrate for PEP carboxylase in the cytosol.

The importance of the PEPPT protein for plant development became obvious by the analysis of corresponding mutants of *Arabidopsis thaliana*. The recently described CUE (chlorophyll *a/b* binding protein underexpressed) mutants are defective in the *peppt* gene (Li *et al*., 1995; J. Chory and S. Streatfield, personal communication). These mutant plants under-express genes for chloroplast components, both in the light and in response to a light pulse. The seedlings are not able to establish photoautotrophic growth and die unless they are germinated on sucrose. The paraveinal regions of the mutant leaves are still green but the interveinal regions are pale green, resulting in a reticulate pattern. Future work will elucidate how this severe phenotype is linked to the role of the PEPPT in plant metabolism and development.

5. Conclusions

Until recently, it was believed that the transport of phosphate, triose phosphates, PGA, PEP and hexose phosphates is mediated by a single transport system with a broad substrate specificity. The recent findings clearly show that plastids contain instead a set of phosphate translocators with different structures but overlapping specificities. Such a system is able to catalyse the uptake of individual phosphorylated substrates even in the presence of high concentrations of other phosphorylated metabolites, which would otherwise compete for the binding site of a single transport system.

Acknowledgements

Work in the author's laboratory was funded by the Deutsche Forschungsgemeinschaft, the Fonds der Chemischen Industrie, the Bundesministerium für Bildung und Forschung, and by the European Communities BIOTECH Programme, as part of the Project of Technological Priority 1993–1996.

References

Borchert, S., Harborth, J., Schünemann, D., Hoferichter, P. and Heldt, H.W. (1993) Studies of the enzymatic capacities and transport properties of pea root plastids. *Plant Physiol.* **101**: 303–312.

Caspar, T., Lin, T.-S., Kakefuda, G., Benbow, L., Preiss, J. and Somerville, C. (1991) Mutants of *Arabidopsis* with altered regulation of starch metabolism. *Plant Physiol.* **95**: 1181–1188.

Denyer, K., Dunlap, F., Thorbjörnsen, T., Keeling, P. and Smith, A.M. (1996) The major form of ADP-glucose pyrophosphorylase in maize endosperm is extra-plastidial. *Plant Physiol.* **112**: 779–785.

Fischer, K., Arbinger, B., Kammerer, K., Busch, C., Brink, S., Wallmeier, H., Sauer, N., Eckerskorn, C. and Flügge, U.I. (1994) Cloning and *in vivo* expression of functional triose phosphate/phosphate translocators from C_3- and C_4-plants: evidence for the putative participation of specific amino acids residues in the recognition of phosphoenolpyruvate. *Plant J.* **5**: 215–226.

Fischer, K., Kammerer, B., Gutensohn, M., Arbinger, B., Weber, A., Häusler, R.E. and Flügge, U.I. (1997) A new class of plastidic phosphate translocators: a putative link between primary and secondary metabolism by the phosphoenolpyruvate/phosphate antiporter. *Plant Cell* 9: 453–462.

Fliege, R., Flügge, U.I., Werdan, K. and Heldt, H.W. (1978) Specific transport of inorganic phosphate, 3-phosphoglycerate and triose phosphates across the inner membrane of the envelope in spinach chloroplasts. *Biochim. Biophys. Acta* 502: 232–247.

Flügge, U.I. (1985) Hydrodynamic properties of the Triton X-100 solubilized chloroplast phosphate translocator. *Biochim. Biophys. Acta* 815: 299–305.

Flügge, U.I. (1992) Reaction mechanism and asymmetric orientation of the reconstituted chloroplast phosphate translocator. *Biochim. Biophys. Acta* 1110: 112–118.

Flügge, U.I. (1995) Phosphate translocation in the regulation of photosynthesis. *J. Exp. Bot.* 46: 1317–1323.

Flügge, U.I., Fischer, K., Gross, A., Sebald, W., Lottspeich, F. and Eckerskorn, C. (1989). The triose phosphate–3-phosphoglycerate–phosphate translocator from spinach chloroplasts: nucleotide sequence of a full-length cDNA clone and import of the *in vitro* synthesized precursor protein into chloroplasts. *EMBO J.* 8: 39–46.

Häusler, R.E., Schlieben, N.H., Schulz, B. and Flügge, U.I. (1998) Compensation of decreased triose phosphate/phosphate transport activity by accelerated starch turnover and glucose transport in transgenic tobacco. *Planta* 204: 366–376.

Hedrich, R., Raschke, K. and Stitt, M. (1985) A role for fructose-2,6-bisphosphate in regulating carbohydrate metabolism in guard cells. *Plant Physiol.* 79: 977–982.

Heineke, D., Kruse, A., Flügge, U.I., Frommer, W.B., Riesmeier, J.W., Willmitzer, L. and Heldt, H.W. (1994) Effect of antisense repression of the chloroplast triose-phosphate translocator on photosynthetic metabolism in transgenic potato plants. *Planta* 193: 174–180.

Herrmann, K.M. (1995) The shikimate pathway: early steps in the biosynthesis of aromatic compounds. *Plant Cell* 7: 907–919.

Hill, L.M. and Smith, A.M. (1991) Evidence that glucose 6-phosphate is imported as the substrate for starch biosynthesis by the plastids of developing pea embryos. *Planta* 185: 91–96.

Kammerer, B., Fischer, K., Hilpert, B., Schubert, S., Gutensohn, M., Weber, A. and Flügge, U.I. (1998) Molecular characterization of a carbon transporter in plastids from heterotrophic tissues: the glucose 6-phosphate/phosphate antiporter. *Plant Cell* 10: 105–117.

Kampfenkel, K., Möhlmann, T., Batz, O., Van Montagu, M., Inzé, D. and Neuhaus, H.E. (1995) Molecular characterization of an *Arabidopsis thaliana* cDNA encoding a novel putative adenylate translocator of higher plants. *FEBS Lett.* 74: 351–355.

Kang, F. and Rawsthorne, S. (1996) Metabolism of glucose-6-phosphate and utilization of multiple metabolites for fatty acid synthesis by plastids from developing oilseed rape embryos. *Planta* 199: 321–327.

Li, H.-M., Culligan, K., Dixon, R.A. and Chory, J. (1995) CUE1: a mesophyll cell-specific positive regulator of light-controlled gene expression in Arabidopsis. *Plant Cell* 7: 1599–1610.

Loddenkötter, B., Kammerer, B., Fischer, K. and Flügge, U.I. (1993) Expression of the functional mature chloroplast triose phosphate translocator in yeast internal membranes and purification of the histidine-tagged protein by a single metal-affinity chromatography step. *Proc. Natl Acad. Sci. USA* 90: 2155–2159.

Miernyk, J.A. and Dennis, D.T. (1992) A developmental analysis of the enolase isoenzymes from *Ricinus communis*. *Plant Physiol.* 99: 748–750.

Neuhaus, H.E., Thom, E., Batz, O. and Scheibe, R. (1993) Purification of highly intact plastids from various heterotrophic plant tissues. Analysis of enzyme equipment and precursor dependency for starch biosynthesis. *Biochem. J.* 296: 395–401.

Overlach, S., Diekmann, W. and Raschke, K. (1993) Phosphate translocator of isolated guard-cell chloroplasts from *Pisum sativum* L. transports glucose-6-phosphate. *Plant Physiol.* 101: 1201–1207.

Riesmeier, J.W., Flügge, U.I., Schulz, B., Heineke, D., Heldt, H.W., Willmitzer, L. and Frommer, W.B. (1993) Antisense repression of the chloroplast triose phosphate translocator affects carbon partitioning in transgenic potato plants. *Proc. Natl. Acad. Sci. USA* **90**: 6160–6164.

Schünemann, D. and Borchert, S. (1994). Specific transport of inorganic phosphate and C_3- and C_6-sugar-phosphates across the envelope membranes of tomato (*Lycopersicon esculentum*) leaf-chloroplasts, tomato fruit-chloroplasts and fruit-chromoplasts. *Bot. Acta* **107**: 461–467.

Schünemann, D., Borchert, S., Flügge, U.I. and Heldt, H.W. (1993) ADP/ATP translocator from pea root plastids: comparison with translocators from spinach chloroplasts and pea leaf mitochondria. *Plant Physiol.* **103**: 131–137.

Stitt, M. and ap Rees, T. (1979) Capacities of pea chloroplasts to catalyse the oxidative pentose phosphate pathway and glycolysis. *Phytochemistry* **18**: 1905–1911.

Sullivan, T. and Kaneko, Y. (1995) The maize brittle1 gene encodes amyloplast membrane polypeptides. *Planta* **196**: 477–484.

Sullivan, T.D., Strelow, L.I., Illingworth, C.A., Phillips, C.A. and Nelson, O.E. (1991) Analysis of the maize *brittle-1* alleles and a defective *Suppressor-mutator*-induced mutable allele. *Plant Cell* **3**: 1337–1348.

Tetlow, I.J., Bowsher, C.G. and Emes, M.J. (1996) Reconstitution of the hexose phosphate translocator from the envelope membranes of wheat endosperm amyloplasts. *Biochem. J.* **319**: 717–723.

Thorbjørnsen, T., Villand, P., Denyer, K., Olsen, O.-A. and Smith, A.M. (1996) Distinct isoforms of ADPglucose pyrophosphorylase occur inside and outside the amyloplasts in barley endosperm. *Plant J.* **10**: 243–250.

Trethewey, R.N. and ap Rees, T. (1994) A mutant of *Arabidopsis thaliana* lacking the ability to transport glucose across the chloroplast envelope. *Biochem. J.* **301**: 449–454.

Walker, J.E. and Runswick, M.J. (1993) The mitochondrial transport protein superfamily. *J. Bioenergetic Biomembranes* **25**: 435–446.

18

Starch synthesis in plant storage tissues: precursor dependency and function of the plastidic ATP/ADP transporter

Torsten Möhlmann, Joachim Tjaden and H. Ekkehard Neuhaus

1. Function of starch in leaves and storage tissues

Starch represents by far the most prominent plant storage product. With a few exceptions (e.g. onions or *Agave*) all higher plants contain starch in autotrophic as well as in heterotrophic organs (Preiss, 1982). One obvious difference between starch and all other types of storage carbohydrates is that the former is water-insoluble making it comparatively easy to extract and to enrich this polysaccharide from plant homogenates.

Starch is a widely used source for various products generated during industrial or food processing (Kleinhanß, 1988). The physico-chemical properties of starch are strongly influenced by the ratio of amylose and amylopectin, the two constituents of starch. Amylose consists of linearly arranged glucose molecules, connected via α-1,4 bonds, and reaching lengths up to 1000 units (Preiss, 1991). In contrast, amylopectin is a branched molecule since in addition to α-1,4 bonds, at every 20–30 glucose units an α-1,6 bond is introduced (Preiss, 1991; Smith *et al.*, 1997). This additional branching allows synthesis of extremely large amylopectin molecules. Starch synthesis is catalysed by the action of the enzymes ADPglucose pyrophosphorylase (AGPase), starch synthase (SS) and starch branching enzyme (SBE). As the last two enzyme activities are exclusively present in the higher plant plastids, the water-insoluble product is confined to the stroma of plastids.

Starch, of course, fulfils an important biological function in addition to its use in industrial processing. It is well known that artificially created perturbations of starch

Plant Carbohydrate Biochemistry, edited by J.A. Bryant, M.M. Burrell and N.J. Kruger.
© 1999 BIOS Scientific Publishers Ltd, Oxford.

metabolism in autotrophic or heterotrophic plant tissues can cause severe negative alterations of plant metabolism. Using *Arabidopsis thaliana* L. as a model system, Casper *et al.* (1986) demonstrated that inhibition of starch synthesis in leaves correlated with strongly reduced growing rates when mutants were kept under short-day conditions. Under conditions of permanent illumination, where even in wild-type plants no diurnal changes of starch levels occur, the growth of mutants is not different to corresponding wild types (Casper *et al.*, 1986). Interestingly, limitation of starch synthesis in heterotrophic storage tissue also correlates with lowered yields. By using an *Agrobacterium*-mediated genetic transformation where a cDNA, coding for the small subunit of the enzyme AGPase (responsible for the synthesis of ADPglucose), is introduced in antisense orientation into the genome from potato it was possible to show that decreased activities of this enzyme correlate with reduced levels of tuber starch, higher sugar concentrations and a significantly increased number of tubers (Müller-Röber *et al.*, 1992).

Both forms of starch, transitory starch in leaves and storage starch in heterotrophic tissues, fulfil various and specific functions. During photosynthetic starch synthesis newly generated hexose phosphates are partly converted to ADPglucose leading to starch synthesis. The water-insoluble property of starch allows green tissue to accumulate large amounts of organic carbon while avoiding high osmotic pressures. To understand the functions of transitory starch we have first to remember that net degradation of transitory starch occurs only at night (Ziegler and Beck, 1989). However, even in the night, chloroplasts still synthesize various compounds requiring intermediates provided by starch degradation. For example, chloroplasts are the site of nucleotide biosynthesis (Doremus and Jagendorf, 1985) making it necessary to keep ribose-5-phosphate (a precursor of nucleotide biosynthesis) at sufficient concentration. This pentose phosphate is an intermediate of the plastidic oxidative pentose phosphate pathway (OPPP, for review see, Emes and Neuhaus, 1998) which is fuelled by a product of starch degradation, namely glucose-6-phosphate (G6P). In addition, the OPPP provides reducing power in the form of NADPH. These reducing equivalents contribute to the nocturnal fatty acid synthesis, which is of course slower in the dark than in the light, but represents an anabolic pathway requiring significant amounts of NADPH (Stumpf, 1980). Moreover, despite the fact that cytosolic nitrate reductase is a highly regulated enzyme, covalently modified by protein phosphorylation, it is still active at night (Kaiser and Brendle-Behnisch, 1991). This leads to synthesis of nitrite, which is a quite toxic compound. By using reducing equivalents liberated from products of starch degradation it is possible to maintain nocturnal nitrite reduction and synthesis of ammonia (Thom and Neuhaus, 1995). Finally, it should be kept in mind that most of the carbon liberated during starch degradation enters the cytosol as glucose, maltose, or as phosphorylated intermediates (3-phosphoglycerate (PGA) or dihydroxyacetone phosphate) (Stitt and Heldt, 1981). Obviously, significant amounts of starch become converted to sucrose and are then subsequently exported via the phloem (Fondy and Gieger, 1982).

In heterotrophic storage tissues starch also fulfils various functions. The most obvious function is the provision of carbon moieties for sucrose, lipid and amino acid synthesis during germination of seeds or during sprouting of, for example, tubers. One major difference in starch mobilization between these two types of tissues is that in the latter case the starch-harbouring plastids generally stay intact throughout the process of starch mobilization. This means that products of starch breakdown have to be transported out of the amyloplast into the surrounding cytosol via processes catalysed by membrane-bound carrier proteins.

2. Precursor dependency of starch biosynthesis in storage plastids

Generally, plastids of higher plants are classified as autotrophic chloroplasts or heterotrophic plastids, the latter comprising, for example, chromoplasts, colourless leucoplasts and amyloplasts. However, the term storage plastid might also cover green chloroplasts in instances in which they are located in storage tissues such as fruits or barks of various plants. One major function of these plastids is the re-assimilation of respiratory carbon dioxide, liberated during mitochondrial metabolism (Blanke and Lenz, 1989). As these tissues contain high levels of starch, especially when they are green, it is important to identify the metabolic precursor for starch biosynthesis in these plastids.

To get an insight into the precursor dependency of starch biosynthesis in a plant tissue it is possible to expose fruits to radioactively labelled carbon dioxide or sucrose and to quantify subsequently the rate of starch biosynthesis measurable as incorporation of radioactivity. However, this approach cannot identify which precursor is used by fruit chloroplasts for starch biosynthesis. The metabolism of such compounds by cytosolic enzymes would lead to the production of a wide range of molecules of which only one (or a few) might act as precursors for starch.

Therefore, we chose another, more simple, experimental approach to study the precursor dependency of starch biosynthesis in these plastids. For this approach, we developed a method allowing the enrichment of chloroplasts from bell pepper (*Capsicum annuum* L.) fruits (Batz et al., 1995). These chloroplast preparations are characterized by very low contamination by other cellular components, which is a prerequisite for analysing the precursor efficiency of starch biosynthesis. Using such enriched bell pepper fruit chloroplasts we were able to demonstrate that exogenous G6P was the most efficient precursor for starch biosynthesis (*Table 1* and *Figure 1*; Batz et al., 1995). This observation was very interesting as the experiments provided for the first time evidence that fruit chloroplasts can import exogenous G6P and use

Table 1. Precursor dependency of starch biosynthesis in various types of storage plastids

Precursor	Rate of starch synthesis (% of most efficient precursor)			
	Bell pepper fruit chloroplasts[b]	Cauliflower bud amyloplasts[c]	Barley etioplasts[d]	Maize endosperm amyloplasts[e]
G6P	100.0	100.0	8.0[a]	13.9
G1P	7.1	15.5	n.m.	11.0
Glucose	31.5	5.7	4.0[a]	n.m.
Fructose	12.0	3.8[a]	2.0[a]	n.m.
DHAP	9.6	13.5	100.0	0.4
PGA	8.0	n.m.	3.0[a]	n.m.
ADPglucose	n.m.	18.1	n.m.	100.0

n.m., not measured; G6P, glucose-6-phosphate; G1P, glucose-1-phosphate; DHAP, dihydroxyacetone phosphate; PGA, 3-phosphoglycerate.
[a]Unpublished.
[b]Data are taken from Batz et al. (1995).
[c]Data are taken from Batz et al. (1994) and Neuhaus et al. (1993).
[d]Data are taken from Batz et al. (1992).
[e]Data are taken from Neuhaus et al. (1993) and Möhlmann et al. (1996).
The most efficient precursor is given as 100%.

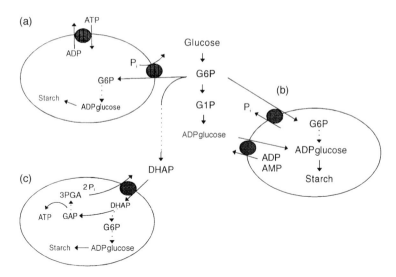

Figure 1. *Pathway of starch biosynthesis in various types of heterotrophic plastids.*
(a) Dicotyledonous amyloplast fruit plastid. (b) Cereal amyloplast.(c) Etioplast. For details see
Batz et al., 1992; Batz et al., 1994; Möhlmann et al., 1997. G6P, glucose-6-phosphate; G1P,
glucose-1-phosphate; GAP, glyceraldehyde-3-phosphate; DHAP, dihydroxyacetone phosphate;
PGA, 3-phosphoglycerate.

the glucose moiety for anabolic reactions. Interestingly, the observation that specific
types of chloroplasts possess a transport system for G6P has also been made before.
Overlach *et al.* (1993) and Rutter and Cobb (1983) clearly demonstrated that chloroplasts enriched from guard cells of pea leaves and from the alga, *Codium fragile*, transport G6P in counter-exchange for orthophosphate. Fruit chloroplasts resemble, in
many respects, leaf chloroplasts as they contain a functional thylakoid system and all
enzymes required for CO_2 fixation. Therefore, it was important to examine whether
illumination of isolated organelles influences starch biosynthesis. We demonstrated
that illumination of fruit chloroplasts stimulates G6P-dependent starch biosynthesis
by a factor of about three and established that this stimulation is mainly due to a stromal synthesis of ATP via the thylakoid-bound ATPase (Batz *et al.*, 1995). Of the various radioactively labelled compounds tested only G6P acted as a precursor for starch
biosynthesis at appreciable rates (*Table 1*) which means that a quite specific transport
protein is located in the inner chloroplast envelope membrane. Previous studies clearly
demonstrated that spinach chloroplasts contain a transport protein that catalyses the
export of dihydroxyacetone phosphate or PGA in strict counter-exchange for inorganic phosphate (P_i), but not G6P (Fliege *et al.*, 1978). However, chloroplasts from
bell pepper import exogenous G6P in counter-exchange for internal P_i. The observation that the latter type of chloroplast resides in a typical sink organ (fruit tissue)
whereas the spinach chloroplast resides in a source tissue indicates that the carbohydrate status of a tissue might influence the transport properties of the corresponding
plastids. To test this hypothesis we fed glucose into detached leaves from spinach
plants via the transpiration stream. This approach has been developed by Krapp *et al.*
(1991) and it has been well demonstrated that elevated glucose levels in leaves switch
their metabolism from that of a source tissue to that of a sink, a change that is characterized by a strong decrease of the chlorophyll content, and of the activity of Calvin

cycle enzymes (Krapp *et al.*, 1991). Surprisingly, after purification of chloroplasts from glucose-fed leaves we were able to demonstrate that chloroplasts had developed a new transport property since they were now able to import G6P and use the carbon unit from this molecule for starch biosynthesis (Quick *et al.*, 1995).

From this result it was tempting to propose that the ratio of various types of phosphate transporter changes in plastids during development. This assumption means that it should be possible to observe various types of phosphate transporters, at the same time, in one type of plastid. By purification of chromoplasts from bell pepper fruits and subsequent reconstitution of the envelope-bound transporters in proteoliposomes we were able to demonstrate that this type of plastid indeed contains both a triose phosphate/P_i and a G6P/P_i transporter (Quick and Neuhaus, 1996). That this is true for the majority of plastids in that tissue is indicated by the observation that the method used for chloroplast isolation allows recovery of significant amounts of chromoplasts from bell pepper fruit tissue (Quick and Neuhaus, 1996). In combination with previous results on pea embryo plastids, which also contained two types of phosphate transporters at the same time (Hill and Smith, 1995), we came to the conclusion that great caution is needed when metabolite transport across plastid envelopes is characterized as a 'single carrier-mediated process'.

However, the plastids discussed above do not represent typical storage plastids as they either contain a photosynthetic apparatus (bell pepper fruit chloroplasts) or because they do not contain significant amounts of starch (chromoplasts). In contrast, amyloplasts from cauliflower buds are located in the chlorophyll-free inflorescences of this plant but contain high amounts of starch (Neuhaus *et al.*, 1993). This starch appears in one or several large starch grains per plastid and fills most of the stroma (Journet and Douce, 1985). Using the method developed by Journet and Douce (1985) for purification of these plastids we were able to show that G6P is by far the most efficient precursor for starch biosynthesis in these organelles (Neuhaus *et al.*, 1993a, 1993b). Subsequently, we demonstrated in competition studies with all other potentially metabolic precursors that G6P is the main substrate for starch synthesis in cauliflower bud amyloplasts (Batz *et al.*, 1994).

It must be mentioned here that G6P-dependent starch synthesis in both bell pepper fruit chloroplasts or cauliflower bud amyloplasts is not only dependent upon the availability of G6P but also upon the additional presence of PGA and ATP (Batz *et al.*, 1995; Neuhaus *et al.*, 1993). Both compounds are additionally required as they activate and energize, respectively, the reaction of the plastidic AGPase. In nearly all plant tissues tested this enzyme is allosterically activated by PGA (Preiss, 1993) and of course always depends upon the ATP availability. By changing the concentration of PGA in the external medium we were able to demonstrate that submillimolar concentrations of this effector stimulate G6P-dependent starch synthesis several-fold (Neuhaus *et al.*, 1993).

This observation and the strict ATP dependency of starch synthesis in cauliflower amyloplasts (Neuhaus *et al.*, 1993) prompted us to develop a model in which heterotrophic tissues are able to adjust the rate of starch biosynthesis in response to the availability of carbohydrate. It is well known that most of the carbon fixed in source tissues is exported into sink tissues. Using a *Chenopodium rubrum* cell suspension culture Hatzfeld *et al.* (1990) were able to demonstrate metabolic events occurring during the switch from carbohydrate starvation of a tissue to starch accumulation. After feeding glucose to starved cells a massive increase of respiratory activity is correlated with elevated levels of hexose phosphates, PGA and ATP. Indeed, as mentioned above, all three metabolic compounds are required to observe high rates of starch synthesis. Thus, the high cytosolic sugar concentrations induce a metabolic situation which signals to the amyloplast that sufficient carbohydrate and energy are present to allow storage of excess organic carbon. Interestingly, in green leaves the rate of chloroplastic starch biosynthesis is also regulated by metabolic signals coming from the cytosol (Neuhaus and Stitt, 1990; Stitt, 1990).

3. Starch synthesis in heterotrophic plastids from cereal tissues

Cereals make by far the greatest contribution to world-wide starch production. The majority of starch is present in endosperm where it is deposited to provide the embryo with carbon and energy during germination. One major problem in analysing the precursor dependency of starch biosynthesis in the cereal endosperm amyloplasts is the extremely high starch content, representing a very dense, water-insoluble compound that destroys the intact plastid envelope during purification. This problem was circumvented by the development of relatively gentle methods allowing the enrichment of cereal amyloplast at low **g**-forces (Tetlow *et al.*, 1994; Tyson and ap Rees, 1988).

Tyson and ap Rees (1988) demonstrated that G1P is the most efficient precursor for starch biosynthesis by wheat endosperm amyloplasts. This result contrasted with subsequent work on several other types of heterotrophic plastids which nearly exclusively use exogenous G6P as the starch precursor (Hill and Smith, 1995; Neuhaus *et al.*, 1993a, 1993b). However, the more recent studies by Tetlow *et al.* (1994) clearly confirm the previous results and establish that wheat amyloplasts import both G1P and G6P. Both hexose phosphates are imported in a strict counter-exchange to P_i (Tetlow *et al.*, 1996) providing further evidence for a wide range of different types of phosphate transporters in plants.

Recently, a new isoform of AGPase, residing in the cytosol of cereal endosperm, and contributing to more than 90% of total AGPase activity has been identified in maize and barley (Denyer *et al.*, 1996; Thorbjørnsen *et al.*, 1996). This observation indicated that at least in the endosperm of these cereals a cytosolic synthesis of ADPglucose might contribute to starch biosynthesis. The high negative charge of ADPglucose, and the large size of the molecule makes it necessary to import ADPglucose into the amyloplast via a membrane-bound transporter. This assumption tempted us to analyse how ADPglucose influences starch biosynthesis in isolated maize amyloplasts, and to characterize the uptake mechanism mediating the import of this nucleotide sugar.

In a previous publication we demonstrated that isolated maize endosperm amyloplasts are able to use both G6P and G1P as precursors for starch (Neuhaus *et al.*, 1993). The examination of the concentration dependency of starch synthesis driven by either G6P or G1P led us to conclude that under *in vivo* conditions G6P is the most likely carbon source for starch synthesis (Neuhaus *et al.*, 1993). However, incubation of maize endosperm amyloplast with radioactively labelled ADP[^{14}C]glucose clearly demonstrated that the rates of starch synthesis are about seven times faster with this precursor (*Table 1*) than with saturating concentrations of G6P (Neuhaus *et al.*, 1993). As ADPglucose-driven starch biosynthesis was strongly dependent upon the integrity of the organelles, we concluded that cytosolically synthesized ADPglucose is imported into maize endosperm amyloplasts and the carbon moiety subsequently used for elongation of starch molecules.

The simple demonstration that a radioactively labelled compound leads to starch synthesis in isolated amyloplasts does not explain the uptake mechanism catalysing the transfer of this compound from one side of the membrane to the other. Therefore, we examined the transport of ADPglucose into maize endosperm amyloplasts in more detail. As short-term uptake into isolated organelles catalysed by a counter-exchange transporter depends upon the presence of sufficient counter-exchange substrate on the stromal site of the organelle we chose a proteoliposome approach. In such an approach the envelope-bound transport proteins are functionally reconstituted into artificial membranes consisting of pure phospholipids. The advantage of such a method is that these proteoliposomes can be preloaded with a defined concentration of various potential counter-exchange substrates. Therefore, differences in the subsequent uptake of the radioactively labelled transport molecule give indications about the efficiency of the respective counter-exchange substrate.

As ADPglucose and the adenine nucleotides ATP and ADP exhibit significant structural similarity it was supposed that these compounds are transported via the plastidic ATP/ADP transporter (Pozueta-Romero *et al.*, 1991), previously discovered in spinach chloroplasts (Heldt, 1969). Schünemann *et al.* (1993) clearly revealed that the ATP/ADP transporter from neither spinach chloroplasts nor pea root plastids interacts with ADPglucose. However, as these plastids are located in cells which obviously do not possess a cytosolic AGPase, the uptake of ADPglucose into these organelles is probably not required *in vivo*. As we have described above, isolated maize endosperm amyloplasts are located in cells harbouring a cytosolic AGPase and are, in addition, able to use ADPglucose for starch biosynthesis. Therefore, we examined whether the plastidic ATP/ADP transporter of these organelles is responsible for ADPglucose uptake. By estimating the effect of this nucleotide sugar on ATP uptake into proteoliposomes preloaded with ADP or AMP we clearly proved that the plastidic ATP/ADP transporter from maize endosperm amyloplast, like that from other sources, does not interact with ADPglucose as a substrate. Interestingly, uptake of ADPglucose into proteoliposomes was substantially stimulated by pre-loading of the vesicles with AMP (Möhlmann *et al.*, 1997). Such a nucleotide sugar/nucleotide monophosphate counter-exchange is in accordance with data about the uptake of nucleotide sugars into the endoplasmatic reticulum from animal tissues (for review see Abeijon and Hirschberg, 1992).

By analogy, we propose that starch biosynthesis in barley endosperm is also driven by uptake of ADPglucose synthesized by the cytosolic AGPase (Thorbjørnsen *et al.*, 1996). If that assumption is right, barley would be the first plant exhibiting totally different pathways for starch biosynthesis in various heterotrophic tissues. It is known that etioplasts possess at the early stages of development prominent starch grains filling about 30–60% of the stroma (Bradbeer, 1976). As these types of plastids lack chlorophyll and several enzymes involved in CO_2 fixation, exogenous carbon precursors have to be taken up into the organelle to allow net carbon accumulation.

To reveal which metabolite is used for starch biosynthesis in this specific type of heterotrophic plastid, we germinated and grew barley plants under continuous dark conditions and developed a method for isolation of the corresponding etioplasts (Batz *et al.*, 1992). An examination of the enzymic complement of these organelles showed that this type of plastid contains, in contrast to most other heterotrophic plastids (Entwistle and ap Rees, 1990), all enzymes required for the conversion of triose phosphates to hexose phosphates, including the strictly required fructose-1,6-bisphosphatase (FBPase).

Short-term uptake experiments in which isolated etioplasts have been incubated for some seconds with either radioactively labelled G6P or dihydroxyacetone phosphate (DHAP) revealed that after separation of the organelles by centrifugation through a silicon oil layer, only DHAP is imported into the stroma (Batz *et al.*, 1992).

These observations encouraged us to quantify a possible DHAP-driven starch biosynthesis in barley etioplasts. As shown in *Table 1*, DHAP is the most efficient precursor for starch synthesis which represented the first example of a triose phosphate-driven starch biosynthesis in plastids from a non-photosynthetic tissue (Batz *et al.*, 1992). Subsequently, Zhang and Murphy (1996) demonstrated that etioplasts purified from pinyon also preferentially use DHAP as the carbon source for starch synthesis. Since in neither of these studies was exogenous ATP required for starch synthesis we have to predict that uptake of triose phosphates provides both the carbon moiety for starch and the energy equivalent in the form of ATP, necessary for energization of AGPase.

A model allowing such supply is given in *Figure 1*. According to this scheme three molecules of DHAP enter the etioplast lumen. Two molecules are used for synthesis of fructose-1,6-bisphosphate which is hydrolysed by FBPase, converted to G1P and subsequently used for ADPglucose synthesis. For the generation of ADPglucose one molecule of ATP is necessary which is generated by the oxidation of triose phosphates to PGA. In this respect, etioplasts represent an example of heterotrophic plastids able to synthesize ATP at sufficient rates for a major anabolic reaction via a glycolytic enzyme sequence. In nearly all other types of heterotrophic plastids the internal production of ATP is insufficient to allow significant rates of anabolic metabolism.

4. Molecular analysis of the plastidic ATP/ADP transporter

As mentioned above, ATP uptake into heterotrophic plastids is required for energization of several important anabolic pathways. For example, starch and fatty acid synthesis in various isolated storage plastids are more or less totally dependent upon the additional presence of ATP in the incubation medium (Kang and Rawsthorne, 1994; Neuhaus *et al.*, 1993; Tetlow *et al.*, 1994; Zhang and Murphy, 1996). Moreover, we have demonstrated that both G6P-driven starch biosynthesis and acetate-dependent fatty acid synthesis in isolated cauliflower bud amyloplasts compete for ATP imported into that organelle (Möhlmann *et al.*, 1994). These observations indicated that the plastidic ATP/ADP transporter has a very important function in controlling the rate of end-product synthesis in storage tissues.

Plant cells, in common with all eukaryotic cells, possess an adenylate transport system residing in the inner mitochondrial membrane and catalysing the exchange of ATP and ADP (Vignais, 1976). The function of this highly active transporter is the provision of ATP-consuming reactions with ATP synthesized in the mitochondrial matrix during oxidative phosphorylation (Klingenberg, 1989). As this transporter catalyses a strict counter-exchange of ATP and ADP in a 1:1 stoichiometry (Klingenberg, 1989), it was attractive to assume that an isoform of this carrier protein resides in the inner plastidic envelope membrane. However, several biochemical, physiological and immunological observations indicated that this is not so. Firstly, the direction of transport in plastids is opposite to adenine nucleotide transport in mitochondria. In plastids ATP is imported in counter-exchange to ADP export (Heldt, 1969; Schünemann *et al.*, 1993). Secondly, mitochondrial ADP/ATP transporters are strongly inhibited by carboxyatractyloside or bongkregic acid (Stubbs, 1981). These inhibitors do not affect the plastidic ATP/ADP transport system significantly (Schünemann et al., 1993). Thirdly, polyclonal antibodies raised against the mitochondrial ADP/ATP transporter from Neurospora crassa do not cross-react with proteins in isolated plastid envelope membranes (Neuhaus et al., 1997; Schünemann et al., 1993). These observations encouraged us to search for a new type of adenine nucleotide transporter in higher plant plastids.

Recently, we screened a cDNA library from Arabidopsis thaliana L. for a copper transporter. We found a cDNA representing a hybrid of a cDNA coding for the copper transporter and a cDNA encoding a protein with high similarity to the ATP/ADP transporter from Rickettsia prowazekii. Bacteria of the genus Rickettsia are obligatory, Gram-negative intracellular parasites which can only grow in the cytoplasm of eukaryotic cells. Rickettsia prowazekii causes typhus and exploits the host cell's cytoplasm by uptake of various metabolic intermediates (Winkler, 1976). One of these intermediates is ATP, which is imported from the eukaryotic cytoplasm in counter-exchange with bacterial ADP (Winkler, 1976) on a translocator subsequently identified at the molecular level (Williamson et al., 1989). The protein sequence deduced after sequencing the full-length clone from A. thaliana revealed a highly hydrophobic membrane protein with 12 putative transmembrane domains (Kampfenkel et al., 1995). In the structural part of the protein the plastidic ATP/ADP transport protein (AATP1, At) exhibits more than 66% similarity to the rickettsial translocase (Kampfenkel et al., 1995; Krause et al., 1985).

In contrast to the rickettsial ATP/ADP translocator, the plastidic ATP/ADP transporter possesses an N-terminal transit peptide which exhibits typical features common to four other membrane proteins located in the plastid envelope (see Kampfenkel et al., 1995). Using radioactively labelled AATP1 (At) pre-protein we were able to show that this protein is targeted into the envelope membrane from spinach chloroplasts and the transit peptide is processed during embedding of the structural part of the transporter (Neuhaus et al., 1997). By using a peptide-specific antibody raised against a fragment of a hydrophilic loop of the protein we demonstrated that a cross-reacting protein of about 63 kDa is present in spinach chloroplast envelopes, but not in mitochondrial membranes.

One feature of the rickettsial transporter is a relatively high substrate affinity for adenine nucleotides (Winkler, 1976). This feature also holds true for the authentic plastidic protein (Neuhaus et al., 1993; Schünemann et al., 1993). To prove whether the gene product of AATP1 (At) also has a high affinity of adenine nucleotides we expressed AATP1 (At) heterologously in Escherichia coli and in yeast (Saccharomyces cerevisiae), reconstituted the hydrophobic membrane protein in proteoliposomes, and measured uptake of [^{32}P]ATP after preloading the vesicles with ADP. As expected, AATP1 (At) encoded an adenine nucleotide transporter exhibiting an apparent K_m for ATP of about 20 μM (Neuhaus et al., 1997).

A Southern blot analysis of genomic DNA purified from two Arabidopsis ecotypes, namely Landsbergia and Columbia, showed that the plastidic ATP/ADP transporter exists in two isoforms in both ecotypes. This observation prompted us to re-screen the A. thaliana cDNA library from which we isolated the cDNA coding for the second isoform (AATP2, At; Möhlmann et al., 1998). The deduced amino acid sequence exhibits 77% identity when compared to AATP1 (At), and 38% to the rickettsial transporter (Möhlmann et al., 1998).

The presence of isoforms of enzymes or transport proteins might have several functions. For example, different biochemical properties of the two plastidic ATP/ADP transporters might be required in various types of plastids: it is known that different types of plastid fulfil very specific functions (Emes and Neuhaus, 1997). Another reason for the existence of isoforms of proteins might be that specific expression of a gene encoding a particular isoform is required during development of a plant. This suggestion is reinforced by the observation that, for example, two isoforms of the mitochondrial ADP/ATP transporter are present in maize (Winning *et al.*, 1991), while three isoforms of this carrier are present in yeast (Kalorov *et al.*, 1990). In order to ascertain whether the two plastidic ATP/ADP transporters from *A. thaliana* have different biochemical properties it is necessary to conduct uptake experiments using the heterologously expressed carrier proteins. As our previously developed method for expressing AATP1 (At) in yeast or *E. coli* necessitated the subsequent reconstitution of membrane proteins in proteoliposomes, pre-loading with counter-exchange substrates and uptake experiments using anion exchange columns, we decided to search for alternative methods which are simpler and quicker to use. Recently, Miroux and Walker (1996) have developed a new *E. coli* strain (C43) that exhibits superior abilities to express heterologous membrane proteins of eukaryotic origin. Using this strain we tried to express AATP1 (At) heterologously with the aim of obtaining functional integration of this carrier protein in the cytoplasmic membrane of the bacterium. This approach was encouraged by the observation that heterologous expression of the rickettsial ATP/ADP transporter exhibiting about 66% similarity to AATP1 (At) can result in a functional adenine nucleotide transport system in the *E. coli* cytoplasmic membrane (Tjaden *et al.*, 1998). Using this approach we have been able to show that both AATP1 and AATP2 (At) are adenine nucleotide carriers exhibiting similar low K_m values (i.e. high affinities) for both ATP and ADP (between 20 and 25 μM) (Möhlmann *et al.*, 1998; Tjaden *et al.*, 1998). These data demonstrate that the basic biochemical properties of the heterologously expressed carriers are comparable to the authentic transporters analysed in isolated plastids (Neuhaus *et al.*, 1993; Schünemann *et al.*, 1993).

To reveal the function of the plastidic ATP/ADP transporter for the synthesis of metabolic storage products like starch it is necessary to change the activity of the carrier. Such changes can be induced by creating transgenic plants with altered levels of mRNA coding for a homologous protein from another species ('sense plants'), and 'antisense plants' transformed with the endogenous cDNA cloned in the antisense orientation. Using an *Agrobacterium tumefaciens*-mediated transfer system we have created potato (*Solanum tuberosum*) plants with altered levels of mRNA coding for plastidic ATP/ADP transporters. For sense plants we chose the heterologous AATP1 (At) cDNA to prevent co-suppression. For antisense plants we first identified a potato homologue (AATP1, St) and cloned it in an antisense orientation. Both constructs were placed under the control of the constitutive cauliflower mosaic virus 35S-promoter leading to a permanent expression of the cDNAs in all types of tissues.

After demonstrating the successful transformation and expression of the cDNAs by northern blot analysis it became obvious that there were no significant changes in the morphological appearance of the green potato tissues. However, several antisense lines showed a strongly increased number of tubers (J. Tjaden *et al.*, unpublished results). In some strong antisense lines, the total number of tubers did not change but the morphology of the tubers became finger-like. Obviously, the reduction of plastid ATP/ADP transporter activity induced a significant change in tuber metabolism. Metabolic studies on leaves and tubers of sense and antisense plants will be carried out in the near future to gain a more detailed understanding of heterotrophic plant metabolism.

References

Abeijon, C. and Hirschberg, C.B. (1992) Topography of glycosylation reactions in the endoplasmic reticulum. *Trends Biol. Sci.* **17**: 32–36.

Batz, O., Scheibe, R. and Neuhaus, H.E. (1992) Transport processes and corresponding changes in metabolite levels in relation to starch synthesis in barley (*Hordeum vulgare* L.) etioplasts. *Plant Physiol.* **100**: 184–190.

Batz, O., Scheibe, R. and Neuhaus, H.E. (1994) Glucose- and ADPGlc-dependent starch synthesis in isolated cauliflower-bud amyloplasts. Analysis of the interaction of various potential precursors. *Biochim. Biophys. Acta* **1200**: 148–154.

Batz, O., Scheibe, R. and Neuhaus, H.E. (1995) Purification of chloroplasts from fruits of green-pepper (*Capsicum annuum* L.) and characterization of starch synthesis. Evidence for a functional hexose-phosphate translocator. *Planta* **196**: 50–57.

Blanke, M.M. and Lenz, F. (1989) Fruit photosynthesis. *Plant Cell Environ.* **12**: 31–46.

Bradbeer, J.W. (1976) Chloroplast development in greening leaves In: *Perspectives in Experimental Biology*, Vol. 2 *Botany* (ed. N. Sutherland). Pergamon Press, Oxford, pp. 131–143.

Casper, T., Huber S.C. and Sommerville, C.R. (1986) Alterations in growth, photosynthesis and respiration in a starchless mutant of *Arabidopsis thaliana* deficient in chloroplast phosphoglucose mutase activity. *Plant Physiol.* **79**: 1–7.

Denyer, K., Dunlap, F., Thorbjønsen, T., Keeling, P. and Smith, A.M. (1996) The major form of ADP-glucose pyrophosphorylase in maize endosperm is extra-plastidial. *Plant Physiol.* **112**: 779–785.

Doremus, H.D. and Jagendorf, A.T. (1985) Subcellular localization of the pathway of *de novo* pyrimidine synthesis in pea leaves. *Plant Physiol.* **79**: 856–861.

Emes, M.J. and Neuhaus, H.E. (1997) Metabolism and transport in non-photosynthetic plastids. *J. Exp. Bot.* **48**: 1995–2005.

Entwistle, G. and ap Rees, T. (1990) Lack of fructose-1,6-bisphosphatase in a range of higher plants that store starch. *Biochem. J.* **271**: 467–472.

Fliege, R., Flügge, U.I., Werdan, K. and Heldt, H.W. (1978) Specific transport of inorganic phosphate, 3-phosphoglycerate and triose phosphates across the inner membrane of the envelope in spinach chloroplasts. *Biochim. Biophys. Acta* **502**: 232–247.

Fondy, B.R. and Geiger, D.R. (1982) Diurnal pattern of translocation and carbohydrate metabolism in source leaves of *Beta vulgaris* L. *Plant Physiol.* **70**: 671–676.

Hatzfeld, W.-D., Dancer, J. and Stitt, M. (1990) Fructose-2,6-bisphosphate, metabolites and 'coarse' control of pyrophosphate: fructose-6-phosphate phosphotransferase during triose-phosphate cycling in heterotrophic cell-suspension cultures of *Chenopodium rubrum*. *Planta* **180**: 205–211.

Heldt, H.W. (1969) Adenine nucleotide translocation in spinach chloroplasts. *FEBS Lett.* **5**: 11–14.

Hill, L.M. and Smith, A.M. (1991) Evidence that glucose-6-phosphate is imported as the substrate for starch synthesis by the plastids of developing pea embryos. *Planta* **185**: 91–96.

Hill, L.M. and Smith, A.M. (1995) Coupled movements of glucose 6-phosphate and triose phosphate across the envelopes of plastids from developing embryos of pea (*Pisum sativum* L.). *J. Plant Physiol.* **146**: 411–417.

Journet, E.-P. and Douce, R. (1985) Enzymic capacities of purified cauliflower bud plastids for lipid synthesis and carbohydrate metabolism. *Plant Physiol.* **79**: 458–467.

Kaiser, W.M. and Brendle-Behnisch, E. (1991) Rapid modulation of spinach leaf nitrate reductase by photosynthesis I. Modulation in vivo by CO_2 availability. *Plant Physiol.* **96**: 363–367.

Kalorov, J., Kalorova, N. and Nelson, N. (1990) A third ADP/ATP translocator gene in yeast. *J. Biol. Chem.* **265**: 12711–12716.

Kampfenkel, K., Möhlmann, T., Batz, O., van Montagu, M., Inzé, D. and Neuhaus, H.E. (1995) *FEBS Lett.* **374**: 351–355.

Kang, F. and Rawsthorne, S. (1994) Starch and fatty acid synthesis in plastids from developing embryos of oilseed rape (*Brassica napus* L.). *Plant J.* **6**: 795–805.

Kleinhanß, W. (1988) Produktion und Nutzungsmöglichkeiten nachwachsender Rohstoffe. Schriftenreihe des Bundesministeriums für Ernährung, Landwirtschaft and Forsten. Reihe A: Angewandte Wissenschaft, Heft 353, Landwirtschaftverlag GmbH, Münster-Hiltrup, Germany.

Klingenberg, M. (1989) Molecular aspects of adenine nucleotide carrier from mitochondria. *Arch. Biochem. Biophys.* **270**: 1–14.

Krapp, A., Quick, W.P. and Stitt, M. (1991) Ribulose1,5bisphosphate carboxylase-oxygenase, other photosynthetic enzymes and chlorophyll decrease when glucose is supplied to mature spinach leaves via the transpiration stream. *Planta* **186**: 58–69.

Krause, D.C., Winkler, H.H. and Wood, D.O. (1985) Cloning and expression of the *Rickettsia prowazekii* ADP/ATP translocator in *Escherichia coli*. *Proc. Natl Acad. Sci. USA* **82**: 3015–3019.

Miroux, B. and Walker, J.E. (1996) Over-production of proteins in *Escherichia coli*: mutant hosts that allow synthesis of some membrane proteins and globular proteins at high levels. *J. Mol. Biol.* **260**: 289–298.

Möhlmann, T., Scheibe, R. and Neuhaus, H.E. (1994) Interaction between starch synthesis and fatty-acid synthesis in isolated cauliflower-bud amyloplasts. *Planta* **194**: 492–497.

Möhlmann, T., Tjaden, J., Henrichs, G., Quick, P.W., Häusler, R. and Neuhaus, H.E. (1997) ADPglucose drives starch synthesis in isolated maize-endosperm amyloplasts. Characterisation of starch synthesis and transport properties across the amyloplastic envelope. *Biochem. J.* **324**: 503–509.

Möhlmann, T., Tjaden, J., Schwöppe, C., Winkler, H.H., Kampfenkel, K. and Neuhaus, H.E. (1998) Occurrence of two plastidic ATP/ADP transporters in *Arabidopsis thaliana*: molecular characterisation and comparative structural analysis of homologous ATP/ADP translocators from plastids and *Rickettsia prowazekii*. *Eur. J. Biochem.* **252**: 353–359.

Müller-Röber, B., Sonnewald, U. and Willmitzer, L. (1992) Inhibition of the ADP-glucose pyrophosphorylase in transgenic potatoes leads to sugar-storing tubers and influences tuber formation and expression of tuber storage protein genes. *EMBO J.* **11**: 1229–1238.

Neuhaus, H.E. and Stitt, M. (1990) Control analysis of photosynthate partitioning. Impact of reduced activity of ADP-glucose pyrophosphorylase or plastid phosphoglucomutase on the fluxes to starch and sucrose in *Arabidopsis thaliana* (L.) Henyh. *Planta*. **182**: 445–454.

Neuhaus, H.E., Henrichs, G. and Scheibe, R. (1993a) Characterization of glucose-6-phosphate incorporation into starch by isolated intact cauliflower-bud plastids. *Plant Physiol.* **101**: 573–578.

Neuhaus, H.E., Thom, E., Batz, O. and Scheibe, R. (1993b) Purification of highly intact plastids from various heterotrophic plant tissues. Analysis of enzymic equipment and precursor dependency for starch biosynthesis. *Biochem J.* **296**: 396–401.

Overlach, S., Dickmann, W. and Raschke, K. (1993) Phosphate translocator of isolated guard-cell chloroplasts from *Pisum sativum* transports glucose-6-phosphate. *Plant Physiol.* **101**: 1201–1207.

Pozueta-Romero, J., Frehner, M., Viale, A.M. and Akazawa, T. (1991) Direct transport of ADPglucose by an adenylate translocator is linked to starch biosynthesis in amyloplasts. *Proc. Natl Acad. Sci. USA* **88**: 5769–5773.

Preiss, J. (1982) Regulation of the biosynthesis and degradation of starch. *Annu. Rev. Plant Physiol.* **33**: 431–454.

Preiss, J. (1991) *Oxford Surveys of Plant Molecular & Cell Biology*, Vol. 7. Oxford University Press, Oxford, pp. 59–114.

Quick, P. and Neuhaus, H.E. (1996) Evidence for two types of phosphate translocators in pepper (*Capsicum annuum* L.) fruit chromoplasts. *Biochem. J.* 320: 7–10.

Quick, P., Scheibe, R. and Neuhaus, H.E. (1995) Induction of a hexose-phosphate translocator activity in spinach chloroplasts. *Plant Physiol.* 109: 113–121.

Rutter, C.J. and Cobb, A.H. (1983) Translocation of orthophosphate and glucose-6-phosphate in *Codium fragile* chloroplasts. *New Phytol.* 95: 559–568.

Schünemann, D., Borchert, S., Flügge, U.I. and Heldt, H.W. (1993) ATP/ADP translocator from pea root plastids. Comparison with translocators from spinach chloroplasts and pea leaf mitochondria. *Plant Physiol.* 103: 131–137.

Smith, A.M., Denyer, K. and Martin, C. (1997) The synthesis of the starch granule. *Annu. Rev. Plant Physiol. Plant Mol. Biol.* 48: 67–87.

Stitt, M. (1990) Fructose-2, 6-bisphosphate in plants. *Annu. Rev. Plant Physiol. Plant Mol. Biol.* 41: 153–185.

Stitt, M. and Heldt, H.W. (1981) Physiological rates of starch breakdown in isolated intact spinach chloroplasts. *Plant Physiol.* 68: 755–761.

Stubbs, M. (1981) Inhibitors of the adenine nucleotide translocase. *Int. Encycl. Pharmacol. Ther.* 107: 283–304.

Stumpf, P.K. (1980) Biosynthesis of saturated and unsaturated fatty acids. In: *The Biochemistry of Plants*, Vol. 4 (eds P.K. Stumpf and E.E. Conn). Academic Press, New York, pp. 177–199.

Tetlow, I.J., Blisset, K.J. and Emes, M.J. (1994) Starch synthesis and carbohydrate oxidation in amyloplasts from developing wheat endosperm. *Planta* 194: 454–460.

Tetlow, I.J., Bowsher, C.G. and Emes, M.J. (1996) Reconstitution of the hexose phosphate translocator from the envelope membranes of wheat endoperm amyloplasts. *Biochem. J.* 319: 717–723.

Thom, E. and Neuhaus, H.E. (1995) Oxidation of imported or endogenous carbohydrates by isolated chloroplasts from green pepper fruits. *Plant Physiol.* 109: 1421–1425.

Thorbjørnsen, T., Villand, P., Denyer, K., Olsen, O.-A. and Smith, A.M. (1996) Distinct isoforms of ADPglucose pyrophosphorylase occur inside and outside the amyloplasts in barley endosperm. *Plant J.* 10: 243–250.

Tjaden, J., Schwöppe, C., Möhlmann, T. and Neuhaus, H.E. (1998) Expression of the plastidic ATP/ADP transporter gene in *Escherichia coli* lead to the presence of a functional adenine nucleotide transport system in the bacterial cytoplasmic membrane. *J. Biol. Chem.* 273: 9630–9636.

Tyson, R.H. and ap Rees, T. (1988) Starch synthesis by isolated amyloplasts from wheat endosperm. *Planta* 175: 33–38.

Vignais, P.V. (1976) Molecular and physiological aspects of adenine nucleotide transport in mitochondria. *Biochim. Biophys. Acta* 456: 1–38.

Williamson, L.R., Plano, G.V., Winkler, H.H., Krause, D.C. and Wood, D.O. (1989) Nucleotide sequence of the *Rickettsia prowazekii* ATP/ADP translocase-encoding gene. *Gene* 80: 269–278.

Winkler, H.H. (1976) Rickettsial permeability: an ADP-ATP transport system. *J. Biol. Chem.* 251: 389–396.

Winning, B.M., Day, C.D., Sarah, C.J. and Leaver, C.J. (1991) Nucleotide sequence of two cDNAs encoding the adenine nucleotide translocator from *Zea mays* L. *Plant Mol. Biol.* 17: 305–307.

Zhang, F. and Murphy, J.B. (1996) A phosphate translocator is present in photosynthetically inactive plastids of dark-germinated pinyon seedlings. *Plant Physiol.* 111(Suppl.): 108.

Ziegler, P. and Beck, E. (1989) Biosynthesis and degradation of starch in higher plants. *Annu. Rev. Plant Physiol. Plant Mol. Biol.* 40: 95–117.

19

The folate status of plant mitochondria

Fabrice Rébeillé and Roland Douce

1. Introduction

One of the most spectacular properties of leaf mitochondria is their involvement in the photorespiratory process. Photorespiration is initiated at the level of the Rubisco. The carboxylation reaction leads to the production of two molecules of PGA whereas the oxygenation reaction leads to one molecule of PGA and one molecule of 2-phosphoglycolate. In order to avoid depletion of the Benson–Calvin cycle, 2-phosphoglycolate (a two-carbon molecule) must be converted into PGA (a three-carbon molecule) and reintroduced in the reductive pentose phosphate cycle. This is the main function of the phosphorespiratory metabolism which involves three different cell compartments: the chloroplast, the peroxisome and the mitochondrion (for a review, see Bourguignon *et al.*, 1998). The conversion of a two-carbon molecule into a three-carbon molecule is carried out in mitochondria through the formation of serine from glycine, involving the glycine cleavage system (GDC) and serine hydroxymethyltransferase (SHMT). Firstly, glycine is broken down by a complex of proteins (GDC) which through their concerted activities, catalyse the oxidative decarboxylation and deamination of glycine with the formation of CO_2, NH_3 and the concomitant reduction of NAD^+ to NADH. The remaining carbon, the methylene carbon of glycine, is then transferred to tetrahydrofolate (H_4FGlu_n) to form methylene tetrahydrofolate ($CH_2H_4FGlu_n$). The latter compound reacts with a second molecule of glycine to form serine, a reaction catalysed by SHMT (*Figure 1a*). During the greening of leaves, the amounts of these two enzymatic systems increase considerably in the matrix space of mitochondria where they represent up to 40% of the soluble proteins (*Figure 1b*). These proteins require the presence of various co-factors: lipoate, pyridoxal phosphate and tetrahydrofolate, and it is likely that their synthesis increases during the greening of leaves in order to keep pace with the accumulation of GDC and SHMT. The origin and availability of these co-factors is still an open question. In this chapter we present evidence indicating that mitochondria play a major role in the synthesis of one of these co-factors, tetrahydrofolate.

Plant Carbohydrate Biochemistry, edited by J.A. Bryant, M.M. Burrell and N.J. Kruger.
© 1999 BIOS Scientific Publishers Ltd, Oxford.

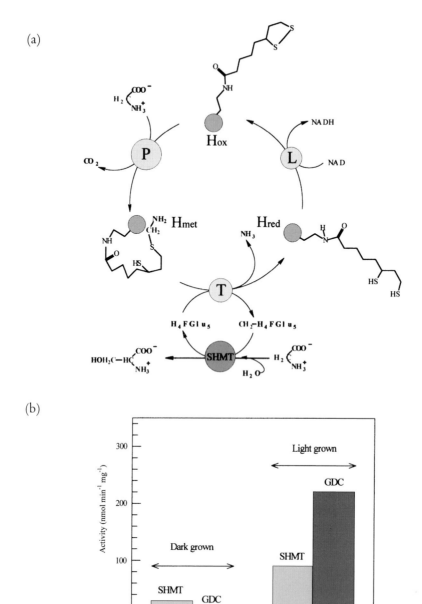

Figure 1. *Schematic representation of the GDC and SHMT coupled reactions (a) and effect of illumination upon the GDC and SHMT content (b). The lipoamide arm of the H protein interacts successively with the P, the T and the L proteins. Once charged in methylamine, the lipoamide arm rotates to come in contact with hydrophobic residues of a cavity opened at the surface of the protein, thus protecting the methylamine group against nucleophilic attack. The rate of GDC is about three times higher than the rate of SHMT, thus maintaining a high level of methylene tetrahydrofolate which drives the SHMT reaction toward the production of serine. Recycling of tetrahydrofolate through the SHMT reaction is the limiting step of the whole reaction.*

2. Folate content of plant mitochondria

There is now strong evidence that folate is present in various subcellular compartments. Several authors (Chen et al., 1997; Okinaka and Iwai, 1970b) have reported
that the largest pool of folate is found in a soluble (cytosol-enriched) fraction. The
size of the mitochondrial folate pool has been estimated by several groups. Analysis of
mitochondria from 4-day-old pea cotyledons suggests that they contain 3.5–4% of the
total pool of folate (Clandinin and Cossins, 1972). However, the folate level in mitochondria might increase significantly during the course of pea development (Coffin
and Cossins, 1986). Another report indicated that mitochondria from 1-day-old pea
cotyledons contain up to 20% of the total folate pool (Okinaka and Iwai, 1970b). The
subcellular distribution of folate in 2-week-old pea leaves has been investigated using
Percoll-purified mitochondria (Neuburger et al., 1996). On a protein basis, these
mitochondria have a folate content of about 400 pmol mg^{-1} protein, in good agreement with other reports (Besson et al., 1993; Chen et al., 1997). This is almost three
times higher than the whole leaf content (Chen et al., 1997), suggesting that mitochondria, which represent only 3–5% of the cytoplasmic (cytosol plus organelles) volume, have most probably the highest folate concentration. Assuming a mitochondrial
volume of 2 μl mg^{-1} protein, it can be estimated that, in pea leaves, the mitochondrial
folate concentration is approximately 0.2 mM. By comparison, the folate content of
chloroplasts is approximately 150 times lower on a protein basis (Neuburger et al.,
1996).

It is now clear that the principal forms of folate in living cells are the polyglutamyl
derivatives of tetrahydrofolate (Imeson et al., 1990; McGuire and Coward, 1984). In
plants, there is considerable variation in the glutamate chain length, depending on the
species (Zheng et al., 1992), although highly conjugated forms predominate. For
example, pea seedlings contain a large proportion of tetra- and pentaglutamate folates
(Imeson et al., 1990). Likewise, the major forms of folate recovered in the matrix space
of pea leaf mitochondria were tetra- (25%) and pentaglutamate (55%) derivatives
(Besson et al., 1993). Considering the variability of the polyglutamate chain, one may
question its physiological role. In pea leaf mitochondria, for example, it was observed
that the affinity of tetrahydrofolate for the T-protein of the GDC increases with the
number of glutamate residues (Rébeillé et al., 1994). There is strong evidence that the
negatively charged α-COO$^-$ groups of the polyglutamate tail could bind at specific
points such as basic groups of the protein. A close association between the tetrahydrofolate co-factor and the protein may have several advantages. Firstly, it can increase
the efficiency of the reaction, lowering the K_m values, and most folate-dependent
enzymes exhibit a higher affinity for polyglutamyl-conjugated substrates than for the
corresponding monoglutamate (McGuire and Coward, 1984; Rébeillé et al., 1994;
Strong et al., 1990). Secondly, folylpolyglutamate binding to folate-dependent
enzymes may contribute significantly to the protection of this readily oxidizable molecule (Rébeillé et al., 1994). If this holds true in vivo, it is possible that tetrahydrofolate derivatives are not free in the matrix space. On the basis of the matrix
concentrations of SHMT and the T-protein of GDC and taking into account that the
tetrameric SHMT binds one folate molecule per subunit and the monomeric T-protein
binds one folate molecule, it was calculated that these two proteins are sufficiently
abundant to bind all folate in the mitochondria (Rébeillé et al., 1994). Clearly, the
optimal glutamate chain length must be the result of a compromise between a strong

affinity for folate enzymes which limits the diffusion in the bulk medium and the ability for the molecule to move from one catalytic site to another. The importance of polyglutamate derivatives is highlighted by the fact that mutations affecting the generation of conjugated folates result in auxotrophy for products of one-carbon metabolism such as methionine (Cossins and Chan, 1984).

3. Interaction between the T-protein and the SHMT through a common pool of tetrahydrofolate

It has been proposed that the binding of folylpolyglutamates increases the efficiency of sequential folate-dependent proteins by enhancing the 'channelling' of intermediates between the active sites (MacKenzie, 1984; McGuire and Bertino, 1981). In other words, the two enzymes must be close enough that the reaction intermediate (i.e. the product of the first reaction and substrate of the second) can be directly transferred from one site to the other. Supporting this view, it has been shown that for formiminotransferase-cyclodeaminase (Paquin et al., 1985) the specificity of channelling was optimum for H_4FGlu_5, the pentaglutamate derivatives being one of the predominant polyglutamate forms of folate in animals (Kisliuk, 1981) and plants (Besson et al., 1993; Imeson et al., 1990). Another example of folate channelling between two catalytic domains is provided by the bifunctional dihydrofolate reductase/thymidylate synthase (DHFR/TS). Indeed, the crystallographic structure of DHFR/TS indicates that transfer of dihydrofolate between the two active sites does not occur by a diffusional pathway through the medium but rather by an electrostatic channelling at the surface of the protein (Knighton et al., 1994). However, it is not certain that such situations exist with all sequential folate-dependent proteins. We have investigated this question in mitochondria from leaf tissues where the SHMT and the T-protein of the GDC interact through a common pool of polyglutamyltetrahydrofolates (*Figure 1*). If CH_2-H_4FGlu_n, product of the GDC reaction and substrate of the SHMT reaction was channelled between the two catalytic sites, it would not equilibrate with the bulk solvent but would be directly transferred from one site to another. In a reconstituted system containing GDC and SHMT, it was observed that, during the steady-state course of glycine oxidation, the pentaglutamate form of CH_2-H_4FGlu_n accumulated in the bulk medium (Rébeillé et al., 1994). In other words, CH_2-H_4FGlu_n is released into solution and must reassociate with SHMT before being converted into H_4FGlu_n. This indicates that, in these *in vitro* experiments, the folate co-factor was not channelled between the T-protein and SHMT. However, it must be kept in mind that the *in vivo* situation might be different since the very high concentrations of T-protein and SHMT in the matrix space (approximately 200 mg ml^{-1}), together with the relatively low tetrahydrofolate level (0.2 mM), could lead to a situation in which the diffusional pathway of folate compounds is considerably reduced (see Prabhu et al., 1996).

4. The synthesis of tetrahydrofolate

Tetrahydrofolate biosynthesis requires, first, the generation of pterin compounds from GTP, a pathway initiated by GTP cyclohydrolase (Brown, 1985). Pterins are a family of molecules that are found in pigments of some organisms (the name pterin originates from the Greek name *pteron*, wings, because these compounds were found on the wings of some butterflies), but are also involved in the metabolism of aromatic amino acids (Kaufman and Kaufman, 1985; Shiman, 1985). One of these pterins, the 6-hydroxymethyl dihydropterin, is the starting point of the tetrahydrofolate biosynthetic pathway. As shown in *Figure 2*, this pathway requires the sequential operation of five reactions. This process is specific to plants and microorganisms, because the three first reactions are absent in animals. Interestingly, all these activities are detectable in Percoll-purified mitochondria (Neuburger *et al.*, 1996). In marked contrast, no significant activities could be detected in the other cell fractions (chloroplasts, cytosol) suggesting that mitochondria play a major role in tetrahydrofolate synthesis (*Table 1*).

4.1 *The bifunctional dihydropterin pyrophosphokinase/dihydropteroate synthase (HPPK/DHPS)*

In higher plants, the two first steps of tetrahydrofolate synthesis are catalysed by a bifunctional protein. This situation is different to that in prokaryotes where two separate enzymes are involved (Lacks *et al.*, 1995; Talarico *et al.*, 1992). In eukaryotes, the related proteins studied so far are always multifunctional. In the protozoa *Plasmodium falciparum* and *Toxoplasma gondii* the enzyme is bifunctional, containing HPPK and DHPS activities (Allegra *et al.*, 1990; Triglia and Cowman, 1994) and in *Pneumocystis carinii* it is a trifunctional enzyme containing dihydroneopterin aldolase, HPPK and DHPS activities (Volpe *et al.*, 1993). The DHPS domain is the target of sulphonamide

Table 1. *Enzyme activities involved in tetrahydrofolate and thymidylate synthesis in the different pea leaf cell compartments. Fumarase, PRK and PEP carboxylase were marker enzymes respectively for mitochondria, chloroplasts and cytosol. In the cytosol-enriched fraction less than 1% of the proteins were from mitochondria and 30–45% were from plastids. The small activities observed in the cytosol fraction were most probably of mitochondrial origin. Although nuclei were purified on a Percoll gradient, about 25% of the proteins were from chloroplasts and 1% were from cytosol*

	Activity in specific compartment in pea leaves			
Enzyme	Mitochondria	Chloroplasts	Cytosol	Nuclei
		($nmol h^{-1} mg^{-1}$ protein)		
Fumarase	$51\,000 \pm 3000$	n.d.	420 ± 120	n.d.
PRK	n.d.	$27\,000 \pm 2400$	$10\,200 \pm 2400$	7200 ± 1200
PEP carboxylase	n.d.	n.d.	6000 ± 800	60 ± 10
HPPK + DHPS	1.8 ± 0.3	n.d.	n.d.	n.d.
DHPS	16 ± 3	n.d.	0.08 ± 0.04	n.d.
DHFS	1.6 ± 0.3	n.d.	n.d.	n.d.
DHFR	180 ± 50	n.d	1.5 ± 0.5	n.d.
FPGS	3.5 ± 0.5	n.d.	0.04 ± 0.03	n.d.
TS	7 ± 2	n.d.	n.d.	n.d.

n.d.: not detected. PRK, phosphoribulokinase; PEP carboxylase, phosphoenolpyruvate carboxylase; HPPK, dihydropterin pyrophosphokinase; DHPS, dihydropteroate synthase; DHFS, dihydrofolate synthetase; DHFR, dihydrofolate reductase; FPGS, folylpolyglutamate synthetase; TS, thymidylate synthase.
Modified from Neuburger *et al.* (1996) Mitochondria are a major site for folate and thymidylate synthesis in plants. *The Journal of Biological Chemistry*, vol. 271, pp. 9466–9472. Reprinted by permission of The American Society for Biochemistry and Molecular Biology.

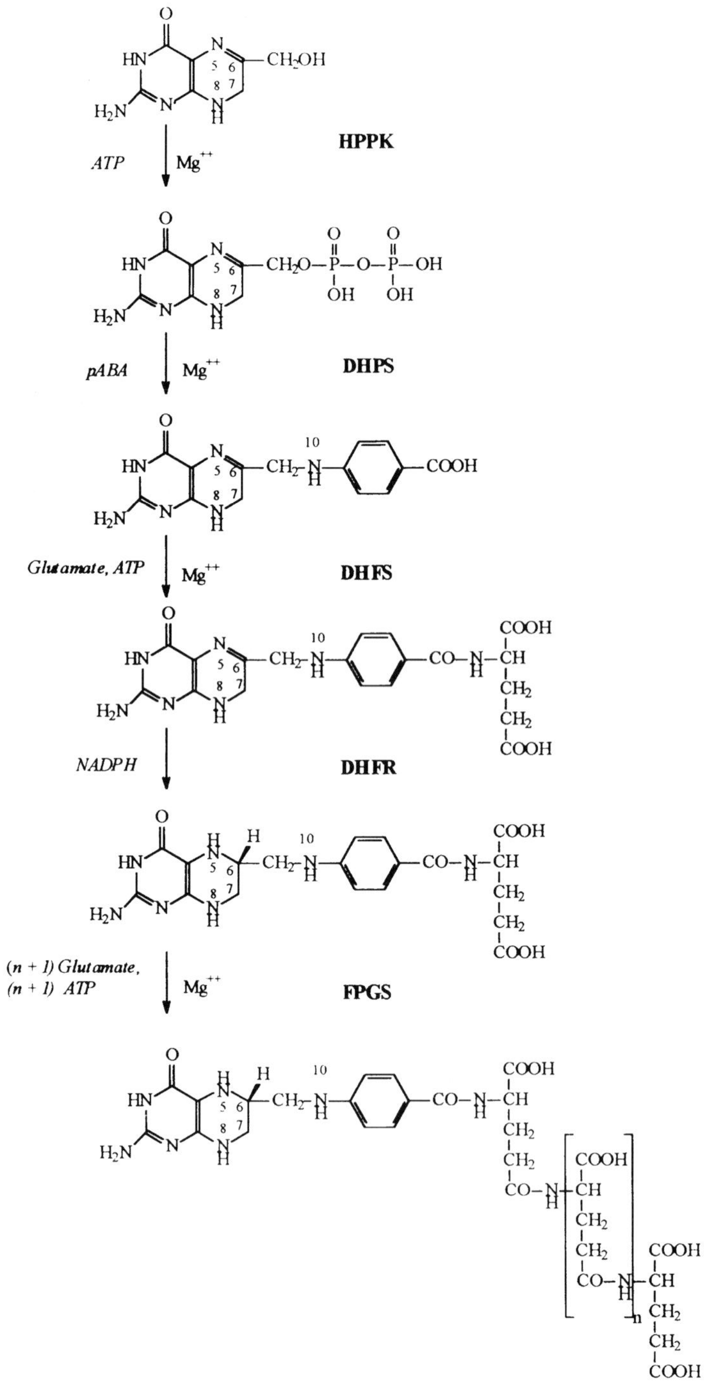

Figure 2. *The biosynthetic pathway of tetrahydrofolate synthesis from 6-hydroxymethyl dihydropterin. HPPK, dihydroxymethyl pyrophosphokinase; DHPS, dihydropteroate synthase; DHFS, dihydrofolate synthetase; DHFR, dihydrofolate reductase; FPGS, folylpolyglutamate synthetase; pABA, acid para-aminobenzoic.*

drugs, which are *p*-aminobenzoic acid analogues that are recognized as alternative substrates (Shiota, 1984). In higher plants, sulphonamide compounds are potential herbicides since they also block DHPS activity (Okinaka and Iwai, 1970a) and inhibit plant growth (Iwai *et al.*, 1962). The bifunctional enzyme, purified from pea leaf mitochondria, represents 0.06% of the matrix protein, has a native molecular mass of 280–300 kDa and is possibly constituted of six identical subunits of 53 kDa. Kinetic studies of the reaction catalysed by the DHPS domain of the protein suggest a random bi-reactant system strongly inhibited by dihydropteroate, its product ($K_i = 10$ μM) (Rébeillé *et al.*, 1997). The related cDNA encodes a polypeptide of 515 residues containing a 28 residue NH_2-terminal extension that resembles the import sequence of mitochondrial proteins. In addition, Southern blot experiments suggest that a single-copy gene codes for the enzyme. This last result, together with the facts that the protein is synthesized with a mitochondrial transit peptide and that the activity is only detected in mitochondria, strongly supports the view that mitochondria are the unique site of 7,8-dihydropteroate synthesis in higher plant cells. When compared to other HPPK and DHPS, the plant enzyme shows highly conserved domains, possibly involved in catalysis and substrate binding, and has the appearance of a mere fusion of the separate bacterial HPPK and DHPS enzymes.

4.2 *Dihydrofolate syntase (DHFS)*

Like HPPK and DHPS, DHFS is absent from animals and is, therefore, a potential target for herbicides. In bacteria, this enzyme is bifunctional, supporting also folylpolyglutamate synthetase (FPGS) activity (Bognar *et al.*, 1987). In yeast, the enzyme is monofunctional and appears to be a monomeric protein of 52 kDa catalysing a Mg^{2+}- and K^+-dependent reaction (McDonald *et al.*, 1995). Little is known about higher plant DHFS. The mitochondrial protein has an apparent native molecular mass of about 54 kDa (Neuburger *et al.*, 1996). It has been purified from pea seedlings and, like its yeast counterpart, requires Mg^{2+} and K^+ for catalysis (Iwai and Ikeda, 1975). DHFS has a high affinity for dihydropteroate (the K_m value is about 1 μM) a situation that limits accumulation of dihydropteroate and feedback inhibition of the DHPS activity.

4.3 *Dihydrofolate reductase (DHFR)*

This enzyme is the target of various anti-cancer drugs, such as methotrexate, used in chemotherapy. In bacteria, yeast and vertebrates the enzyme is a monomer of about 20 kDa. In protozoa and higher plants, as well as in green algae, the situation is different because DHFR is part of a bifunctional protein, also containing TS activity, and appears in most studies as an homodimer of 50–60 kDa (Beverley *et al.*, 1986; Cella *et al.*, 1988; Neuburger *et al.*, 1996). TS is involved in the synthesis of dTMP from dUMP according to the following equation:

$$CH_2\text{-}H_4FGlu_n + dUMP \rightarrow H_2FGlu_n + dTMP.$$

In this reaction, methylene tetrahydrofolate ($CH_2\text{-}H_4FGlu_n$) serves not only as a one-carbon donor but also as an electron donor. The resulting dihydrofolate (H_2FGlu_n) must be recycled back to tetrahydrofolate (H_4FGlu_n) by DHFR then to $CH_2\text{-}H_4FGlu_n$ by the SHMT to ensure the continuous operation of the reaction. Compared to the monofunctional enzyme, the bifunctional protein exhibits distinct biochemical properties such as metabolic channelling from TS to DHFR (Knighton *et al.*, 1994). The plant enzyme has been purified to homogeneity from isolated pea leaf mitochondria (Neuburger *et al.*, 1996). Interestingly, the plant DHFR exhibits a high affinity for both the monoglutamate and the polyglutamate forms of dihydrofolate whereas TS, as generally observed for most folate enzymes (McGuire and Bertino, 1981), shows a high affinity for only the polyglutamate forms of the co-factor (Neuburger *et al.*, 1996). This illustrates the multiple roles of the mitochondrial DHFR. Indeed, the natural substrate of DHFR is the monoglutamate form of dihydrofolate when it is involved in tetrahydrofolate synthesis, and the polyglutamate form of dihydrofolate when it is coupled to TS for CH_2-H_4FGlu_n recycling.

In animal cells, TS is a monofunctional enzyme localized in the cytosol and/or nuclei (Brown *et al.*, 1965). Thus, the question arises whether this monofunctional enzyme also exists in higher plant cells. However, no activity can be detected in cell fractions other than mitochondria (*Table 1*) where, as described above, it is associated with the DHFR activity. These biochemical findings make it tempting to postulate that mitochondria are also a major site for thymidylate synthesis. However, two genes coding for the DHFR/TS have been reported for *Arabidopsis thaliana* (Lazar *et al.*, 1993). In addition, immunocytochemical studies of carrot DHFR/TS indicate a plastidial and, in some tissues, a nuclear location (Luo *et al.*, 1993). Clearly, the question of the subcellular localization of the plant DHFR/TS remains open.

4.4 *Folylpolyglutamate synthetase (FPGS)*

In bacteria, FPGS also displays DHFS activity and there is strong evidence indicating that a single catalytic site is involved for both activities (Kimlova *et al.*, 1991). The bacterial bifunctional protein is monomeric with a molecular mass of about 47–53 kDa (Bognar *et al.*, 1985). In eukaryotes, FPGS is a monofunctional protein of 60 kDa in animals (Garrow *et al.*, 1992) and 65–70 kDa in *Neurospora crassa* and plants (Chan *et al.*, 1991; Imeson and Cossins, 1991a; Neuburger *et al.*, 1996). The folate substrate specificity varies with the source of the enzyme. Indeed, although tetrahydrofolate monoglutamate is always an effective substrate, some species display a higher affinity for 10-formyltetrahydrofolate (*Escherichia coli*) or 5,10-methylene tetrahydrofolate (*Corynebacterium* SP; *N. crassa*) (Chan *et al.*, 1991; McGuire and Coward, 1984). In higher plants, 5,10-methylene tetrahydrofolate is also an effective substrate for FPGS: the enzyme displays a similar affinity as for tetrahydrofolate (Imeson and Cossins, 1991b). *In vitro* analyses of *Neurospora* and higher plant FPGS indicate that diglutamates are the major product of the reaction during the first 2 h (Chan *et al.*, 1991; Imeson and Cossins, 1991b), but longer incubation periods result in more highly conjugated derivatives. In mammalian cells, FPGS is mainly localized in the cytosol (McGuire and Coward, 1984) but the enzyme has also been detected in purified mitochondria (Lin *et al.*, 1993). In *N. crassa*, approximately 50% of the FPGS activity is cytosolic and 50% mitochondrial (Cossins and Chan, 1984). In higher plants, FPGS has been detected only in mitochondria (Neuburger *et al.*, 1996), a situation which, if it holds true, raises the problem of folylpolyglutamate transport across the membranes of the different cell compartments. Indeed, the glutamate chain of folate derivatives is negatively charged, and highly conjugated folates are generally retained within the compartment where they are localized (McGuire and Bertino, 1981). Thus, it would be logical to think that folate derivatives must be deconjugated before being transported and conjugated again having arrived at their final destination, a condition that would require the presence of FPGS in all the cell compartments. However, in animal cells, folylpolyglutamates formed within the mitochondria have been observed to be released into the cytosol without prior hydrolysis (Kim and Shane, 1994), a situation that might also exist in higher plants where mitochondria appear as a major site for folate synthesis. Clearly, the mechanisms involved in folate transport and folate traffic within the cell remain to be discovered.

5. Conclusions

Are mitochondria the only site for folate synthesis in plants? All the enzymes involved in the pathway described above are present in purified mitochondria, which clearly indicates that these organelles have the capacity for folate synthesis. These activities could not be detected in the other cell compartments, but it is, however, difficult to ascertain negative results. The literature contains some apparently contradictory results about the cellular localization of DHFR/TS. Indeed, it has been reported that DHFR/TS in *Daucus carota* cell suspension cultures is localized in plastids, but not in mitochondria (Luo *et al.*, 1993), a conclusion based on the analysis of the corresponding gene and on immunogold labelling studies. This claim is very surprising because DHFR/TS has been purified from isolated leaf mitochondria (Neuburger *et al.*, 1996). However, the possibility remains that the subcellular distribution of DHFR/TS in differentiated tissues, such as leaves, differs from that in non-differentiated cells such as those grown in suspension cultures. The molecular biology data obtained for HPPK/DHPS strongly reinforce the idea that the two first steps of the pathway, at least, are only present in mitochondria. Taken as a whole, higher plant mitochondria appear to be a major, if not the unique, site for folate synthesis.

Is the mitochondrial localization of folate synthesis a unique feature of plants? The subcellular distribution of tetrahydrofolate-synthesizing enzymes in other eukaryotic organisms is not clear. In the protozoa *Plasmodium falciparum* (Triglia and Cowman, 1994) and *Pneumocystis carinii* (Volpe *et al.*, 1993), the reported open reading frame sequences coding for HPPK/DHPS do not indicate the presence of a mitochondrial transit peptide, although this assumption should be confirmed by the analysis of the N-terminal parts of the proteins. In contrast, the DNA sequence coding for the bifunctional DHFR/TS in *Leishmania major* predicts an amino acid extension from the N-terminal part that resembles the import sequence of mitochondrial proteins (Beverley *et al.*, 1986). However, immunogold localization experiments indicate a cytosolic distribution for the protein (Swafford *et al.*, 1990). In animal cells, DHFR is believed to be primarily localized in the cytosol (Brown *et al.*, 1965). In these cells, FPGS is also mainly localized in the cytosol (McGuire and Coward, 1984) but the enzyme has also been detected in purified mitochondria (Lin *et al.*, 1993). Finally, in *N. crassa*, approximately 50% of the FPGS activity is cytosolic and 50% mitochondrial (Cossins and Chan, 1984).

If mitochondria are a major site for folate synthesis in plants, tetrahydrofolate molecules synthesized within mitochondria must be exported to the cytosol and chloroplasts. Release of mitochondrial folylpolyglutamates into the cytosol has been observed in animal cells (Kim and Shane, 1994), but there is, to date, no evidence that, mitochondrial tetrahydrofolate is transported to the cytosol in plants. It is certain that the demonstration of this process, and the comprehension of how the cell regulates its folate requirements between the cytosol, the mitochondria and the chloroplasts, are great challenges for future research in plant folate metabolism.

These observations emphasize the idea that plant mitochondria have a metabolic role that is not simply restricted to respiration and energetic metabolism. Indeed, these organelles are not only involved in tetrahydrofolate biosynthesis but also in the production of biotin (Baldet *et al.*, 1997) and octanoic acid, a precursor of lipoate (Wada *et al.*, 1997). All these compounds are of primary importance for cell metabolism and cell division and are required in all cell compartments. The physiological reason for the biosynthesis of these 'vitamins' in plant mitochondria is an open question.

Acknowledgement

In my post-California years, one of us (R.D.) has felt greatly indebted to Professor Tom ap Rees for his fascinating explanations of plant cell metabolism.

References

Allegra, C.J, Boarman, D., Kovacs, J.A., Morrison, P., Beaver, J., Chabner, B.A. and Masur, H. (1990) Interaction of sulfonamide and sulfone compounds with *Toxoplasma gondii* dihydropteroate synthase. *J. Clin. Invest.* **85**: 371–379.

Baldet, P., Alban, C. and Douce, R. (1997) Biotin synthesis in higher plants: purification and characterization of bioB gene product equivalent from *Arabidopsis thaliana* overexpressed in *Escherichia coli* and its subcellular localization in pea leaf cells. *FEBS Lett.* **419**: 206–210.

Besson, V., Rébeillé, F., Neuburger, M., Douce, R. and Cossins, E.A. (1993) Effects of tetrahydrofolate polyglutamates on the kinetic parameters of serine hydroxymethyltransferase and glycine decarboxylase from pea leaf mitochondria. *Biochem. J.* **292**: 425–430.

Beverley, S.M., Ellenberg, T.E. and Cordingley, J.S. (1986) Primary structure of the gene encoding the bifunctional dihydrofolate reductase-thymidylate synthase of *Leishmania major*. *Proc. Natl Acad. Sci. USA* **83**: 2584–2588.

Bognar, A.L., Osborne, C., Shane, B., Singer, S.C. and Ferone, R. (1985) Folylpoly-γ-glutamate synthetase. Cloning and high expression of the *Escherichia coli folC* gene and purification and properties of the gene product. *J. Biol. Chem.* **260**: 5625–5630.

Bognar, A.L., Osborne, C. and Shane, B. (1987) Primary structure of the *Escherichia coli folC* gene and its folylpolyglutamate synthetase-dihydrofolate synthetase product and regulation by an upstream gene. *J. Biol. Chem.* **262**: 12337–12342.

Bourguignon, J., Rébeillé, F. and Douce, R. (1998) Serine and glycine metabolism in higher plants. In: *Plant Amino Acids* (ed. B. Singh). Marcel Dekker, New York, pp. 111–146.

Brown, G.M. (1985) Biosynthesis of pterins. In: *Folates and Pterins*, Vol. 2 (ed. R.L. Blakley and S.J. Benkovic). Wiley Interscience, New York, pp. 115–154.

Brown, S.S., Neal, G.E. and Williams, D.C. (1965) Subcellular distribution of some folic acid-linked enzymes in rat liver. *Biochem. J.* **97**: 34c–36c.

Cella, R., Nielsen, E. and Parisi, B. (1988) *Daucus carota* cells contain a dihydrofolate reductase: thymidylate synthase bifunctional polypeptide. *Plant Mol. Biol.* **10**: 331–338.

Chan, P.Y., Dale, P.L. and Cossins, E.A. (1991) Purification and properties of *Neurospora* folylpolyglutamate synthetase. *Phytochemistry* **30**: 3525–3531.

Chen, L., Chan, S. and Cossins, E.A. (1997) Distribution of folate derivatives and enzymes for synthesis of 10-formyltetrahydrofolate in cytosolic and mitochondrial fractions of pea leaves. *Plant Physiol.* **115**: 299–309.

Clandinin, M.T. and Cossins, E.A. (1972) Localization and interconversion of tetrahydropteroylpolyglutamates in isolated pea mitochondria. *Biochem. J.* **128**: 29–40.

Coffin, J.W. and Cossins, E.A. (1986) Mitochondrial folates and methionyl-tRNA transformy-lase activity during germination and early growth of seeds. *Phytochemistry* 25: 2481–2487.

Cossins, E.A. and Chan, P.Y. (1984) Folypolyglutamate synthetase activities of *Neurospora*. *Phytochemistry* 23: 965–971.

Garrow, T.A., Admon, A. and Shane, B. (1992) Expression cloning of a human cDNA encod-ing folylpoly(γ-glutamate) synthetase and determination of its primary structure. *Proc. Natl Acad. Sci. USA* 89: 9151–9155.

Imeson, H.C. and Cossins, E.A. (1991a) Higher plant folylpolyglutamate synthetase. I. Purification, stability and reaction requirements of the enzyme from pea seedlings. *J. Plant Physiol.* 138: 476–482.

Imeson, H.C. and Cossins, E.A. (1991b) Higher plant folylpolyglutamate synthetase. II. Some major catalytic properties of the enzyme from *Pisum sativum* L. *J. Plant Physiol.* 138: 483–488.

Imeson, H.C., Zheng, L. and Cossins, E.A. (1990) Folylpolyglutamate derivatives of *Pisum sativum* L. Determination of polyglutamate chain lengths by high performance liquid chro-matography following conversion to p-aminobenzoylpolyglutamates. *Plant Cell Physiol.* 31: 223–231.

Iwai, K. and Ikeda, M. (1975) Purification and properties of the dihydrofolate synthetase from pea seedlings. *J. Nutr. Sci. Vitaminol.* 21: 7–18.

Iwai, K., Nakagawa, S. and Okinaka, O. (1962) The growth inhibition of the germinating seeds by sulfonamides and its reversal by folic acid analogues. *J. Vitaminol.* 8: 20–29.

Kaufman, S. and Kaufman, E.E. (1985) Tyrosine hydroxylase. In: *Folates and Pterins*, Vol. 2 (eds R.L. Blakley and S.J. Benkovic). Wiley Interscience, New York, pp. 251–352.

Kim, J.-S. and Shane, B. (1994) Role of folylpolyglutamate synthetase in the metabolism and cytotoxicity of 5-deazaacyclotetrahydrofolate, an antipurine drug. *J. Biol. Chem.* 269: 9714–9720.

Kimlova, L.J., Pyne, C., Keshavjee, K., Huy, J., Beebakhee, G. and Bognar, A.L. (1991) Mutagenesis of the *folC* gene encoding folylpolyglutamate synthetase-dihydrofolate syn-thetase in *Escherichia coli*. *Arch. Biochem. Biophys.* 284: 9–16.

Kisliuk, R.L. (1981) Pteroylpolyglutamates. *Mol. Cell. Biochem.* 39: 331–345.

Knighton, D.R., Kan, C.C., Howland, E., Janson, C.A., Hostomska, Z., Welsh, K.M. and Matthews, D.A. (1994) Structure of and kinetic channelling in bifunctional dihydrofolate reductase-thymidylate synthase. *Nat. Struct. Biol.* 1: 186–194.

Lacks, S.A., Greenberg, B. and Lopez, P. (1995) A cluster of four genes encoding enzymes for five steps in the folate biosynthetic pathway of *Streptococcus pneumoniae*. *J. Bacteriol.* 177: 66–74.

Lazar, G., Zhang, H. and Goodman, H.M. (1993) The origin of the bifunctional dihydrofolate reductase-thymidylate synthase isogenes of *Arabidopsis thaliana*. *Plant J.* 3: 657–668.

Lin, B.F., Huang, R.F.S. and Shane, B. (1993) Regulation of folate and one-carbon metabolism in mammalian cells. III: role of mitochondrial folylpoly-γ-glutamate synthetase. *J. Biol. Chem.* 268: 21674–21679.

Luo, M., Piffanelli, P., Rastelli, L. and Cella, R. (1993) Molecular cloning and analysis of a cDNA coding for the bifunctional dihydrofolate reductase-thymidylate synthase of *Daucus carota*. *Plant Mol. Biol.* 22: 427–435.

MacKenzie, R.E. (1984) Biogenesis and interconversion of substituted tetrahydrofolates. In: *Folates and Pterins*, Vol. 1 (eds R.L. Blakley and S.J. Benkovic). Wiley Interscience, New York, pp. 255–306.

McDonald, D., Atkinson, I.J., Cossins, E.A. and Shane, B. (1995) Isolation of dihydrofolate and folylpolyglutamate synthetase activities from *Neurospora*. *Phytochemistry* 38: 327–333.

McGuire, J.J. and Bertino, J.R. (1981) Enzymatic synthesis and function of polyglutamates. *Mol. Cell. Biochem.* 38: 19–48.

McGuire, J.J. and Coward, J.K. (1984) Pteroylpolyglutamates: biosynthesis, degradation and function. In: *Folates and Pterins*, Vol. 1 (eds R.L. Blakley and S.J. Benkovic). Wiley Interscience, New York, pp. 135–190.

Neuburger, M., Rébeillé, F., Jourdain, A., Nakamura, S. and Douce, R. (1996) Mitochondria are a major site for folate and thymidylate synthesis in plants. *J. Biol. Chem.* **271:** 9466–9472.

Okinaka, O. and Iwai, K. (1970a) The biosynthesis of folic acid compounds in plants. IV. Purification and properties of the dihydropteroate-synthesizing enzyme from pea seedlings. *J. Vitaminol.* **16:** 201–209.

Okinaka, O. and Iwai, K. (1970b) The biosynthesis of folic acid compounds in plants. III. Distribution of the dihydropteroate synthesizing enzyme in plants. *J. Vitaminol.* **16:** 196–200.

Paquin, J., Baugh, C.M. and MacKenzie, R.E. (1985) Channeling between the active sites of formiminotransferase-cyclodeaminase. Binding and kinetic studies. *J. Biol. Chem.* **260:** 14925–14931.

Prabhu, V., Chatson, K.B., Abrams, G.D. and King, J. (1996) ^{13}C nuclear magnetic resonance detection of interactions of serine hydroxymethyltransferase with C1-tetrahydrofolate synthase and glycine decarboxylase complex activities in *Arabidopsis. Plant Physiol.* **112:** 207–216.

Rébeillé, F., Neuburger, M. and Douce, R. (1994) Interaction between glycine decarboxylase, serine hydroxymethyltransferase and tetrahydrofolate polyglutamates in pea leaf mitochondria. *Biochem. J.* **302:** 223–228.

Rébeillé, F., Macherel, D., Mouillon, J.M., Garin, J. and Douce, R. (1997) Folate biosynthesis in higher plants: purification and molecular cloning of a bifunctional 6-hydroxymethyl-7,8-dihydropterin pyrophosphokinase/7,8-dihydropteroate synthase localized in mitochondria. *EMBO J.* **16:** 947–957.

Shiman, R. (1985) Phenylalanine hydroxylase and dihydropterin reductase. In: *Folates and Pterins,* Vol. 2 (eds R.L. Blakley and S.J. Benkovic). Wiley Interscience, New York, pp. 115–154.

Shiota, T. (1984) Biosynthesis of folate from pterin precursors. In: *Folates and Pterins,* Vol. 1 (eds R.L. Blakley and S.J. Benkovic). Wiley Interscience, New York, pp. 121–134.

Strong, W.B., Tendler, S.J., Seither, R.L., Goldman, I.D. and Schich, V. (1990) Purification and properties of serine hydroxymethyltransferase and C$_1$-tetrahydrofolate synthetase from L1210 cells. *J. Biol. Chem.* **265:** 12149–12155.

Swafford, J.R., Beverley, S.N., Kan, C.C. and Caulfield, J.P. (1990) Distribution of dihydrofolate reductase-thymidylate synthase polypeptide in methotrexate-resistant and wild type Leishmania major: evidence for cytoplasmic distribution despite a hydrophilic leader sequence. In: *Proceedings of the XIIth Congress for Electron Microscopy* (eds L.D. Peachey and D.B. Williams). San Francisco Press, San Francisco, pp. 610–611.

Talarico, T.L., Ray, P.H., Dev, I.K., Merill, B. and Dallas, W.S. (1992) Cloning, sequence analysis, and overexpression of Escherichia coli folK, the gene coding for 7,8-dihydro-6-hydroxymethylpterin-pyrophosphokinase. *J. Bacteriol.* **174:** 5971–5977.

Triglia, T. and Cowman, A.F. (1994) Primary structure and expression of the dihydropteroate synthase gene of *Plasmodium falciparum. Proc. Natl Acad. Sci. USA* **91:** 7149–7153.

Volpe, F., Ballantine, S.P. and Delves, C.J. (1993) The multifunctional folic acid synthesis *fas* gene of *Pneumocystis carinii* encodes dihydroneopterin aldolase, hydroxymethyldihydropterin pyrophosphokinase and dihydropteroate synthase. *Eur. J. Biochem.* **216:** 449–458.

Wada, H., Shintan, D. and Ohlrogge, J. (1997) Why do mitochondria synthesize fatty acids? Evidence for lipoic acid production. *Proc. Natl Acad. Sci. USA* **94:** 1591–1596.

Zheng, L.L., Lin, Y., Lin, S. and Cossins, E.A. (1992) The polyglutamate nature of plant folates. *Phytochemistry* **31:** 2277–2282.

20

Compartmentation of tetrapyrrole synthesis in plant cells

Alison G. Smith, Johanna E. Cornah, Jennifer M. Roper and Davinder Pal Singh

1. Introduction

Tetrapyrroles are essential molecules in plants, as in all eukaryotes and most prokaryotes. Plants have four major types of tetrapyrrole: chlorophyll, the molecule responsible for the harvesting and trapping of light during photosynthesis; haem, the co-factor for proteins involved in electron transfer (cytochromes), oxygen metabolism (catalase, peroxidase, oxidases and hydroxylases) and oxygen binding (leghaemoglobin); sirohaem, a co-factor for sulphite and nitrite reductases; and phytochromobilin, the chromophore of the red light detector phytochrome. The molecules are made up of four pyrrole rings (a five-membered ring containing a N atom) linked by methene bridges, with various substituents on the pyrrole rings. Chlorophyll, haem and sirohaem are circular, with a metal atom co-ordinated by the four nitrogen atoms: magnesium in chlorophyll, and iron in haem and sirohaem. Phytochromobilin is a linear molecule and does not co-ordinate a metal atom. Within the plant cell, the different tetrapyrroles are located in different compartments. Chlorophyll is obviously confined to the chloroplast, and since sulphite and nitrite reductase are both exclusively plastidic (Crawford, 1995; Leustek, 1996), the same is presumably true for sirohaem. Haem, however, is found not just in chloroplasts (for photosynthetic cytochromes), but also in mitochondria (respiratory cytochromes), endoplasmic reticulum (cytochromes P450 and cytochrome b_5) and peroxisomes (catalase, peroxidase), as well as outside the cell in cell wall-associated peroxidases. Phytochrome is located in the cytosol (Terry et al., 1993). In this article, the significance of this distribution for the compartmentation and regulation of the tetrapyrrole synthesis pathway is considered, with particular emphasis on recent work on ferrochelatase, the last enzyme of haem biosynthesis.

Plant Carbohydrate Biochemistry, edited by J.A. Bryant, M.M. Burrell and N.J. Kruger.
© 1999 BIOS Scientific Publishers Ltd, Oxford.

2. Pathway of tetrapyrrole synthesis

Tetrapyrroles are synthesized by a common, branched pathway (*Figure 1a*). In plants, algae and most bacteria [including *Escherichia coli*, cyanobacteria and archaebacteria (Beale, 1990)], the first committed precursor 5-aminolaevulinic acid (ALA) is synthesized from the intact carbon skeleton of glutamate in three steps, involving the chloroplast-encoded tRNA^{Glu}. This is known as the C_5 pathway (Kannangara *et al.*, 1988). In animals, fungi and the α-group of the proteobacteria, ALA is synthesized by a single enzyme, ALA synthase, from succinyl CoA and glycine (Beale, 1990). Two molecules of ALA are condensed by ALA dehydratase to form the pyrrole porphobilinogen. The concerted action of two enzymes, porphobilinogen deaminase and uroporphyrinogen III synthase then results in the formation of uroporphyrinogen III (urogen III) from four porphobilinogen molecules. Urogen III is the first circular tetrapyrrole of the pathway and is the central template from which all biologically functional tetrapyrroles are made (Warren and Scott, 1990). Methylations followed by chelation of an iron atom produces sirohaem. Alternatively, urogen III undergoes a series of decarboxylations and oxidations by urogen decarboxylase, coproporphyrinogen III oxidase (coprogen oxidase) and protoporphyrinogen IX oxidase (protogen oxidase) to form protoporphyrin IX, which has a completely conjugated ring system, and so is the first coloured intermediate. It is also now able to co-ordinate a metal atom. Ferrochelatase inserts Fe^{2+} into the ring to form protohaem, while a separate enzyme, Mg-chelatase, chelates Mg^{2+} into the ring to form Mg-protoporphyrin IX. This is then converted to chlorophyll by a series of reactions involving methylation, formation of the isocyclic ring, reduction of one of the bonds to form a chlorin, and finally addition of the isoprenoid phytol tail. Phytochromobilin is synthesized from protohaem by the action of haem oxygenase, which produces biliverdin IXa after cleavage of the tetrapyrrole ring and release of both the iron atom and a molecule of carbon monoxide. Phytochromobilin synthase converts biliverdin to the 3Z-isomer of phytochromobilin, which must be converted to the 3E-isomer before assembly with the apoprotein of phytochrome (Terry *et al.*, 1993).

As discussed above (Section 1), tetrapyrroles are found in several compartments within the plant cell. The question arises, therefore, where are all the tetrapyrroles synthesized? Is there a single pathway which provides all the end-products, or is there duplication of some or all of the common enzymes in more than one cellular compartment? If there is a single pathway, how is the flux through the different branches controlled, particularly when the relative proportions of the different end-products are so different? For example, in a leaf cell there is approximately 2.5 μmol g^{-1} fresh weight chlorophyll, compared to 2 nmol g^{-1} fresh weight for mitochondrial haem (Werck-Reichardt *et al.*, 1988). The rates of tetrapyrrole synthesis can also vary considerably depending on the tissue or environmental conditions: in an expanding green leaf, chlorophyll synthesis has been estimated to be 3.3 nmol h^{-1} g^{-1} fresh weight, whereas it can be up to 10 times greater during the greening of an etiolated leaf. Similarly, haem synthesis is greatly increased during the formation of the root nodule in legumes, to provide the haem prosthetic group for leghaemoglobin (Santana *et al.*, 1998).

At present, our understanding of these questions is incomplete. However, through a combination of biochemical and molecular approaches, primarily from the investigation of the subcellular location of the enzymes of the pathway, it is becoming possible to address the question of the compartmentation of tetrapyrrole synthesis, as described in the following section.

(a)

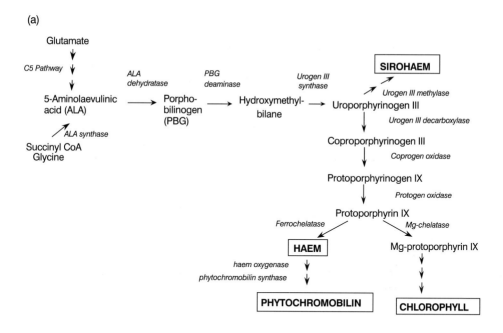

(b)

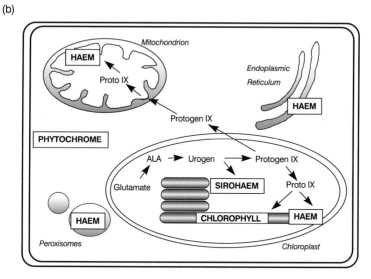

Figure 1. *(a) Pathway of tetrapyrrole synthesis in higher plants, showing the major end products (boxed) and intermediates and enzymes mentioned in this review. In plants, algae and most bacteria, the first committed precursor ALA is synthesized from glutamate in the C_5 pathway in three steps. In animals, fungi and the α-group of purple bacteria, ALA is synthesized in a single reaction from succinyl CoA and glycine. (b) Schematic representation of the subcellular distribution of tetrapyrroles (boxed) and their biosynthetic intermediates within a photosynthetic plant cell. All the enzymes up to and including coprogen oxidase are confined to the chloroplast, while protogen oxidase and ferrochelatase are also found in the mitochondrion. This necessitates the export of protogen IX from the chloroplast to the mitochondrion. Phytochromobilin is synthesized from haem in the chloroplast, but nothing is known of where haem in the ER or peroxisomes is synthesized. Part (b) reproduced with permission from Smith et al., 1993,* Biochemical Journal, *vol. 292, pp. 503–508. © Biochemical Society.*

3. Subcellular location of tetrapyrrole synthesis

3.1 *Location of chlorophyll and plastidic haem synthesis*

The most abundant tetrapyrrole in plants by far is chlorophyll, and as a consequence the majority of the work has been focussed on its synthesis. As might be expected, it is made entirely within the chloroplast, as shown by the fact that isolated chloroplasts readily synthesize chlorophyll from [^{14}C]ALA or [^{14}C]glutamate (Fuesler *et al.*, 1984; Gomez-Silva *et al.*, 1985), and so presumably contain all the requisite enzymes. Haem is also synthesized by chloroplasts, since they can make protoporphyrin IX (also an intermediate for chlorophyll), and ferrochelatase activity is readily demonstrable in chloroplasts, etioplasts (Jones, 1968; Little and Jones, 1976) and root plastids (see Section 4.2).

3.2 *Location of sirohaem synthesis*

Sulphite and nitrite reductases are confined to the plastid (Crawford, 1995; Leustek, 1996), so it is likely that the synthesis of their sirohaem co-factor occurs in the same compartment. A cDNA for urogen III methylase (the first enzyme for sirohaem synthesis from urogen III) from *Arabidopsis thaliana* has been identified. When the cDNA was transcribed and translated *in vitro*, the precursor protein it encoded could be imported and processed to the mature size by isolated chloroplasts, but not by mitochondria (Leustek *et al.*, 1997). However, to date nothing is known in higher plants about the other enzyme(s) needed for the dehydrogenation and ferrochelation steps to make sirohaem.

3.3 *Location of phytochromobilin synthesis*

Although phytochrome itself is cytosolic, phytochromobilin synthesis is thought to occur in plastids, since these contain measurable amounts of haem oxygenase and phytochromobilin synthase (Terry and Lagarias, 1991; Terry *et al.*, 1995). Assembly of holophytochrome occurs spontaneously when the bilin is incubated with the apoprotein *in vitro* (Lagarias and Lagarias, 1989) suggesting that the apoprotein itself catalyses the formation of the thiol linkage which ligates the chromophore to a cysteine within the protein. This reaction is likely to occur in the cytosol (Terry *et al.*, 1993).

In animals, haem oxygenase is present in the endoplasmic reticulum (ER), where it plays an essential role not just in haem turnover, but also in the production of CO (which acts as a neurotransmitter through activation of guanyl cyclase) and in response to oxidative stress. However, mammalian haem oxygenase comprises an NADPH-dependent cytochrome P450 reductase, and a membrane-associated haem binding oxygenase (Maines, 1988). This is quite unlike the chloroplast-located enzyme, which is soluble and uses reducing power supplied by ferredoxin. It is conceivable that plants also have a microsomal haem oxygenase, but there is currently no information on this.

3.4 *Location of mitochondrial haem synthesis*

The synthesis of plant mitochondrial haem was originally considered to follow a similar pathway to that in animals and yeast. In animals, ALA is made from succinyl CoA and glycine by ALA synthase (*Figure 1a*) in the mitochondrial matrix. It is then exported to the cytosol and converted to coprogen III, which is then reimported into the mitochondrion for the last three enzyme reactions (Dailey, 1990). The situation in yeast is identical, except that coprogen oxidase is in the cytosol (Camadro *et al.*, 1986), although associated with the outer mitochondrial membrane, rather than in the inter-membrane space of the mitochondrion as in animals (Elder and Evans, 1978). However, it is unlikely that plant mitochondrial haem is synthesized in the same manner as in animals and yeast.

Firstly, ALA synthase activity cannot be detected in plant mitochondria (A.G.S. unpublished; D. Werck-Reichhardt, personal communication), and a careful labelling study by Schneegurt and Beale (1986) indicated that in etiolated maize epicotyl sections glutamate labels haem *a* (confined to mitochondria) to the same extent as protohaem (found in chloroplasts and ER as well). From this work, it can be concluded that ALA for all cellular tetrapyrroles is synthesized from glutamate in the C_5 pathway in the plastids. Since the C_5 pathway requires chloroplast-encoded tRNA[Glu], this might be one reason for the retention of a plastid genome (Howe and Smith, 1991), even in non-photosynthetic parasitic plants (dePamphilis and Palmer, 1990) or bleached algal mutants (Siemeister and Hachtel, 1989). The only possible exception to this is the *albostrians* mutant of barley, which induces plastid mutations. The mutant tissue has no plastid ribosomes, and severely reduced levels of chlorophyll and other chloroplast components, but does contain respiratory cytochromes and photoactive phytochrome. On northern blots there is no detectable chloroplast tRNA[Glu] (Hess *et al.*, 1992), suggesting that the C_5 pathway cannot be operating, and the authors conclude that there is an alternative method to make ALA in these cells. It remains possible, nevertheless, that low levels of tRNA[Glu] were present, and were sufficient to support haem and phytochromobilin synthesis.

The distribution of the other enzymes of haem biosynthesis in plants has been established using both biochemical and molecular methods. In our laboratory, subcellular fractionation studies followed by the measurement of the activity of both marker enzymes and tetrapyrrole synthesis enzymes indicated that ALA dehydratase, porphobilinogen deaminase and coprogen oxidase activities are confined to the plastid both in etiolated pea leaves, where the major biosynthetic capacity of the tetrapyrrole pathway is for chlorophyll, and in the spadices of *Arum* species, where the major tetrapyrrole synthesized is likely to be mitochondrial haem (Smith, 1988; Smith *et al.*, 1993). No evidence for activity of these enzymes in either mitochondria or the cytosol was obtained. Higher plant cDNAs which have been identified for all the enzymes up to and including coprogen oxidase encode proteins with N-terminal extensions possessing the characteristics of chloroplast transit peptides, and in many instances have been experimentally demonstrated to function as such (e.g. Kruse *et al.*, 1995; Lim *et al.*, 1994; Mock *et al.*, 1995). Immunogold electron microscopy has provided further evidence for an exclusively plastidic location for porphobilinogen deaminase (Witty *et al.*, 1996) and coprogen oxidase (M.A. Santana and A.G.S., manuscript in preparation).

In contrast, activity of the penultimate enzyme of haem synthesis, protogen oxidase, was readily detectable in mitochondria from both pea leaves and *Arum* spadices (Smith *et al.*, 1993). Protogen oxidase could also be measured in mitochondria from several other higher plants including barley, potato and maize (Camadro *et al.*, 1991; Jacobs and Jacobs, 1987). In tobacco, two separate genes which encode protogen oxidase have been identified. One encodes a protein which is imported into isolated chloroplasts, whereas the other encodes a mitochondrially targeted isoform (Lermontova *et al.*, 1997). Similarly, ferrochelatase activity has been demonstrated in mitochondria from potato tubers (Porra and Lascelles, 1968) and etiolated barley leaves (Little and Jones, 1976). Furthermore, in *Arabidopsis* two ferrochelatase genes have been identified, which encode proteins with different subcellular locations: ferrochelatase-II is chloroplast located, while ferrochelatase-I is imported into both chloroplasts and mitochondria (see Section 4.1 below).

Although the studies on the subcellular distribution of haem biosynthesis enzymes have been from a number of different higher plants, the results are all consistent with the model shown in *Figure 1b*, in which all the enzymes up to and including coprogen oxidase are confined to the plastid, even in non-photosynthetic tissues. However, protogen oxidase and ferrochelatase are present also in plant mitochondria. For these organelles to synthesize haem therefore, the transport of the intermediate protogen IX (or conceivably proto IX) from the plastids to the mitochondria is required. Jacobs and Jacobs (1993) demonstrated that barley etioplasts were capable of exporting both protogen IX and protoporphyrin IX when incubated with ALA.

A possible consequence of the transfer of protogen IX through the cell, is that if its uptake into the mitochondria is prevented in some way, it may accumulate in the cytosol, where some or all is likely to oxidize to protoporphyrin IX. The oxidation may be non-enzymic, since this occurs spontaneously in light and oxygen. More likely, it is an enzymic reaction since protogen oxidase activity has been detected in plasma membrane preparations from barley root (Jacobs *et al.*, 1991) and in the microsomal fraction of etiolated maize seedlings (Retzlaff and Böger, 1996). If protoporphyrin IX cannot pass into the mitochondria or plastids, it cannot be used as a substrate by ferrochelatase or Mg-chelatase, and so would be a metabolic dead-end. This would provide an explanation for the unusual effect of diphenylether herbicides such as acifluorfen methyl or oxyfluorfen. These compounds are potent inhibitors of mitochondrial and plastidic protogen oxidases, but do not affect the chelatase enzymes (Matringe *et al.*, 1989; Witowski and Halling, 1988), and the microsomal and plasma membrane-associated protogen oxidases are also relatively insensitive (Jacobs *et al.*, 1991; Reztlaff and Böger, 1996). In herbicide-treated plants it is protoporphyrin IX (the product of protogen oxidase) and not protogen IX (the substrate) which accumulates. The protoporphyrin IX is sequestered in plant cell membranes and causes photo-oxidative damage in the light (Matringe and Scalla, 1988). Interestingly, an equivalent situation is seen in human variegate porphyria, where protogen oxidase activity is congenitally deficient. Again it is protoporphyrin IX which accumulates, presumably because once formed it cannot pass into the mitochondrial matrix, towards which the active site of ferrochelatase is orientated (Harbin and Dailey, 1985; Jones and Jones, 1969).

3.5 *Location of synthesis of haem for other subcellular locations*

Although the major haemoproteins, at least in a leaf cell, are the cytochromes found in the chloroplasts and mitochondria, there are several other haemoproteins found elsewhere in the cell, most notably in the ER (cytochromes P450 and cytochrome b_5) and peroxisomes (peroxidase and catalase). In addition, there are cell wall-associated peroxidases (for lignin and suberin biosynthesis) and in the specialized cell type of the root nodule, oxygen-binding leghaemoglobin is found in the cytosol (Appleby, 1984).

However, nothing is known about the subcellular source of haem for these haemoproteins. It is possible that the haem is provided by the plastid, as it is for phytochromobilin (Section 3.3). Chloroplasts readily export haem (Thomas and Weinstein, 1990). Alternatively, the haem may come from the mitochondrion, as it presumably does in non-photosynthetic organisms such as animals or yeast, although again little is known about the precise mechanisms. Finally, recent studies have detected protogen oxidase activity in enriched plasma membrane fractions from barley (Jacobs et al., 1991) and maize ER (Retzlaff and Böger, 1996), and ferrochelatase activity in plasma membrane (Jacobs and Jacobs, 1995), so it is conceivable that haem in these different subcellular locations is synthesized in situ, from protogen IX supplied by the plastids.

To date, most of the work on the synthesis of haemoproteins other than cytochromes has concentrated on the apoprotein (e.g. catalase, peroxidase, leghaemoglobin), although more recently, the source of the haem co-factor for leghaemoglobin has started to be addressed (Madsen et al., 1993; Santana et al., 1998). In our laboratory, we have been studying the expression of genes for ferrochelatase from Arabidopsis, and found evidence for enhanced levels of expression in response to environmental stresses, which would be expected to induce catalase and peroxidase levels. This therefore is a potential system for studying the synthesis of their haem co-factors. In the last part of this chapter (Section 4), our recent results on ferrochelatase are summarized.

4. Role of ferrochelatase

4.1 Isolation of ferrochelatase genes from Arabidopsis thaliana

We used the technique of functional complementation of a yeast mutant deficient in ferrochelatase to identify Arabidopsis cDNAs encoding ferrochelatase (Chow et al., 1998; Smith et al., 1994). A total of 27 independent cDNAs were isolated, which could be classified into two groups on the basis of restriction maps, plasmid Southern analysis and sequence data. These two groups were clearly the product of two different genes, which were subsequently mapped to different locations on the Arabidopsis genome by restriction fragment length polymorphism analysis of recombinant inbred lines. Ferrochelatase-I and ferrochelatase-II isoforms are 69% identical to each other at the amino acid level, but interestingly, ferrochelatase-II is more similar to the cyanobacterial ferrochelatase than it is to ferrochelatase-I. On the other hand, Arabidopsis ferrochelatase-I is more similar to ferrochelatases identified from cucumber and barley (Miyamoto et al., 1994) than it is to Arabidopsis ferrochelatase-II. The ferrochelatase-II gene encodes a precursor protein which is imported and processed to the mature size by isolated chloroplasts but not mitochondria. From northern analysis, the gene is expressed in photosynthetic tissue, but no transcripts were detectable in root RNA. Expression of the gene appeared to be enhanced by light, but was still present in the dark. The ferrochelatase-I gene, on the other hand, is expressed in all tissues of the plant. Its expression is enhanced by light, although again it occurs in etiolated leaves, and is induced by growth on 3% sucrose, which has no effect on ferrochelatase-II transcript levels (Chow et al., 1998). Interestingly, the ferrochelatase-I precursor protein can be imported into both chloroplasts and mitochondria (Chow et al., 1997). This is an unusual phenomenon, since it is generally assumed that the recognition signals for organelle targeting are very specific. However, there are increasing numbers of dual-targeted proteins reported in the literature, including glutathione reductase from peas (Creissen *et al.*, 1995), and *Arabidopsis* histidyl and methionyl-tRNA synthetases (Small *et al.*, 1998). Furthermore, both the barley and cucumber ferrochelatases (Miyamoto *et al.*, 1994) behave in a similar fashion to *Arabidopsis* ferrochelatase-I (D.P.S. and A.G.S., manuscript in preparation). At present we do not have any information on whether there is dual-targeting of ferrochelatase *in vivo*, but this is being investigated by generating chimeric fusion proteins with reporter proteins, and by raising specific antibodies to the proteins.

Assuming our data reflect the situation *in vivo*, we can conclude that the mitochondrial isoform of ferrochelatase is encoded by the ferrochelatase-I gene, and so too is the plastidic isoform in non-photosynthetic cells. However, in chloroplasts there may be ferrochelatase-II or ferrochelatase-I, or both. If there are two isoforms of the enzyme in the same chloroplast, it may be that they have different roles to play. For instance, ferrochelatase-II may be responsible for the synthesis of photosynthetic cytochromes, while ferrochelatase-I may provide haem for other tetrapyrroles, such as phytochromobilin, or haem in the endomembrane system or peroxisomes.

4.2 *Activity of ferrochelatase in different organelles*

To investigate the roles of the two ferrochelatase isoforms further, we have studied the levels of ferrochelatase in different organelles, so that this can be correlated with levels of different haemoproteins. Ferrochelatase is notoriously difficult to assay, particularly using the natural substrates protoporphyrin IX and ferrous ion: protoporphyrin IX tends to aggregate in aqueous solution (and it is only the monomer which is a substrate), and ferrous ion oxidizes readily to ferric ion (which again is not a substrate), so the assay must be carried out under strictly anaerobic conditions. Fortunately, it is possible to substitute synthetic porphyrins and other divalent metal cations, which are often much better substrates for ferrochelatase *in vitro*. We have developed a novel assay using Co^{2+} and deuteroporphyrin, which gives high rates of ferrochelatase activity with plant tissue (J.M.R., J.E.C., D.P.S. and A.G.S., manuscript in preparation).

Using this method, we determined levels of ferrochelatase activity in plastids and

Table 1. *Activity of ferrochelatase in different organelles isolated from pea tissue. Green leaf organelles were isolated as described by Chow* et al. *(1997) and those from etiolated leaf and roots as described by Smith (1988). Ferrochelatase was assayed using a novel Co^{2+}–deuteroporphyrin assay (see text for details)*

Tissue	Enzyme activity in plastids		Enzyme activity in mitochondria	
	nmol min^{-1} mg protein^{-1}	nmol min^{-1} g FW^{-1}	nmol min^{-1} mg protein^{-1}	nmol min^{-1} g FW^{-1}
Green leaves	0.548	3.6	0.473	0.157
Etiolated leaves	0.192	0.384	n.d.	n.d.
Roots	0.151	0.03	0.119	0.009

n.d., not detectable.

mitochondria from different tissues of peas (*Table 1*). In green leaf tissue and in roots, the specific activity of ferrochelatase is the same order of magnitude in the two organelles, although it is lower overall in roots. However, when the activity is expressed per gram fresh weight, there is clearly much more activity in plastids than in mitochondria: a 20-fold increase in green leaves, and a three-fold increase in roots. This implies a greater capacity for haem synthesis in plastids than in mitochondria. In leaves, this is likely to be due in part both to the presence of the two ferrochelatase iso-forms, and to the need to make large amounts of haem for photosynthetic cytochromes. Nonetheless, there is a possibility that this greater capacity is for the synthesis of non-plastidic haem (e.g. for the ER, peroxisomes), while the mitochon-drial activity is purely for mitochondrial haem.

4.3 *Regulation of expression of ferrochelatase-I gene*

Until we have a means to distinguish between ferrochelatase-I and ferrochelatase-II directly, for instance with specific antibodies, the presence of the two isoforms com-plicates the study of the role of each within the cell, and indeed within the plant as a whole. We therefore decided to investigate expression of the ferrochelatase-I gene separately, by fusing different lengths of promoter sequence to the reporter gene *uidA* which encodes β-glucuronidase (GUS), and introducing the constructs into tobacco. Different tissues were then assayed for GUS activity (D.P.S., J.E.C. and A.G.S., man-uscript in preparation). Although the absolute levels of GUS activity varied between constructs, essentially similar patterns were seen with all the promoter fragments (the shortest being –358 to +144 relative to the transcription start size), implying that the important response elements are contained within the first 350 bp of the promoter. *Figure 2* summarizes the results which we obtained with the longest promoter frag-ment (–1615 to +144). The promoter is active in all tissues, as expected from the northern results (Chow *et al.*, 1998), but it is greatly elevated in flowers. Although the promoter activity may not be a true reflection of enzyme levels in the tissue, we have found that, in a number of different higher plants, ferrochelatase activity is much higher in flowers than in leaves (J.E.C. and A.G.S., unpublished observations). Why there should be such high levels of ferrochelatase in flowers is unclear, but this may reflect the requirement for mitochondrial haem to sustain the 20–40-fold increase in mitochondria and associated increase in respiration which has been reported in anther tissue during sporogenesis (Huang *et al.*, 1994). It will therefore be of great interest to establish whether the increased enzyme activity is in both plastids and mitochondria, or is confined to the latter organelle.

The ferrochelatase-I gene promoter is also more active in tissue grown in higher sucrose levels, again as observed by northern analysis (Chow *et al.*, 1998). Since fer-rochelatase-I provides the mitochondrial isoform, this can be explained by the need for more respiratory cytochromes to respire the supplied sugar. More unexpected is the observation that there is stimulation of gene expression by wounding of stems. This was investigated further in a more detailed experiment, in which leaf discs were removed from a tobacco leaf, and then the wounded leaf was assayed for GUS activity after 24 h (*Figure 3a*). Compared to GUS levels in an unwounded leaf from the same plant, there is up to three-fold stimulation in the wounded tissue. Again this might reflect the need for increased respiratory cytochromes, but another intriguing possi-bility is that the increase is to supply haem for other haemoproteins such as catalase or

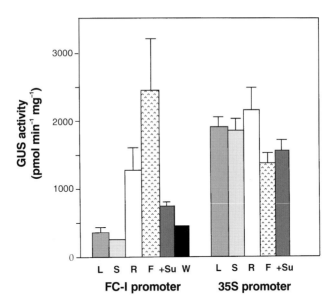

Figure 2. *Activity of the reporter enzyme GUS in transgenic tobacco plants containing the ferrochelatase-I (FC-I) or CaMV 35S (35S) promoters fused to the uidA gene encoding GUS. The FC-I promoter was the region –1615 to +144 relative to the transcription start site, but essentially identical profiles were obtained with all promoter constructs tested. The values represent the mean ± SE of several independent primary transformants. L, leaf; S, stem; R, root; F, flower; +Su, leaf from plants grown in 3% sucrose; W, wounded stem.*

peroxidase. For example, the activity of anionic peroxidases, which play an integral role in secondary cell wall biosynthesis, was shown to increase after wounding, reaching a maximum after 72 h (Lagrimini and Rothstein, 1987).

The production of lignin and suberin is also important not just in wound repair, but also to protect against pathogen attack (Roberts *et al.*, 1988). When the FC-I: GUS tobacco plants, which carried the N resistance gene, were challenged with tobacco mosaic virus, they underwent a hypersensitive response, characterized by an oxidative burst at the site of inoculation, causing host cell death and a distinctive necrotic lesion. In the tissue surrounding the lesion, there was a dramatic increase of up to 30-fold in GUS activity (*Figure 3b*). This induction of the ferrochelatase-I promoter is unlikely to be merely a response to wounding at the inoculation site, since the the up-regulation was so much greater than seen in the wounded plants (*Figure 3a*).

5. Conclusion

In this review, we have tried to give an overview of the current understanding of tetrapyrrole synthesis in higher plant cells. However, there remain many unanswered questions. It is to be hoped that in the future, the combination of biochemical and molecular techniques described here will be brought to bear to address some or all of them. Perhaps one of the most intriguing at present is the assembly of the tetrapyrrole with the apoprotein to form a functional protein. Recent advances in molecular genetics have led to considerable insight into the biogenesis of c-type cytochromes in

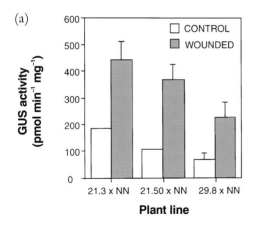

(a)

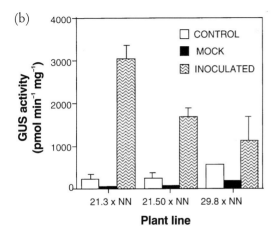

(b)

Figure 3. Effect of wounding and viral infection on the expression of FC-I promoter in leaves of transgenic tobacco. Primary transformants containing the uidA gene fused to different FC-I promoter constructs were crossed with wild-type tobacco homozygous for the N gene conferring resistance to tobacco mosaic virus. The progeny of the cross was then used in the following experiments: 21.3 and 21.50 were primary transformants with the –1615 to +144 promoter region (transcriptional construct), while 29.8 was a primary transformant with a –1615 to +144 promoter region (translational construct). (a) Tobacco leaves were wounded by removing leaf discs with a cork borer. The wounded leaves were assayed for GUS activity 24 h later. The unwounded control was leaf material from the same plant. (b) Tobacco leaves were abraded with carborundum and then either 5 μg ml^{-1} tobacco mosaic virus (inoculated) or water (mock) was applied. Leaves were assayed after 5 days, when necrotic lesions, indicative of a hypersensitive response, had appeared on the inoculated plants.

chloroplasts, mitochondria and bacteria (Kranz *et al.*, 1998). Similar approaches to study other haemoprotein assembly are likely to be seminal in influencing our understanding of the compartmentation and regulation of tetrapyrrole synthesis within the plant cell.

References

Appleby, C.A. (1984) Leghaemoglobin and *Rhizobium* respiration. *Annu. Rev. Plant Physiol.* **35:** 443–478.

Beale, S.I. (1990) Biosynthesis of the tetrapyrrole pigment precursor, delta-aminolevulinic acid, from glutamate. *Plant Physiol.* **93:** 1273–1279.

Camadro, J.-M., Chambon, H., Jolles, J. and Labbe, P. (1986) Purification and properties of coproporphyrinogen oxidase from the yeast *Saccharomyces cerevisiae. Eur. J. Biochem.* **156:** 579–587.

Camadro, J.-M., Matringe, M., Scalla, R. and Labbe, P. (1991) Kinetic-studies on protoporphyrinogen oxidase inhibition by diphenyl ether herbicides. *Biochem. J.* **277:** 17–21.

Chow, K.-S., Singh, D.P., Roper, J.M. and Smith, A.G. (1997) A single precursor protein for ferrochelatase-I from *Arabidopsis* is imported *in vitro* into both chloroplasts and mitochondria. *J. Biol. Chem.* **272:** 27565–27571.

Chow, K.S., Singh, D.P., Roper, J.M., Walker, A.R. and Smith, A.G. (1998) Two different genes encode ferrochelatase in Arabidopsis: mapping, expression and subcellular targeting of the precursor proteins. *Plant J.* **15**: 531–541.

Crawford, N.M. (1995) Nitrate – nutrient and signal for plant-growth. *Plant Cell* **7**: 859–868.

Creissen, G., Reynolds, H., Xue, Y. and Mullineaux, P. (1995) Simultaneous targeting of pea glutathione reductase and of a bacterial fusion protein to chloroplasts and mitochondria in transgenic tobacco. *Plant J.* **8**: 167–175.

Dailey, H.A. (1990) Conversion of coproporphyrinogen to protoheme in higher eukaryotes and bacteria: terminal three enzymes. In: *Biosynthesis of Heme and Chlorophylls* (ed. H.A. Dailey). McGraw-Hill, New York, pp. 123–161.

dePamphilis, C.W. and Palmer, J.D. (1990) Loss of photosynthetic and chlororespiratory genes from the plastid genome of a parasitic flowering plant. *Nature* **348**: 337–339.

Elder, G.H. and Evans, J.O. (1978) Evidence that the coproporphyrinogen oxidase activity of rat liver is situated in the intermembrane space of mitochondria. *Biochem. J.* **172**: 345–347.

Fuesler, T.P., Castelfranco, P.A. and Wong, Y.-S. (1984) Formation of Mg-containing chlorophyll precursors from protoporphyrin IX, 5-aminolevulinic acid, and glutamate in isolated, photosynthetically competent, developing chloroplasts. *Plant Physiol.* **74**: 928–933.

Gomez-Silva, B., Timko, M.P. and Schiff, J.A. (1985) Chlorophyll biosynthesis from glutamate or 5-aminolevulinate in intact *Euglena* chloroplasts. *Planta* **165**: 12–22.

Harbin, B.M. and Dailey, H.A. (1985) Orientation of ferrochelatase in bovine liver mitochondria. *Biochemistry* **24**: 366–370.

Hess, W.R., Schendel, R., Rudiger, W., Fieder, B. and Borner, T. (1992) Components of chlorophyll biosynthesis in a barley albina mutant unable to synthesize δ-aminolevulinic-acid by utilizing the transfer-RNA for glutamic-acid. *Planta* **188**: 19–27.

Howe, C.J. and Smith, A.G. (1991) Plants without chlorophyll. *Nature* **349**: 109.

Huang, J., Struck, F., Matzinger, D.F. and Levings, C.S. (1994) Flower-enhanced expression of a nuclear-encoded mitochondrial respiratory protein is associated with changes in mitochondrion number. *Plant Cell* **6**: 439–448.

Jacobs, J.M. and Jacobs, N.J. (1987) Oxidation of protoporphyrinogen to protoporphyrin, a step in chlorophyll and haem biosynthesis. Purification and partial characterization of the enzyme from barley organelles. *Biochem. J.* **244**: 219–224.

Jacobs, J.M. and Jacobs, N.J. (1993) Porphyrin accumulation and export by isolated barley (*Hordeum vulgare*) plastids – effect of diphenylether herbicides. *Plant Physiol.* **101**: 1181–1187.

Jacobs, J.M. and Jacobs, N.J. (1995) Terminal enzymes of heme biosynthesis in the plant plasma membrane. *Arch. Biochem. Biophys.* **323**: 274–278.

Jacobs, J.M., Jacobs, N.J., Sherman, T.D. and Duke, S.O. (1991) Effect of diphenyl ether herbicides on oxidation of protoporphyrinogen to protoporphyrin in organellar and plasma-membrane enriched fractions of barley. *Plant Physiol.* **97**: 197–203.

Jones, O.T.G. (1968) Ferrochelatase of spinach chloroplasts. *Biochem. J.* **107**: 113–119.

Jones, M.S. and Jones, O.T.G. (1969) The structural organisation of haem synthesis in rat liver mitochondria. *Biochem. J.* **113**: 507–514.

Kannangara, C.G., Gough, S.P., Bruyant, P., Hoober, J.K., Kahn, A. and von Wettstein, D. (1988) tRNA^Glu as a cofactor in delta-aminolevulinate biosynthesis: steps that regulate chlorophyll synthesis. *Trends Biochem. Sci.* **13**: 139–143.

Kranz, R., Lill, R., Goldman, B., Bonnard, G. and Merchant, S. (1998) Molecular mechanisms of cytochrome *c* biogenesis: three distinct systems. *Mol. Microbiol.* **29**: 383–396.

Kruse, E., Mock, H.P. and Grimm, B. (1995) Coproporphyrinogen-III oxidase from barley and tobacco – sequence analysis and initial expression studies. *Planta* **196**: 796–803.

Lagarias, J.C. and Lagarias, D.M. (1989) Self-assembly of synthetic phytochrome holoprotein *in vitro*. *Proc. Natl Acad. Sci. USA* **86**: 5778–5780.

Lagrimini, L.M. and Rothstein, S. (1987) Tissue specificity of tobacco peroxidase isozymes and their induction by wounding and tobacco mosaic virus infection. *Plant Physiol.* **84**: 438–442.

Lermontova, I., Kruse, E., Mock, H.P. and Grimm, B. (1997) Cloning and characterization of a plastidal and a mitochondrial isoform of tobacco protoporphyrinogen IX oxidase. *Proc. Natl Acad. Sci. USA* **94**: 8895–8900.

Leustek, T. (1996) Molecular genetics of sulfate assimilation in plants. *Physiol. Plant.* **97**: 411–419.

Leustek, T., Smith, M., Murillo, M., Singh, D.P., Smith, A.G., Woodcock, S.C., Awan, S.J. and Warren, M.J. (1997) Siroheme biosynthesis in higher plants: analysis of an S-adenosyl-L-methionine-dependent uroporphyrinogen III methyltransferase from *Arabidopsis thaliana. J. Biol. Chem.* **272**: 2744–2752.

Lim, S.H., Witty, M., Wallace-Cook, A.D.M., Ilag, L.I. and Smith, A.G. (1994) Porphobilinogen deaminase is encoded by a single gene in *Arabidopsis thaliana* and is targeted to the chloroplasts. *Plant Mol. Biol.* **26**: 863–872.

Little, H.N. and Jones, O.T.G. (1976) The subcellular localization and properties of the ferrochelatase of etiolated barley. *Biochem. J.* **156**: 309–314.

Madsen, O., Sandal, L., Sandal, N. and Marcker, K.A. (1993) A soybean coproporphyrinogen oxidase gene is highly expressed in root nodules. *Plant Mol. Biol.* **23**: 35–43.

Maines, M.D. (1988) Heme oxygenase – function, multiplicity, regulatory mechanisms, and clinical-applications. *FASEB J.* **2**: 2557–2568.

Matringe, M. and Scalla, R. (1988) Studies on the mode of action of acifluorofen-methyl in nonchlorophyllous soybean cells. Accumulation of tetrapyrroles. *Plant Physiol.* **86**: 619–622.

Matringe, M., Camadro, J.M., Labbe, P. and Scalla, R. (1989) Protopophyrinogen oxidase as a molecular target for diphenyl ether herbicides. *Biochem. J.* **260**: 231–235.

Miyamoto, K., Tanaka, R., Teramoto, H., Masuda, T., Tsuji, H. and Inokuchi, H. (1994) Nucleotide sequence of cDNA clones encoding ferrochelatase from barley and cucumber. *Plant Physiol.* **105**: 769.

Mock, H.P., Trainotti, L., Kruse, E. and Grimm, B. (1995) Isolation, sequencing and expression of cDNA sequences encoding uroporphyrinogen decarboxylase from tobacco and barley. *Plant Mol. Biol.* **28**: 245–256.

Porra, R.J. and Lascelles, J. (1968) Studies on ferrochelatase: the enzymic formation of haem in proplastids, chloroplsts and plant mitochondria. *Biochem. J.* **108**: 343–348.

Retzlaff, K. and Böger, P. (1996) An endoplasmic reticulum plant enzyme has protoporphyrinogen IX oxidase activity. *Pest. Biochem. Physiol.* **54**: 105–114.

Roberts, E., Kutchan, T. and Kolattukudy, P.E. (1988) Cloning and sequencing of a cDNA for a highly anionic peroxidase from potato and the induction of its mRNA in suberising potato tubers and tomato fruit. *Plant Mol. Biol.* **11**: 15–26.

Santana, M.A., Pihakaski-Maunsbach, K., Sandal, N., Marcker, K.A. and Smith, A.G. (1998) Evidence that the plant host synthesizes the heme moiety of leghemoglobin in root nodules. *Plant Physiol.* **116**: 1259–1269.

Schneegurt, M.A. and Beale, S.I. (1986) Biosynthesis of protoheme and heme *a* from glutamate in maize. *Plant Physiol.* **81**: 965–971.

Siemeister, G. and Hachtel, W. (1989) Organization and nucleotide-sequence of ribosomal-RNA genes on a circular 73 kbp DNA from the colorless flagellate *Astasia longa. Curr. Genet.* **15**: 435–441.

Small, I.D., Wintz, H., Akashi, K. and Mireau, H. (1998) Two birds with one stone: genes that encode products targeted to two or more compartments. *Plant Mol. Biol.* **38**: 265–277.

Smith, A.G. (1988) Subcellular localization of two porphyrin-synthesis enzymes in *Pisum sativum* (pea) and *Arum* (cuckoo-pint) species. *Biochem. J.* **249**: 423–428.

Smith, A.G., Marsh, O. and Elder, G.H. (1993) Investigation of the subcellular location of the tetrapyrrole-biosynthesis enzyme coproporphyrinogen oxidase in higher plants. *Biochem. J.* **292**: 503–508.

Smith, A.G., Santana, M.A., Wallace-Cook, A.D.M., Roper, J.M. and Labbe-Bois, R. (1994) Isolation of a cDNA encoding chloroplast ferrochelatase from *Arabidopsis thaliana* by functional complementation of a yeast mutant. *J. Biol. Chem.* **269**: 13405–13413.

Terry, M.J. and Lagarias, J.C. (1991) Holophytochrome assembly – coupled assay for phytochromobilin synthase in organello. *J. Biol. Chem.* **266**: 22215–22221.

Terry, M.J., Maines, M.D. and Lagarias, J.C. (1993) Inactivation of phytochrome- and phycobiliprotein-chromophore precursors by rat liver biliverdin reductase. *J. Biol. Chem.* **268**: 26099–26106.

Terry, M.J., McDowell, M.T. and Lagarias, J.C. (1995) (3Z)-phytochromobilin and (3E)-phytochromobilin are intermediates in the biosynthesis of the phytochrome chromophore. *J. Biol. Chem.* **270**: 11111–11118.

Thomas, J. and Weinstein, J.D. (1990) Measurement of heme efflux and heme content in isolated developing chloroplasts. *Plant Physiol.* **94**: 1414–1423.

Warren, M.J. and Scott, A.I. (1990) Tetrapyrrole assembly and modification into the ligands of biological functional cofactors. *Trends Biochem. Sci.* **15**: 486–491.

Werck-Reichhart, D., Jones, O.T.G. and Durst, F. (1988) Haem-synthesis during cytochrome P450 induction in higher plants. 5-ALA synthesis through a 5-carbon pathway in *Helianthus tuberosus* tuber tissues aged in the dark. *Biochem. J.* **249**: 473–480.

Witowski, D.A. and Halling, B.P. (1988) Accumulation of photodynamic tetrapyrroles induced by acifluorofen-methyl. *Plant Physiol.* **87**: 632–637.

Witty, M., Jones, R.M., Robb, M.S., Shoolingin Jordan, P.M. and Smith, A.G. (1996) Subcellular location of the tetrapyrrole synthesis enzyme porphobilinogen deaminase in higher plants: an immunological investigation. *Planta* **199**: 557–564.

21

What's a nice enzyme like you doing in a place like this? A possible link between glycolysis and DNA replication

John A. Bryant and Louise E. Anderson

1. Introduction

At first sight, the inclusion of a chapter dealing with aspects of DNA replication in a book concerned primarily with plant carbohydrate metabolism may seem strange. It could be argued of course that the two are linked via the synthesis of deoxyribose and via the provision of reducing potential and energy. However, we propose in this chapter that there may be a more direct link between carbohydrate metabolism, particularly respiration, and DNA replication. The key to this linkage is the enzyme phosphoglycerate kinase (PGK). In respiratory metabolism it catalyses the first ATP-generating step in glycolysis, as illustrated in *Figure 1*. The same enzyme is involved in

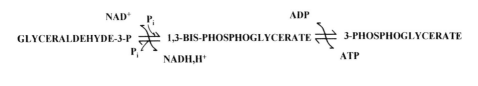

Figure 1. *The reactions catalysed by glyceraldehyde phosphate dehydrogenase and phosphoglycerate kinase (PGK) in glycolysis. For assay of PGK, the two reactions are driven in the opposite direction to that shown here and the activity is determined by the decrease in the amount of NADH, measured by absorbance at 340 nm.*

Plant Carbohydrate Biochemistry, edited by J.A. Bryant, M.M. Burrell and N.J. Kruger.
© 1999 BIOS Scientific Publishers Ltd, Oxford.

gluconeogenesis and a separate form of PGK occurs in the chloroplast where it participates in the Calvin cycle (Bringloe *et al.*, 1996; Longstaff *et al.*,1989; Macioszek *et al.*,1990). In addition to these well-known roles there is more recent evidence that PGK is also directly involved in DNA replication. In order to present this evidence it is first necessary to describe certain features of DNA replication, and in particular, the synthesis of the lagging strand.

Figure 2a illustrates the early stages of DNA replication at a single replication origin. It is unlikely that the initiation of the copying of the two template strands happens as symmetrically within the origin site as is shown in the diagram but the key features of the process are nevertheless clear. Because of the anti-parallel configuration of the double helix, and because DNA replication uses precursors of only one polarity, that is deoxyribonucleoside-5'-triphosphates, synthesis of one of the strands on either side of the initiation site proceeds discontinuously via the formation of short nascent pieces of DNA known as Okazaki fragments. The whole process of Okazaki fragment synthesis is shown in *Figure 2b*. Synthesis of the Okazaki fragments is brought about in eukaryotes by a particular DNA polymerase, DNA polymerase-α (recently reviewed by Foiani *et al.*, 1997). In common with all DNA polymerases, polymerase-α is unable to initiate a DNA strand *de novo* but requires the prior synthesis of an RNA primer which is later removed (*Figure 2b*). The enzyme which synthesizes the primers, primase, is closely associated with polymerase-α.

2. DNA polymerase-α in pea

The DNA polymerase-α of pea (*Pisum sativum*) has been extensively characterized over a period of several years (Bryant *et al.*,1992). Like its counterpart in animals it is closely associated with primase and is more loosely associated with other activities

(a)

Origin

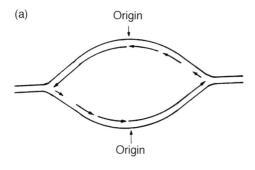

Origin

(b)

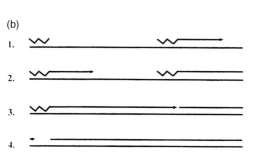

Figure 2. (a) Diagram of the initiation of DNA replication at an origin of replication. Note that on the 'lagging strand' the DNA is made initially as short pieces, the Okazaki fragments. (b) Diagram illustrating the synthesis of Okazaki fragments. The first step is the synthesis of an RNA primer (indicated by the short 'wavy' lines). The primers are about 11–17 nucleotides long and are synthesized by a primase associated with DNA polymerase-α. The RNA primer is then extended by DNA polymerase-α; in the meantime, the RNA primer of the previously initiated Okazaki fragment is degraded by RNase-H. DNA polymerase-α continues synthesis up to the 5' end of the previously synthesized fragment and the fragments are joined by DNA ligase.

involved in synthesis of Okazaki fragments, to form what has been termed the lagging strand replication complex. Amongst these associated activities are ribonuclease-H, which removes the RNA primers, and DNA-binding activity (*Figure 3a*), the function of which was initially not clear. DNA polymerase itself has DNA binding activity but the major DNA binding activity in the pea lagging strand complex is in fact associated with a 42 kDa polypeptide (*Figure 3b*) that may be separated from the polymerase by chromatography on heparin-agarose (Al-Rashdi and Bryant, 1994).

(a)

(b)

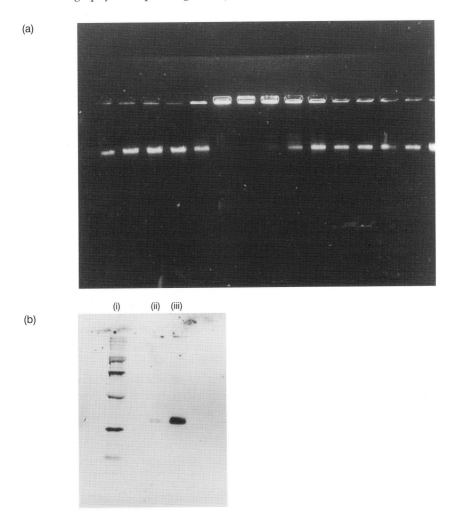

Figure 3. *(a) Assay of DNA-binding activity (detected by retardation of linearized plasmid DNA at the top of an agarose electrophoresis gel) in fractions of DNA polymerase-α eluted from a column of phospho-cellulose. (b) South-western blot of purified DNA-binding protein. After purification, the protein was electrophoresed in an SDS-polyacrylamide gel and then blotted into a nitrocellulose membrane. The blot was probed with radioactive DNA and then autoradiographed, revealing a single band of DNA-binding activity at 42 kDa. From left to right, the lanes are (i) radioactive protein molecular mass markers, (ii) 0.6 μg DNA-binding protein, (iii) 6.0 μg DNA-binding protein. From Al-Rashdi and Bryant (1994). Purification of a DNA-binding protein from a multi-protein complex associated with DNA polymerase-α in pea.* Journal of Experimental Botany, *vol. 45, pp. 1867–1871. Reprinted with permission of Oxford University Press.*

3. Identification of a primer-recognition protein in pea

The 42 kDa DNA-binding protein referred to above has been characterized by electron microscopy and by gel mobility assays in respect of its binding to different types of DNA. Gel mobility assays, of the type illustrated in *Figure 3a*, clearly show that the protein does not bind to either completely single-stranded or completely double-stranded DNA (Burton *et al.*, 1997). For example, the protein will not bind to the circular single-stranded DNA of bacteriophage M13. However, it will bind to the M13 DNA if a short oligonucleotide (e.g. a 15-nucleotide sequencing primer) is annealed to the single-stranded DNA prior to the binding assay (*Figure 4*). This suggests that discontinuities in the DNA double helix such as double-strand/single-strand junctions are the targets for binding. This feature is seen even more clearly when binding is studied by electron microscopy (Burton *et al.*, 1997). *Figure 5* shows the binding of the DNA to circular plasmids in which single-strand breaks have been induced by γ-irradiation and to the cohesive termini of a restriction fragment cut from a plasmid by the restriction enzymes *Eco*RI and *Hind*III. Binding to the restriction fragments is abolished if the cohesive termini are removed by S_1 nuclease prior to the binding assay (Burton *et al.*, 1997).

The preference of the DNA-binding protein for double-strand/single-strand junctions, particularly its ability to bind to a primed M13 DNA template and its association with DNA polymerase-α lead to the suggestion that it may be involved in recognition of primers during the synthesis of the lagging strand. A primed template (as illustrated in *Figure 2b*) clearly provides a double-strand/single-strand junction, albeit with the junction being provided by the presence of a short RNA molecule. To test this idea, pea DNA polymerase-α was assayed in the presence or absence of the DNA-binding protein using different types of primed template. These were (a) linear DNA with very frequent primers generated by 'gapping' the DNA with deoxyribonuclease, (b) M13 bacteriophage DNA with one oligonucleotide primer annealed per bacteriophage DNA molecule and (c) M13 DNA which was self-primed with an RNA primer (or primers) by allowing the polymerase's own primase activity to undertake primer synthesis. The latter, of course, corresponds more closely to the *in vivo* situation. The results (*Table 1*) show very clearly that the DNA-binding protein stimulates significantly the activity of DNA polymerase using the primed M13 DNA templates

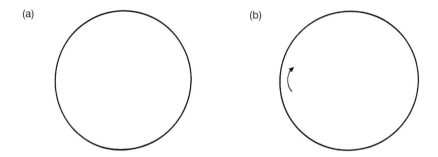

Figure 4. *(a) Diagram illustrating an unprimed M13 DNA molecule. (b) Diagram illustrating an M13 DNA molecule primed by annealing to it an oligodeoxyribonucleotide (not to scale). The DNA binding protein binds to the primed but not to the unprimed molecule.*

(a)

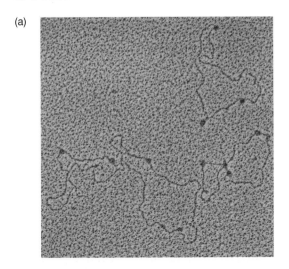

(b)

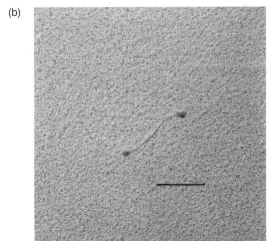

Figure 5. (a) *Binding of the polymerase-α-associated DNA-binding protein to plasmids nicked by exposure to γ-radiation.* (b) *Binding of the polymerase-α-associated DNA-binding protein to the cohesive termini generated by restriction endonucleases EcoR1 and HindIII. The scale bar represents 200 nm. From Burton et al. (1997). Novel DNA-binding properties of a protein associated with DNA polymerase-α in pea.* Plant Journal, *vol. 12, pp. 357–365. Reprinted with permission of Blackwell Science Ltd.*

Table 1. *Effect of adding the 42 kDa DNA-binding protein back to the DNA polymerase-α-primase complex with different template-primer systems*

	Template-primer system		
	'Gapped' DNA[a]	Primed M13 DNA[b]	'Self-primed' M13 DNA[c]
Stimulation of polymerase activity (%)	25	110	400

[a]Gapped DNA: calf thymus DNA treated with deoxyribonuclease.
[b]Primed M13 DNA: M13 DNA to which an oligodeoxyribonucleotide-sequencing primer had been annealed.
[c]Self-primed DNA: M13 DNA primed by the polymerase's own primase activity (non-radioactive ribonucleotides were provided as substrates for the primase).

and especially the self-primed template. However, with the 'gapped' DNA template, the stimulation is very much less. In other words, when the primers are very closely spaced along the template (as in gapped DNA) the DNA-binding protein has little effect on DNA polymerase activity but when the primers are infrequent and/or when the template has been self-primed by primase, there is a major effect on polymerase activity. These results lead us to conclude that the 42 kDa protein facilitates the correct use of primers by DNA polymerase-α. This feature, considered alongside the protein's preferred DNA configurations for efficient DNA-binding is consistent with the view that the 42 kDa is a primer-recognition protein which works as an accessory to DNA polymerase-α in synthesis of the lagging strand.

4. Primer-recognition protein and phosphoglycerate kinase

Although this is the first description of a primer-recognition protein in plants, a similar protein has been described in mammalian cells. It was initially discovered as a protein which stimulated the activity of DNA polymerase-α rather than as a DNA-binding protein (Pritchard et al., 1983). Subsequently, it was shown that the extent of the stimulation was inversely related to the frequency of primers on the DNA template and it was deduced that the protein aided 'productive' binding of the polymerase to the template (i.e. binding adjacent to the 3′ hydroxy terminus of a primer). The protein was thus termed a primer-recognition protein. Analysis of the primer-recognition protein in HeLa cells showed that it consisted of two polypeptides of 36 and 42 kDa (in contrast to the protein from pea, described above, which consists of only one polypeptide of 42 kDa). Purification and sequence analysis of the 42 kDa protein in HeLa cells led to the surprising conclusion that it was identical to a glycolytic enzyme, PGK (Jindal and Vishwanatha, 1990).

  Although such a conclusion may be considered unlikely, evidence is accumulating that the 42 kDa primer-recognition protein of pea is also identical to PGK. Antibodies raised against purified chloroplastic PGK from pea recognize antigen in the nuclei of pea leaves, as well as the expected location in chloroplasts and cytosol (Anderson et al., 1995). The location of PGK-like antigen in the nucleus is consistent with a role for the enzyme in nuclear metabolism in addition to its established roles in glycolysis, gluconeogenesis and the Calvin cycle. This idea is further supported by the finding that DNA polymerase-α preparations from pea exhibit PGK activity (*Figure 6*) and that the PGK coincides with the primer-recognition protein. To this must be added the finding that the anti-PGK antibodies (referred to above) recognize purified primer-recognition protein in western blots (*Figure 7*) and that pre-incubation of the primer-recognition protein with anti-PGK antibodies leads to loss of DNA-binding activity (*Figure 8*). This immuno-depletion of activity does not happen if the primer-recognition protein is pre-incubated with antibodies raised against MCM3, a DNA replication-origin-activation protein. Finally, it has been shown recently that stimulation of mammalian DNA polymerase-α activity on primed M13 DNA templates may be achieved by the addition of purified PGK which mimics completely the activity of primer-recognition protein (Popanda et al., 1998). This result calls to question the earlier suggestion (see above) that the mammalian primer-recognition protein consists of two completely different polypeptides but it is consistent with the data from pea which indicate that primer-recognition protein consists only of a PGK-like polypeptide.

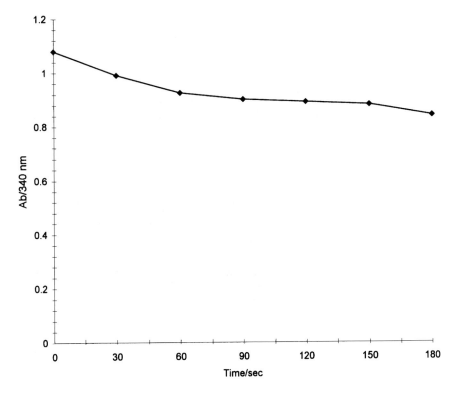

Figure 6. Phosphoglycerate kinase activity assayed in a preparation of DNA polymerase-α from pea. The same protein fractions also exhibited primer-recognition activity.

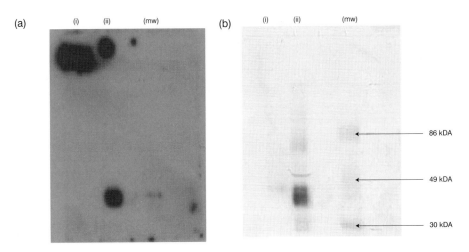

Figure 7. South-western blot of DNA-binding protein. (a) Autoradiograph of blot probed with radioactive DNA to detect DNA-binding activity. (b) The same blot probed with anti-phosphoglycerate kinase antibodies. Lane (i): polymerase-α complex from which DNA binding activity has been removed. Lane (ii): purified DNA-binding protein. mw=molecular weight markers.

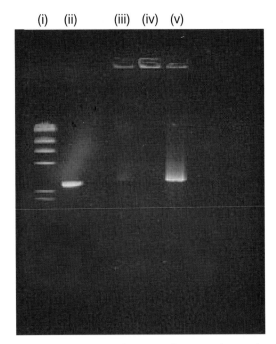

Figure 8. *Depletion of DNA-binding activity by pre-incubation with anti-phosphoglycerate kinase (PGK) antibodies. DNA-binding activity was detected by retardation of the DNA at the origin of an agarose electrophoresis gel. From left to right the lanes contained: (i) bacteriophage-λ DNA, digested with the restriction endonuclease HindIII (mobility marker); (ii) linearized plasmid DNA (pUC18); (iii) linearized pUC18 incubated with DNA-binding protein; (iv) linearized pUC18 incubated with DNA-binding protein that had been pre-incubated with antibodies raised against MCM3 (an origin-activating protein) – this did not diminish the DNA-binding activity; (v) linearized pUC18 incubated with DNA-binding protein that had been pre-incubated with anti-PGK antibodies. The ability of binding protein to retard the DNA at the top of the gel was strongly inhibited.*

5. Phosphoglycerate kinase in the nucleus: how and why?

The discovery that PGK has a role in DNA replication within the nucleus, a role which is apparently unrelated to its roles in the cytosol and chloroplast, raises a number of interesting questions. Two of these questions will be considered here. First, how does PGK get into the nucleus? Sequence analysis of animal and plant PGKs, including the chloroplastic PGKs, shows that they possess near the N-terminus a bipartite nuclear localization signal, or a very near match to it (Anderson *et al.*, 1995, 1996; Fleming and Littlechild, 1997). Such signals consist of two adjacent basic amino acid residues separated by 10 residues from a motif in which three out of five residues are basic (Dingwall and Laskey, 1991; *Figure 9*). Such nuclear localization signals are recognized by the

Tobacco cytosol	**MAVKKSVGSLKEADLKGKR**
Wheat cytosol	**MATKRSVGTLGEADLKGKK**
Tobacco c`plast	**ASMAKKSVGDLTAAELKGKK**
Spinach c`plast	**ASMAKKSVGDLTSADLKGKK**
Wheat c`plast	**ATMAKKSVGDLTAADLEGKR**

Figure 9. *Bi-partite nuclear localization signals in plant phosphoglycerate kinases.*

nuclear import protein, importin, which may be shown to effect nuclear uptake *in vitro*, as recently reported for the *Arabidopsis* importin (Smith *et al*., 1997). However, the possession of a nuclear localization signal is clearly not sufficient for nuclear uptake, otherwise all the cell's PGK would be taken up into the nucleus. Indeed, several proteins that are solely located in the cytosol possess nuclear uptake signals (R. A. Laskey, personal communication) and thus other factors must contribute to nuclear uptake. Any suggestion about the selection of a particular sub-population of PGK for nuclear uptake must be speculative but it is possible that some form of protein modification, such as phosphorylation, may be involved. A subsidiary question is which of the two plant PGKs is it that enters the nucleus? Again there is no clear answer to this but the fact that the same situation prevails in mammalian cells which do not possess a chloroplastic PGK suggests that it is a sub-population of the cytosolic PGK which enters the nucleus.

The possession of two such completely different activities in one polypeptide is very unusual, leading to the second major question: what is the functional significance? Again, any answer must be purely speculative. The position of PGK in glycolysis, catalysing the first ATP-generating step in respiration, places the enzyme in a position to 'sense' the state of the cell's ATP-generating pathways. If that sensing were combined with phosphorylation to facilitate nuclear uptake, then PGK could provide a link between ATP generation and the expenditure of energy in DNA replication.

Clearly there is much research still to be done before we reach an understanding of this fascinating biochemical situation, a situation that would surely have intrigued Tom ap Rees, to whose memory this chapter, like the rest of the volume, is dedicated.

Acknowledgements

We are grateful to past and present members of our research teams, Jamila Al-Rashdi, David Brice, Sara Burton, Paul Fitchett, Amina Ibrahim and Wang Xing-Wu, for their participation in this project and to Jack Van't Hof for his collaboration and continued interest. We thank the BBSRC, Exeter University, the National Science Foundation, the University of Illinois-Chicago and NATO for research grants. John Bryant acknowledges with gratitude the excellent research training he had as a PhD student in Tom ap Rees's laboratory. Louise Anderson profited greatly from spending a year at the CSIRO Plant Physiology Unit, University of Sydney, when Tom ap Rees was a staff member there.

References

Al-Rashdi, J. and Bryant, J.A. (1994) Purification of a DNA-binding protein from a multi-protein complex associated with DNA polymerase-α in pea. *J. Exp. Bot.* **45**: 1867–1871.

Anderson, L.E., Wang, X. and Gibbons, J.T. (1995) Three enzymes of carbon metabolism or their antigenic analogs in pea leaf nuclei. *Plant Physiol.* **108**: 659–667.

Anderson, L.E., Gibbons, J.T. and Wang, X. (1996) Distribution of ten enzymes of carbon metabolism in pea (*Pisum sativum*) chloroplasts. *Int. J. Plant Sci.* **157**: 525–538.

Bringloe, D.H., Rao, S.K., Dyer, T.A., Raines, C.A. and Bradbeer, J.W. (1996) Differential gene expression of chloroplast and cytosolic phosphoglycerate kinase in tobacco. *Plant Mol. Biol.* **30**: 637–640.

Bryant, J.A., Fitchett, P.N., Hughes, S.G. and Sibson, D.R. (1992) DNA polymerase-α in pea is part of a large multi-protein complex. *J. Exp. Bot.* **43**: 31–45.

Burton, S.K., Van't Hof, J. and Bryant, J.A. (1997) Novel DNA-binding properties of a protein associated with DNA polymerase-α in pea. *Plant J.* **12**: 357–365.

Dingwall, C. and Laskey, R.A. (1991) Nuclear targeting sequences: a consensus. *Trends Biochem. Sci.* **16**: 478–481.

Fleming, T. and Littlechild, J.A. (1997) Sequence and structural comparison of thermophilic phosphoglycerate kinases with a mesophilic equivalent. *Comp. Biochem. Physiol.* **A118**: 439–451.

Foiani, M., Lucchini, G. and Plevani, P. (1997) The DNA polymerase-α-primase complex couples DNA replication, cell-cycle progression and DNA-damage response. *Trends Biochem. Sci.* **22**: 424–427.

Jindal, H.K. and Vishwanatha, J.K. (1990) Purification and characterisation of primer-recognition proteins from Hela cells. *Biochemistry* **29**: 4767–4773.

Longstaff, M., Raines, C.A., McMorrow, E.M., Bradbeer, J.W. and Dyer, T.A. (1989) Wheat phosphoglycerate kinase: evidence for recombination between the genes for the chloroplastic and cytosolic enzymes. *Nucl. Acids Res.* **17**: 6569–6580.

Macioszek, J., Anderson, J.B. and Anderson, L.E. (1990) Isolation of chloroplastic phosphoglycerate kinase: kinetics of the two-enzyme phosphoglycerate kinase glyceraldehyde-3-phosphate dehydrogenase couple. *Plant Physiol.* **94**: 291–296.

Popanda, O., Fox, G. and Theilmann, H.W. (1998) Modulation of DNA polymerases α, δ and ε by lactate dehydrogenase and 3-phosphoglycerate kinase. *Biochim. Biophys. Acta* **1397**: 102–117.

Pritchard, C.G., Weavers, D.T., Baril, E.F. and De Pamphilis, M.L. (1983) DNA polymerase-α co-factors C_1, C_2 function as primer-recognition proteins. *J. Biol. Chem.* **258**: 9810–9819.

Smith, H.M.S., Hicks, G.R. and Reikhel, N.V. (1997) Importin alpha from *Arabidopsis thaliana* is a nuclear import receptor that recognises three classes of import signals. *Plant Physiol.* **114**: 411–417.

Index